Studies in Big Data

Volume 166

The series "Studies in Big Data" (SBD) publishes new developments and advances in the various areas of Big Data- quickly and with a high quality. The intent is to cover the theory, research, development, and applications of Big Data, as embedded in the fields of engineering, computer science, physics, economics and life sciences. The books of the series refer to the analysis and understanding of large, complex, and/or distributed data sets generated from recent digital sources coming from sensors or other physical instruments as well as simulations, crowd sourcing, social networks or other internet transactions, such as emails or video click streams and other. The series contains monographs, lecture notes and edited volumes in Big Data spanning the areas of computational intelligence including neural networks, evolutionary computation, soft computing, fuzzy systems, as well as artificial intelligence, data mining, modern statistics and Operations research, as well as self-organizing systems. Of particular value to both the contributors and the readership are the short publication timeframe and the world-wide distribution, which enable both wide and rapid dissemination of research output.

The books of this series are reviewed in a single blind peer review process.

Indexed by SCOPUS, EI Compendex, SCIMAGO and zbMATH.

All books published in the series are submitted for consideration in Web of Science.

Jerzy S. Respondek

Fast Matrix Multiplication with Applications

Springer

Jerzy S. Respondek
Faculty of Automatic Control, Electronics
and Computer Science
Silesian University of Technology
Gliwice, Poland

ISSN 2197-6503 ISSN 2197-6511 (electronic)
Studies in Big Data
ISBN 978-3-031-76929-0 ISBN 978-3-031-76930-6 (eBook)
https://doi.org/10.1007/978-3-031-76930-6

This Springer imprint is published by the registered company Springer Nature Switzerland AG
The registered company address is: Gewerbestrasse 11, 6330 Cham, Switzerland

Contents

Chapter 1
Introduction

The main theme of this book is to present the methods of constructing fast matrix multiplication algorithms and—no less importantly—to show how to profit from their acceleration in other, also non-matrix, problems. Interestingly, today we can apply fast matrix multiplication algorithms not only to other linear algebra problems, but also to current problems in science and engineering, such as:

- artificial intelligence,
- databases,
- parallel computations,
- computational biology,
- pattern recognition,
- compiler construction,
- finding optimal route

and others.

Another interesting aspect of fast matrix multiplication is that these algorithms work on dense matrices, whereas we have developed methods to apply them to sparse matrices.

In Sect. 9.12 we have enumerated all the applications, with the articles in which they are presented. In Chaps. 9 and 10 we could only present a part of them in detail and illustrate them with examples.

Roughly speaking, all the above problems are as fast as the fast matrix multiplication algorithm used. This means that the construction of a faster matrix multiplication algorithm automatically makes it possible to speed up the solution of, among other things, the above problems using the methods described in this book in Chap. 9. The flowchart on Fig. 9.7 illustrates this idea for selected problems.

The concept of a two-dimensional mathematical matrix is a fundamental notion in science and engineering. Let us begin with a discussion of the formal aspects, namely the mathematical matrix definition. The classical monograph Bellman (1997) on page 16 defines the matrix as a rectangular or, in particular, square table of complex

J. S. Respondek, *Fast Matrix Multiplication with Applications*, Studies in Big Data 166,
https://doi.org/10.1007/978-3-031-76930-6_1

numbers. Gantmacher (1960, page 1) states that the entries of the 2D table must be elements of a certain field. Some authors define the matrix notion in a more formal way. Kaczorek (1998) on page 15 defines the matrix as "*the function of two variables which assigns to each pair of natural numbers (i,j), where* $i = 1,\ldots,m$, $j = 1,\ldots,n$, *exactly one real or complex number* a_{ij}".

Although the concept of a matrix was first introduced by Sylvester (1850, page 369), and Sylvester (1851, page 302), many authors consider Cayley (1857) article to be the first comprehensive work dedicated specifically to the topic of matrices. The text, which occupies 21 pages, discusses 58 matrix operations and their properties. In addition to the widely recognized and appreciated monographs by Gantmacher and Bellman, it is also worthwhile to consider earlier works that are still valuable and treat matrix problems in a mature way. The above monographs are (we have deliberately presented them in reverse chronological order), Aitken (1944), preceded by Frazer et al. (1938), Wedderburn (1934), Mac Duffee (1933), Turnbull and Aitken (1932), Turnbull (1928) and Cullis (1913–1925, Vol. I–III). Two notable monographs on determinants are Muir and Metzler (1933) and Scott (1880).

This monograph is dedicated exclusively to the fast matrix multiplication algorithms and their applications, but it is also worth noting that the monograph Golub and Van Loan (2013) is recognised as the most comprehensive work on numerical recipes for matrices. Another seminal work on numerical algorithms for matrices is Householder (1964). As documented e.g. in the monograph by Yong and Se (2004), the concept of Gaussian elimination was already known in ancient times. In Sect. 7.7, pages 145–154, the process of transforming the source system matrix to a triangular form in order to solve three linear equations with three unknowns is described.

We will now proceed to the matrix multiplication operation. This operation plays a pivotal role in matrix algebra algorithms, as it forms the foundation for many other matrix operations, particularly those presented in this monograph. Consequently, the efficiency of the latter depends on the efficiency of the former. Firstly, it is necessary to define the matrix multiplication. In point 11 of the article by Cayley (1857), on pages 20–21, the definition of matrix multiplication in the case of a pair of matrices of dimensionality 3×3 is provided (However, in a notation very different from the contemporary). In the early years of its development, matrix multiplication was also defined indirectly as an auxiliary tool for determinants. In his monograph from 1880, Scott provides an unintentional definition of matrix multiplication in Chap. 4, entitled "*Multiplication of Determinants*". In Sect. 3, pages 46–47, the defining formula is presented in the context of the formula for the determinant of a product of matrices. In contemporary monographs, such as Gantmacher (1960, p 6) and Bellman (1997, p 39), the operation is defined as follows.

1.1 Matrix Multiplication Definition

Let us be given two rectangular matrices $A = [a_{ij}]_{i=1,\ldots,m,\ j=1,\ldots,n}$, $B = [b_{ij}]_{i=1,\ldots,n,\ j=1,\ldots,l}$. *As a result of the multiplication of the A matrix by the B matrix we define the matrix* $C = [c_{ij}]_{i=1,\ldots,m,\ j=1,\ldots,l}$ *with entries expressed by the sum*

$$c_{ij} = \sum_{k=1}^{n} a_{ik} b_{kj}, \quad i = 1, \ldots, m, \ j = 1, \ldots, l \tag{1.1}$$

It is important to note that a product of two rectangular matrices exists if and only if the number of columns of the first matrix is equal to the number of rows of the second. Throughout this monograph the matrix multiplication of dimensionality $m \times n$ by $n \times l$ will be abbreviated as (m, n, l). The matrix product has the following most important properties (assuming the feasibility of the respective operations):

- $A(B + C) = AB + AC, \quad (B + C)D = BD + CD$ (left and right distributivity).
- $\alpha(AB) = (\alpha A)B = (AB)\alpha = A(B\alpha)$ (commutativity with respect to a scalar number).

$$(AB)C = A(BC) \text{(associativity)}. \tag{1.2}$$

$$(AB)^T = B^T A^T \text{(transposition of a product)}. \tag{1.3}$$

$$(AB)^{-1} = B^{-1} A^{-1} \text{(inversion of a product)}. \tag{1.4}$$

$$\det(AB) = \det(A) \cdot \det(B) \text{(determinant of a product)}. \tag{1.5}$$

$$\operatorname{rank}(AB) \leq \min\{\operatorname{rank}(A), \operatorname{rank}(B)\}. \tag{1.6}$$

- $\operatorname{trace}(AB) = \operatorname{trace}(A) \cdot \operatorname{trace}(B)$.

Furthermore, it should be noted that the matrix product operation is, in general, not commutative:

$$AB \neq BA \tag{1.7}$$

Due to Hawkins (1974), the non-commutativity of the matrix product was first identified by Eisenstein in works from 1844 and 1852 (see references therein). This fact is also stated by Cayley (1857) in point 11.

The monograph is structured in the following way:

- In Chap. 2 we explained how to multiply matrices using commutative algorithms. These algorithms cannot be applied in a recursive way. We started by looking at the first Winograd result from 1968. Then we surveyed later results up to 2023. Finally, we presented algorithms for the 3×3 problem in 22 and 21 multiplications.[1]
- Chapter 3 is about non-commutative exact algorithms. We start with the classic Strassen algorithm and show how it can be derived. Then we look at its Winograd variation. We then present the Gastinel generalisation of the Strassen algorithm to an arbitrary dimension. Finally, we look at the Diophantine equation method for finding fast matrix multiplication algorithms. Next we present the Laderman's algorithm for 3×3 problem in 23 multiplication, and its generalisation by Sykora. We also briefly describe permanents and Latin squares, which are used by the Sykora construction. We then present the duality notion of fast matrix multiplication algorithms. Shortly, once we have an algorithm for (k, m, n) dimensionality it shows how to obtain algorithms for the remaining five dimensionalities $(3! - 1 = 5)$ that work in the same number of non-commutative multiplications. We then briefly describe the elementary, general methods for constructing fast algorithms. We then present classical trilinear identity algorithms for even-dimensional matrices. We then introduce the reader to the notion of disjoint matrix multiplication, which allows to multiply more than one matrix product at a time, with less non-commutative multiplications than follows from the definition (1.1). For this purpose, we present the double disjoint algorithm.
- In Chap. 4 we presented some exact, triple disjoint algorithms with application to single matrix multiplication. We presented two algorithms for triple disjoint multiplication and one algorithm for single matrix multiplication, constructed on their basis.
- In Chap. 5 we introduce the reader to the notion of so-called Arbitrary Precision Algorithms (APA), using Schonhage's (1981) result for—again—the same 3×3 problem, but now with 21 multiplications. In short, they allow to obtain an approximate value of a matrix product, but in return in fewer products. We have deliberately chosen this problem, rather than the chronological order in which the papers were published, for the sake of clarity. We then show how to construct an exact algorithm based on a given algorithm of arbitrary precision, and we prove what complexity it has. Next, we go back in time to the notion of partial matrix multiplication, proposed by Bini et al. (1979), who in conjunction with arbitrary precision algorithms, and (already mentioned) duality of matrix multiplication, obtained the algorithm with the best time complexity at the year of publication. Next, we show how to transform, the previously presented in Chap. 3, (exact) double disjoint matrix multiplication algorithm to the corresponding approximate version, saving some non-scalar multiplications in return. Finally we present in detail two approximate algorithms by Schonhage from 1981: the (relatively) simple double disjoint algorithm, and a more sophisticated one that computes three matrix products at once.

[1] Whenever we use the short form $n \times n$, we mean that we are multiplying two square matrices of equal dimensions $n \times n$ both.

- Chapter 6 is entirely devoted to the problem of how to use the disjoint algorithms, which calculate several separate matrix products at once, to calculate a single matrix product. In general, the chapter is divided into two parts: for disjoint algorithms working on matrices of equal dimensionality, and the second, more general part, showing how to use algorithms working on a set of matrix products of arbitrary dimension to obtain a fast algorithm that calculates a single matrix product. In the first part we present a block method by Schonhage (1981), then a recursive method and some of its improvements proposed by the author of this monograph. In the second part, the most general, summarising result is, of course, the famous tau-theorem. In order to make the lecture as clear as possible not only for advanced readers, we present its proof in a (not so much) special case—often encountered in practice—of a disjoint algorithm, working on three matrices with cyclically shifted dimensions. Finally, we use the tau-theorem to determine the complexity of previously presented double and triple disjoint algorithms applied to single matrix multiplication.
- Chapter 7 is entirely devoted to the notion of partial matrix multiplication, originally proposed by Bini et al. (1979). First, we present a particular partial matrix multiplication algorithm proposed by Schonhage (1981). Then we show a general methodology that allows to transform any partial matrix multiplication algorithm into a total one. This was also done by Bini et al. (1979), but in a special case.
- In Chap. 8 we tried to summarise the problems of fast matrix multiplication algorithms. We gave a chronological overview of the most important milestones in the field. Next, using the example of an 5×5 problem, we showed how much research effort is put into saving as many multiplications as possible. Then we showed how the efficiency exponent of multiplication algorithms decreases over the years and discussed some theoretically obtained limits. Finally, we presented a simple but effective and universal method for using matrix multiplication algorithms designed for square matrices for rectangular problems, and reviewed other work on fast multiplication of rectangular matrices. We also introduced the reader to the method of speeding up matrix multiplication by choosing an order of parenthesis in the case of finding a multiple matrix multiplication.
- In Chap. 9 we move on to applications of fast matrix multiplication algorithms. In detail we described a series of linear algebra problems, in particular: inversion of matrices, LU decomposition, solving systems of linear equations, finding determinants, characteristic polynomials, successive powers of a given matrix and value of a matrix polynomial. We also focus on non-matrix problems, such as finding a series of polynomials at once and computing the value of a given polynomial in a series of points. Finally, we surveyed other known applications of fast matrix multiplication algorithms. We distinguished between matrix-related problems, graph problems and others. We presented separately problems using matrix multiplication algorithms for square and rectangular matrices. A flowchart of the linear algebra problems with respect to, particularly fast, matrix multiplication is given.
- Chapter 10 is devoted to the fast multiplication of matrices of a special domain, i.e. with Boolean entries. We have explained how to use the classical, exact, fast

matrix multiplication algorithms presented in the monograph in this case, and have presented a specialised algorithm for this purpose.

- A separate Chap. 11 we devoted to an auxiliary tool in fast matrix multiplication algorithms, i.e. the Vandermonde matrices. They are useful in the title problem at two points: in transforming an arbitrary precision algorithm into an exact one (Chap. 5), and in transforming a partial matrix multiplication algorithm into a total one (Chap. 7). The (classical) Vandermonde matrix is a special case of the confluent generalisation, so the algorithms that work on the latter can also be applied to the former. Thus, we present the problem of the Vandermonde matrix using its confluent version. We have sketched its history back to 1901 (or even 1891), then listed a number of applications of this type of structured (some call it also as 'special') matrix, and described two of them in detail. Since the applications of the Vandermonde matrix in fast matrix multiplication require its inversion, we surveyed articles on the inversion of a confluent Vandermonde matrix and indicated the only one which, in a general case, always works in quadratic time.

We hope that the reader will find this monograph useful and interesting in the field of fast matrix multiplication algorithms and their applications.

References

Aitken AC (1944) Determinants and matrices, 3rd edn. Interscience Publishers, New York, NY

Bellman R (1997) Introduction to matrix analysis, 2nd edn. SIAM, Philadelphia

Bini D, Capovani M, Lotti G, Romani F (1979) O(n^2.7799) complexity for approximate matrix multiplication. Inform Process Lett 8(5):234–235

Cayley (1857) A memoir on the theory of matrices. Philos Trans R Soc Lond 148:17–37

Cullis CE (1913, 1918, 1925) Matrices and determinoids, vol. 1–3. Cambridge University Press, Cambridge

Frazer RA, Duncan WJ, Collar AR (1938) Elementary matrices and some applications to dynamics and differential equations. Cambridge University Press, Cambridge

Gantmacher FR (1960) The theory of matrices. Chelsea Publishing Company, New York, NY

Golub GH, Van Loan ChF (2013) Matrix computations, 4th edn. The Johns Hopkins University Press, Baltimore

Hawkins T (1974) The theory of matrices in the 19th century. Proc Int Cong Math Vancouver 2:561–570

Householder AS (1964) The theory of matrices in numerical analysis, 1st edn. Blaisdell Publishing Company, New York

Kaczorek T (1998) Wektory i macierze w automatyce i elektrotechnice, 2nd edn. WNT, Warsaw (in Polish)

Mac Duffee CC (1933) The theory of matrices. Springer, Berlin

Muir T, Metzler WH (1933) Theory of determinants. Longmans, Green and CO., Bombay, Calcutta, Madras

Schonhage A (1981) Partial and total matrix multiplication. SIAM J Comput 10(3):434–455

Scott RF (1880) Theory of determinants. Cambridge University Press, Cambridge

Sylvester JJ (1850) Additions to the articles in the September number of this journal, "On a new class of theorems," and on Pascal's theorem. Phil Mag 3(37):363–370

Sylvester JJ (1851) On the relation between the minor determinants of linearly equivalent quadratic functions. Phil Mag 4(1):295–305

Turnbull HW, Aitken AC (1932) An introduction to the theory of canonical matrices. Blackie and Son, London, Glasgow and Bombay

Turnbull HW (1928) The theory of determinants, matrices and invariants. Blackie and Son, London, Glasgow

Wedderburn JHM (1934) Lectures on matrices. American Mathematical Society. Colloquium Publications, Providence, Rhode Island

Yong LL, Se AT (2004) Fleeting footsteps. Tracing the conception of arithmetic and algebra in Ancient China. World Scientific Publishing, Singapore

Chapter 2
Commutative Matrix Multiplication Algorithms

2.1 Introduction

The most standard implementation of the algorithm for multiplying of square matrices achieves the time complexity $O(n^3)$, which follows directly from its Definition 1.1, requiring exactly n^3 multiplications. For multiplied matrices of dimensions $(m, n) \times (n, p)$ it requires $m \cdot n \cdot p$ multiplications.

For a dozen of years, this number of multiplications seemed to be a limit not to be beaten until 1968,[1] when the first successful attempt to improve the efficiency of the matrix multiplication was made by Winograd (1968). Let us first give some theoretical background.

The field of fast matrix multiplication algorithms distinguishes two main classes of algorithms, i.e. commutative and non-commutative. The first class of algorithms requires that the elements of the matrices belong to a commutative field with respect to multiplication. This means that the commutative matrix multiplication algorithms cannot be applied recursively. This follows from the fact that during the recursion we come to scalars at the last recursion level, but on the previous but last levels we come to matrices. As we know for matrices in general $AB \neq BA$ [property (1.7)].

The commutative paradigm often—but not always (e.g. see the discussion on the optimality of the Strassen algorithm in Sect. 3.2)—allows to construct algorithms with less number of scalar multiplications compared to the non-commutative algorithm (for the problem of the same dimensionality), but the multiplied elements must only be the scalars (cannot be submatrices themselves). Thus, we obtain an algorithm with a fixed size of the matrices to be multiplied.

The chapter is organized as follows:

[1] In introduction of the article Pan (1982) it is claimed, that at seminar in Moscow in 1966 was presented the first method of non-asymptotic acceleration of matrix multiplication which was created in 1965. It remained unpublished, though.

J. S. Respondek, *Fast Matrix Multiplication with Applications*, Studies in Big Data 166,
https://doi.org/10.1007/978-3-031-76930-6_2

- In Sect. 2.2 we describe the result of Winograd from 1968, for scalar product, with the way it can be applied to matrix multiplication. We also give an example for the case 3×3.
- In Sect. 2.3 we described and compared other methods of non-asymptotic acceleration of matrix multiplication published up to 2023.
- In Sect. 2.4 we presented a commutative algorithm for the 3×3 problem in 22 multiplication, given by Makarov (1986). It is a solution specialised for this dimensionality.
- In Sect. 2.5 we explained how Rosowski (2023) solved the 3×3 problem in 21 multiplications.

2.2 The First Attempt to Improve Efficiency

Winograd (1968) proposed a modified algorithm for computing the inner product of vectors, but at the same time showed how this new algorithm could be applied to matrix multiplication, reducing the number of scalar multiplications to be performed. For two vectors $x = (x_1, \ldots, x_n)$ and $y = (y_1, \ldots, y_n)$, he proposed to calculate the inner products in two steps. In the first step, we compute auxiliary numbers $\xi,\ \eta$:

$$\xi = \sum_{i=1}^{\lfloor n/2 \rfloor} x_{2i-1}x_{2i},\ \eta = \sum_{i=1}^{\lfloor n/2 \rfloor} y_{2i-1}y_{2i}$$

Next, in the second step, using $\xi,\ \eta$, the inner product is given by:

$$x^T y = \begin{cases} x_{even}^T y_{even} = \left[\sum_{i=1}^{\lfloor n/2 \rfloor} (x_{2i-1} + y_{2i})(x_{2i} + y_{2i-1}) - \xi - \eta \right], & \text{for even } n \\ x_{even}^T y_{even} + x_n y_n, & \text{for odd } n \end{cases} \tag{2.1}$$

To compute a **single** inner product by the formula (2.1) we have to perform roughly $(3/2)n$ multiplications, which is worse than $1 \cdot n$ that follows directly from the definition (1.1). But if we want to compute a **series** of t inner products of n-dimensional vectors, of the amount of N, thanks to the formula (2.1) we have to perform $N\lfloor n/2 \rfloor + t\lfloor (n+1)/2 \rfloor$ scalar multiplications, instead of tn in—let's call it—the direct method (Winograd 1968).

In the monograph's title application to the multiplication of matrices, here square, we have $t = n^2$ and $N = 2n$ (n horizontal vectors come from the A matrix in the AB product, and n vertical vectors we take from the B matrix, respectively). Considering that $\lfloor n/2 \rfloor = n/2$ for even n and $\lfloor n/2 \rfloor = (n-1)/2$ for odd n, the required number of multiplications is equal to:

$$2n\left\lfloor \frac{n}{2} \right\rfloor + n^2\left\lfloor \frac{n+1}{2} \right\rfloor = 2n^2 + (n^2 - 2n)\left\lfloor \frac{n+1}{2} \right\rfloor \quad (2.2)$$

Essentially, the efficiency of the method of inner product calculation proposed by Winograd lies in the idea that, for a given matrix product, it is necessary to calculate the variables ξ, η **only once** for a given matrix product, for each of the $2n$ vectors separately. This is possible thanks to the fact, that the ξ, η variables are the linear combinations of their own elements; they do not contain any "mixed" scalar products of the form $x.y.$ (or $y.x.$). Next, we use it n^2 times in the formula (2.1), separately for each $n \times n$ matrix entry, what now requires about $n/2$ multiplications.

Thus, thanks to the new inner product algorithm by Winograd (1968), the total number of scalar multiplications to compute the product of two $n \times n$ matrices was reduced from "standard" n^3 to approximately $n^3/2$ for large n for the first time in history.

Formula (2.1) was the first example of a more general method of **aggregation and correction**. Namely, we are **aggregating** a series of multiplications into a single product of sums [here the product $(x_{2i-1} + y_{2i})(x_{2i} + y_{2i-1})$], then "**adjusting**" the result by the **correction terms**, in (2.1) ξ and η. The key idea is to reduce the number of—costly—multiplications as much as possible. In the following chapters we will show that the cost of multiplication compared to addition/subtraction applied recursively to matrices is much higher than in the scalar case.

2.2.1 Example—3 × 3 Case

Let us use the Winograd's method to find the product of 3 × 3 matrices. First, we find the inner product of 3-dimensional vectors using formula (2.1):

$$\begin{bmatrix} x_1 & x_2 & x_3 \end{bmatrix} \cdot \begin{bmatrix} y_1 & y_2 & y_3 \end{bmatrix}^T = (x_1 + y_2)(x_2 + y_1) - x_1x_2 - y_1y_2 + x_3y_3 \quad (2.3)$$

which clearly leads to the "definitional" inner product $x_1y_1 + x_2y_2 + x_3y_3$. To multiply the 3 × 3 matrices A, B we write them in the form of a block of vectors.

$$AB = \begin{bmatrix} a_1 \\ \hline a_2 \\ \hline a_3 \end{bmatrix} \begin{bmatrix} b_1 \,|\, b_2 \,|\, b_2 \end{bmatrix}, \quad a_i = \begin{bmatrix} a_{i1} & a_{i2} & a_{i3} \end{bmatrix}, \quad b_i = \begin{bmatrix} b_{1i} \\ b_{2i} \\ b_{3i} \end{bmatrix}, \quad i = 1, 2, 3 \quad (2.4)$$

It is now sufficient to apply the scalar product formula (2.3) to the vectors that build the A, B matrices. For better clarity, we have graphically distinguished the common products.

$$\left\{\begin{array}{l}
a_1b_1 = (a_{11}+b_{21})\cdot(a_{12}+b_{11}) - \boxed{a_{11}a_{12}} - b_{11}b_{21} + a_{13}b_{31}\\
a_1b_2 = (a_{11}+b_{22})\cdot(a_{12}+b_{12}) - \boxed{a_{11}a_{12}} - b_{12}b_{22} + a_{13}b_{32}\\
a_1b_3 = (a_{11}+b_{23})\cdot(a_{12}+b_{13}) - \boxed{a_{11}a_{12}} - \boxed{b_{13}b_{23}} + a_{13}b_{33}\\
\hline
a_2b_1 = (a_{21}+b_{21})\cdot(a_{22}+b_{11}) - a_{21}a_{22} - b_{11}b_{21} + a_{23}b_{31}\\
a_2b_2 = (a_{21}+b_{22})\cdot(a_{22}+b_{12}) - a_{21}a_{22} - b_{12}b_{22} + a_{23}b_{32}\\
a_2b_3 = (a_{21}+b_{23})\cdot(a_{22}+b_{13}) - a_{21}a_{22} - \boxed{b_{13}b_{23}} + a_{23}b_{33}\\
\hline
a_3b_1 = (a_{31}+b_{21})\cdot(a_{32}+b_{11}) - a_{31}a_{32} - b_{11}b_{21} + a_{33}b_{31}\\
a_3b_2 = (a_{31}+b_{22})\cdot(a_{32}+b_{12}) - a_{31}a_{32} - b_{12}b_{22} + a_{33}b_{32}\\
a_3b_3 = (a_{31}+b_{23})\cdot(a_{32}+b_{13}) - a_{31}a_{32} - b_{13}b_{23} + a_{33}b_{33}
\end{array}\right. \tag{2.5}$$

So we obtained an algorithm to compute the product of 3×3-size matrices by 24 scalar multiplications, compared to $3^3 = 27$ multiplications from the definition, saving 3 of them.

According to formula (2.2), the number of multiplications in Winograd's inner product algorithm is 8 for $n = 2$, 24 for $n = 3$, 48 for $n = 4$ and 95 for $n = 5$. After Winograd's result from 1968, further progress was made for certain matrix sizes (in a commutative version), especially for $n = 3$. We shall discus this size in Sects. 2.4 and 2.5.

We can see that the inner product (2.2) does not always lead to lower number of multiplications; for $n = 2$ it is just equal to the $2^3 = 8$ of the direct method.

The attentive reader will notice that the Winograd's inner product formula (2.1) contains another concern, which is one of the crucial issues in the subject of the matrix multiplication. There are entries whose correctness, from the point of view of the matrix product definition, depends on the commutativity of the domain of the elements of the multiplied matrices. For example in the expression for a_1b_1 from the aggregated product $(a_{11} + b_{21}) \cdot (a_{12} + b_{11})$ we get a term $b_{21}b_{11}$, which next should be cancelled by $b_{11}b_{21}$. In general formula (2.1) to cancel the term $y_{2i}y_{2i-1}$ by the correction term $y_{2i-1}y_{2i}$ the elements of the matrices must commute under the operation of multiplication.

The Winograd's inner product formula is a classic example of a commutative algorithm, in this case reducing the number of multiplications roughly by half.

2.3 Further Attempts to Reduce the Number of Multiplications

There are further attempts to present faster commutative algorithms for matrix multiplication, both for square and for rectangular matrices. The first of this class is the article Waksman (1970), which presents a commutative algorithm for square matrix multiplication of even dimensions in $n(n^2 + 2n - 1)/2$ multiplications. The number of multiplications (2.2) required by the formula Winograd (1968) in the case of even n is equal to:

$$2n^2 + (n^2 - 2n)\left\lfloor \frac{n+1}{2} \right\rfloor = 2n^2 + (n^2 - 2n)\frac{n}{2} = \frac{n}{2}(n^2 + 2n)$$

because for even n we have $\lfloor (n+1)/2 \rfloor = \lfloor n/2 + 1/2 \rfloor = n/2$. Thus Waksman (1970) gives an improvement over Winograd's algorithm with number of multiplications $n/2$ less, reaching $n(n^2 + 2n - 1)/2$ multiplications for any size, in particular 7 multiplications in the 2×2 case. Nevertheless, the additional cost paid for this by Waksman (1970) are additional divisions by 2.

The next result is given by work Islam (2009) in Chap. 6, pages 56–59. Unlike Waksman (1970), it is designed for rectangular matrices. For square matrices it requires the same number of multiplications as Waksman (1970) and also requires additional divisions by two.

Article Rosowski (2023) gives better results. His algorithms works on rectangular matrices. For even n the Theorem 2, pages 306–307, gives an algorithm with the same multiplication amount as Waksman (1970), but no additional divisions are needed. For the case of odd n a separate Theorem 3, pages 307–309, is given, constructed on the idea of the commutative product of 3×3 matrices in 21 multiplications. It further divides the matrix dimensions parity to m parameters, resulting in general in algorithms that work in less multiplications than those presented by Islam (2009), except for minimal values of p, m parameters. Also, for odd n, no additional divisions by 2 are required. Table 2.1 gives a detailed comparison of all these algorithms, for commutative matrix multiplication paradigm.

One might ask whether it is possible to aggregate more than two summands in the way that formula (2.1) does. Unfortunately, Harter (1972) proved that the Winograd's formula is the best of this form, so further acceleration of matrix multiplication is possible, but not on the basis of such an—in principle simple—idea.

Article Brent (1970) carries out the analysis of the numerical stability of the Winograd formula and proposes a ready to use implementation of the Winograd's result in the LU decomposition.

To sum up, the article by Winograd (1968) and others covered in this section reduced the number of scalar multiplications to multiply matrices, but they did not improve the asymptotic complexity, which is just still $O(n^3)$ for both the definition based algorithm as well as the algorithms covered in this chapter. In the next sections we will survey other examples of commutative algorithms, just by the example of the 3×3-dimensional problem.

Table 2.1 Comparison of algorithms for commutative matrix multiplication

<table>
<tr><th rowspan="2">Work</th><th colspan="2">Multiplication number</th><th rowspan="2" colspan="2">Parity</th><th rowspan="2">Additional divisions by 2</th></tr>
<tr><th>Rectangular case</th><th>Square case</th></tr>
<tr><td>Winograd (1968)</td><td>$(m+p)n+(mp-m-p)\lfloor\frac{n+1}{2}\rfloor$</td><td>$2n^2+(n^2-2n)\lfloor\frac{n+1}{2}\rfloor$</td><td colspan="2">–</td><td>No</td></tr>
<tr><td>Waksman (1970)</td><td>–</td><td>$\frac{n(n^2+2n-1)}{2}$ = [Winograd (1968) – $n/2$]</td><td colspan="2">Even n</td><td>Yes</td></tr>
<tr><td rowspan="2">Islam (2009)</td><td>$\frac{n}{2}(mp+m+p-1)$</td><td>= [Waksman (1970)]</td><td colspan="2">Even n</td><td rowspan="2">Yes</td></tr>
<tr><td>$\frac{n-1}{2}(mp+m+p-1)+mp$</td><td>–</td><td colspan="2">Odd n</td></tr>
<tr><td rowspan="3">Rosowski (2023)</td><td>$\frac{n}{2}(mp+m+p-1)$ = [Waksman (1970)]</td><td>= [Waksman (1970)]</td><td colspan="2">Even n</td><td rowspan="3">No</td></tr>
<tr><td>$\frac{n}{2}(mp+m+p-1)+\frac{p-1}{2}$ = [Islam (2009) – $\frac{(p-1)(m-2)}{2}$]</td><td>–</td><td>Even m</td><td rowspan="2">Odd n</td></tr>
<tr><td>$\frac{n}{2}(mp+m+p-1)$ = [Islam (2009) – $\frac{(p-1)(m-1)}{2}$]</td><td>$\frac{n(n^2+2n-1)}{2}$</td><td>odd m</td></tr>
</table>

2.4 Commutative Algorithm for 3 × 3 Matrices in 22 Multiplications

The fastest solution up to the year 2023 for 3×3 size in a commutative way (Observe the elements of the A and B matrices on both sides of aggregated products) was given in 1986 by Makarov, by 22 following multiplications:

$$
\begin{cases}
m_1 = (a_{13} - a_{23} + b_{13})(a_{11} + b_{31} - b_{32} + b_{33}) \\
m_2 = (a_{12} + a_{22} + b_{12})(a_{21} - b_{21} + b_{22} - b_{23}) \\
m_3 = (a_{12} + a_{32} + b_{12})(a_{31} - b_{21} + b_{22} - b_{23}) \\
m_4 = (-a_{23} - a_{33} + b_{13})(a_{31} - b_{31} + b_{32} - b_{33}) \\
m_5 = a_{11}(-a_{13} + a_{23} + b_{11}) \\
m_6 = a_{21}(a_{12} + a_{22} + b_{11}) \\
m_7 = a_{31}(a_{12} + a_{23} + a_{32} + a_{33} + b_{11}), \\
m_8 = b_{12}(a_{11} + b_{21} - b_{22} + b_{23}) \\
m_9 = b_{13}(a_{21} + b_{31} - b_{32} + b_{33}) \\
m_{10} = a_{12}b_{21} \\
m_{11} = a_{23}b_{31} \\
m_{12} = (a_{13} - a_{23})(a_{11} + b_{31}) \\
m_{13} = (a_{12} + a_{22})(b_{21} - a_{21}) \\
m_{14} = (a_{12} + b_{12})(b_{21} - b_{22} + b_{23}) \\
m_{15} = a_{22}b_{23} \\
m_{16} = (b_{13} - a_{23})(b_{31} - b_{32} + b_{33}) \\
m_{17} = a_{23}b_{32} \\
m_{18} = (-a_{23} + a_{32} - a_{33})b_{23} \\
m_{19} = (-a_{12} + a_{13} - a_{32} + a_{33})b_{32} \\
m_{20} = (a_{12} + a_{32})(-a_{31} + b_{21} + b_{23} + b_{32}) \\
m_{21} = (a_{23} + a_{33})(a_{31} + b_{23} - b_{31} + b_{32}) \\
m_{22} = (-a_{12} + a_{23} - a_{32} + a_{33})(b_{23} + b_{32})
\end{cases}
$$

linearly combined as follows:

$$
\begin{cases}
c_{11} = m_5 + m_{10} + m_{11} + m_{12} \\
c_{12} = m_8 + m_{10} - m_{14} + m_{17} - m_{18} + m_{19} - m_{22} \\
c_{13} = m_1 - m_{11} - m_{12} - m_{16} + m_{17} - m_{18} + m_{19} - m_{22} \\
\\
c_{21} = m_6 - m_{10} + m_{11} + m_{13} \\
c_{22} = m_2 - m_{10} + m_{13} + m_{14} + m_{15} + m_{17} \\
c_{23} = m_9 - m_{11} + m_{15} - m_{16} + m_{17} \\
\\
c_{31} = m_7 - m_{10} - m_{11} + m_{20} - m_{21} + m_{22} \\
c_{32} = m_3 - m_{10} + m_{14} - m_{17} + m_{18} + m_{20} + m_{22} \\
c_{33} = m_4 + m_{11} + m_{16} - m_{17} + m_{18} + m_{21}
\end{cases}
$$

This result remained the leader of commutative algorithms for 3 × 3 dimensionality for 37 years since 1986. In the next section we present an algorithm that has surpassed it.

2.5 Commutative Algorithm for 3 x 3 Matrices in 21 Multiplications

Article Rosowski (2023) in Sect. 2 devised a commutative algorithm for 3×3 dimensionality which works in 21 multiplications. The proposed algorithm is constructed in two stages. In the first stage we consider a 1×3 vector by 3×3 matrix multiplication of the form:

$$\overline{a}_{(i)} \cdot B = \begin{bmatrix} a_i^{(1)} & a_i^{(2)} & a_i^{(3)} \end{bmatrix} \begin{bmatrix} b_{11} & b_{12} & b_{13} \\ b_{21} & b_{22} & b_{23} \\ b_{31} & b_{32} & b_{33} \end{bmatrix}$$

Article Rosowski (2023) proposed to multiply it by the following 9 products:

$$\begin{cases} p_1(\overline{a}_{(i)}) = (a_i^{(2)} + b_{12})(a_i^{(1)} + b_{21}) \\ p_2(\overline{a}_{(i)}) = (a_i^{(3)} + b_{13})(a_i^{(1)} + b_{31}) \\ p_3(\overline{a}_{(i)}) = (a_i^{(3)} + b_{23})(a_i^{(2)} + b_{32}) \\ \\ p_4(\overline{a}_{(i)}) = a_i^{(1)}(b_{11} - b_{12} - b_{13} - a_i^{(2)} - a_i^{(3)}) \\ p_5(\overline{a}_{(i)}) = a_i^{(2)}(b_{22} - b_{21} - b_{23} - a_i^{(1)} - a_i^{(3)}) \\ p_6(\overline{a}_{(i)}) = a_i^{(3)}(b_{33} - b_{31} - b_{32} - a_i^{(1)} - a_i^{(2)}) \\ \cdots\cdots\cdots\cdots\cdots\cdots\cdots\cdots\cdots\cdots\cdots\cdots \\ p_7 = b_{12}b_{21} \\ p_8 = b_{13}b_{31} \\ p_9 = b_{23}b_{32} \end{cases}$$

The 9 products are the same amount that follows from the definition (1.1) for the dimensionality (1, 3, 3). However, we can observe that only the first six depend on the vector $\overline{a}^{(i)}$ (for clarity, we inserted additional line between 6 and 7th product). Thus, in the second step, we can smartly solve the (3, 3, 3)-dimensional problem as follows:

$$A \cdot B = \begin{bmatrix} \overline{a}_{(1)} \\ \hline \overline{a}_{(2)} \\ \hline \overline{a}_{(3)} \end{bmatrix} \cdot B$$

In fact, the obtained commutative algorithm for matrix multiplication with (3, 3, 3)-dimensionality requires $3 \cdot 6 + 3 = 21$ multiplication. It is so far the best known commutative algorithm for this problem.

In addition, the inner product based article by Waksman (1970) described in this section leads to 21 multiplications, but hides additional divisions by 2 and is commutative.

References

Brent R (1970) Error analysis of algorithms for matrix multiplication and triangular decomposition using Winograd's identity. Numer Math 16:145–156

Harter R (1972) The optimality of Winograd's formula. Commun ACM 5(15):352

Islam MN (2009) Optimization techniques for matrix multiplication. Master's thesis. University of Western, ON

Makarov OM (1986) An algorithm for multiplying 3 × 3 matrices. Comput Maths Math Phys 26(1):179–180

Pan V (1982) Fast matrix multiplication without APA-algorithms. Comput Math Appl 8(5):343–366

Rosowski A (2023) Fast commutative matrix algorithms. J Symb Comput 114:302–321

Waksman A (1970) On Winograd's algorithm for inner products. IEEE Trans Comput C19:360–361

Winograd S (1968) A new algorithm for inner product. IEEE Trans Comput 7:693–694

Chapter 3
The Exact Non-commutative Algorithms

The non-commutative class of the algorithms allows the elements to be multiplied to belong to a non-commutative field. This requirement enables to use the non-commutative matrix multiplication algorithms in a recursive way, which supports to apply them to matrices of arbitrary size. The necessity of distinguishing a separate class of exact algorithms will become clear in the next chapter, devoted to so-called arbitrary precision algorithms.

The chapter is organized as follows:

- In Sect. 3.1 we give the classical Strassen's result, show in detail one of the possible ways to derive it, and discuss some practical aspects of the algorithm.
- In Sect. 3.2 we presented Winograd's modification of Strassen's algorithm and reviewed some theoretical properties of the problem. In particular, we surveyed the proven theoretical bounds of this problem, i.e. the minimum number of multiplications and additions, also in the approximate version of the algorithm.
- In Sect. 3.3 we presented the generalisation of Strassen's algorithm to an arbitrary dimensionality of the problem, which reduces the definitional cubic number of non-scalar multiplications by a linear term.
- In Sect. 3.4 we introduced one of the general tools for constructing non-commutative matrix multiplication algorithms, i.e. the Brent Diophantine equations method. The Strassen algorithm is given as an example of the application of this method.
- In Sect. 3.5 we returned to the 3×3 problem, now in a non-commutative version. The classical Laderman's algorithm, which works in 23 non-scalar multiplication, is presented and discussed, also in relation to the Brent method.
- The Sect. 3.6 is devoted to a further generalisation of Strassen's algorithm, which reduces the definitional number of non-scalar multiplications by a quadratic term. The Laderman result for the 3×3 problem is given as a special case of the generalisation.
- Section 3.7 deals with the so-called duality of matrix multiplication. In short, the results contained therein show how to easy obtain 5 remaining permutations of

J. S. Respondek, *Fast Matrix Multiplication with Applications*, Studies in Big Data 166,
https://doi.org/10.1007/978-3-031-76930-6_3

the (k, m, n) dimensional problem, having an existing algorithm multiplying that problem in r steps $(r < kmn)$.

- Section 3.8 introduces some new generic methods for constructing fast matrix multiplication algorithms.
- Section 3.9 describes the classic so-called trilinear identity algorithm. Importantly, this algorithm leads to a worse efficiency exponent than the (earlier) Strassen's algorithm, but presents methodology which is later used in the construction of faster algorithms, in particular the so-called double disjoint matrix multiplication algorithm.
- Section 3.10 is devoted to another concept of double disjoint matrix multiplication. It is a special case of the general s-disjoint matrix multiplication. Simply, it allows more than one matrix multiplication to be performed at once, at a lower cost than performing these products separately.

In the meantime, this chapter has introduced some theoretical concepts and facts about fast matrix multiplication:

- **Scalar and non-scalar multiplications**: Sect. 3.4 explains the difference between these notions.
- **Trilinear representation** of the non-commutative fast matrix multiplication algorithm: Sect. 3.9.
- **Memory complexity** of non-commutative algorithms: Sect. 3.9.

3.1 The First Attempt to Improve the Asymptotic Efficiency

The first result that improved the asymptotic efficiency of matrix multiplication was proposed in 1969 by Volker Strassen in the famous article Strassen (1969), where an algorithm with $O(n^{2.807})$ time complexity was presented. Strassen achieved this result thanks to recursion, which potentially offers great possibilities for improving the efficiency of matrix multiplication. The Strassen (1969) article initiated a long series of works with a common general aim, i.e. to decrease the ω exponent in the time complexity of the algorithm $O(n^{\omega}) = O(M(n))$.

Let C denote the matrix product of $n \times n$ matrices A, B. We divide each of the matrix factors A, B into four submatrices, each with the dimensions of $(n/2) \times (n/2)$, as in the following formula

$$\left[\begin{array}{c|c} C_{11} & C_{12} \\ \hline C_{21} & C_{22} \end{array}\right]_C = \left[\begin{array}{c|c} A_{11} & A_{12} \\ \hline A_{21} & A_{22} \end{array}\right]_A \cdot \left[\begin{array}{c|c} B_{11} & B_{12} \\ \hline B_{21} & B_{22} \end{array}\right]_B \tag{3.1}$$

In general, the particular algorithm constructed by the recursive paradigm, that requires r multiplications and b additions, has a time complexity that can be calculated from the recursive equation of the form

$$T(n) = rT\left(\frac{n}{2}\right) + b\left(\frac{n}{2}\right)^2$$

which solution is $T(n) = O\left(n^{\log_2 r}\right) = O(n^{\omega})$ for $n^2 < n^{\omega} \leq n^3$.

Unfortunately, not every algorithm constructed by this paradigm—even if mathematically correct—guarantees better efficiency than the standard algorithm with cubic time complexity. Let us directly extend the block matrix equality (3.1) [e.g. Gantmacher (1960)[1]]:

$$\left[\begin{array}{c|c} C_{11} & C_{12} \\ \hline C_{21} & C_{22} \end{array}\right]_C = \left[\begin{array}{c|c} A_{11}B_{11} + A_{12}B_{21} & A_{11}B_{12} + A_{12}B_{22} \\ \hline A_{21}B_{11} + A_{22}B_{21} & A_{21}B_{12} + A_{22}B_{22} \end{array}\right] \tag{3.2}$$

We can see that the computation of the auxiliary matrix sub-expressions in (3.2) requires exactly 8 half-size matrix multiplications, giving the recursive equation of the form $T(n) = 8T(n/2) + 4(n/2)^2$. According to the well-known general recursion theorem (Cormen et al. 2001, Sect. 4.3), this equation leads to the algorithm with a time complexity equal to $O\left(n^{\log_2 8}\right) = O\left(n^3\right)$.

We can see that if we want to improve the time efficiency of matrix multiplication by the recursive paradigm we have to find a linear combination of sub-expressions with a number of half-size matrix multiplications smaller than 8. Nevertheless, the number of matrix additions can be arbitrary and, in practice, can be (and is) much larger than 4, not as in formula (3.2).

Basically, in all those algorithms constructed in the general recursive way, the computational work is shifted from the time-consuming matrix multiplications to the much less costly additions[2] and—if necessary—to the transpositions, both of which require only $O(n^2)$ operations.

The Strassen algorithm, as an auxiliary sub-expression, proposes to compute the following 7 matrix terms $m_1, \ldots, m_7$, each of which requires exactly one matrix multiplication:

$$\begin{cases} m_1 = (A_{11} + A_{22})(B_{11} + B_{22}) \\ m_2 = (A_{21} + A_{22})B_{11} \\ m_3 = A_{11}(B_{12} - B_{22}) \\ m_4 = A_{22}(B_{21} - B_{11}) \\ m_5 = (A_{11} + A_{12})B_{22} \\ m_6 = (A_{21} - A_{11})(B_{11} + B_{12}) \\ m_7 = (A_{12} - A_{22})(B_{21} + B_{22}) \end{cases} \tag{3.3}$$

[1] Other appreciated and renowned works on various properties and problems of block matrices are chapter XIV of the monograph Pease (1965) and the articles Egervary (1954), Friedman (1961), Silvester (2000) (together with references therein) and even chapter V of the monograph Scott (1880).

[2] Throughout the monograph, in the context of time complexity, we count a subtraction as an addition.

Formula (3.4) gives the final solution to the problem, i.e. appropriate linear combinations:

$$\left[\begin{array}{c|c} C_{11} & C_{12} \\ \hline C_{21} & C_{22} \end{array}\right]_C = \left[\begin{array}{c|c} m_1 + m_4 - m_5 + m_7 & m_3 + m_5 \\ \hline m_2 + m_4 & m_1 - m_2 + m_3 + m_6 \end{array}\right] \tag{3.4}$$

We leave it as an exercise for the interested reader to check how, with the use of the matrix terms m. Equation (3.3), it is possible to construct the matrix product, as in formula (3.4).

We can see that we now need 7 half-size matrix multiplications and 18 (4.5 times more!) matrix additions, and the recursive complexity equation receives the form:

$$T(n) = 7T\left(\frac{n}{2}\right) + 18\left(\frac{n}{2}\right)^2$$

In the general recursion theorem we have the case where the "recursive term" dominates over a non-recursive function and the algorithm time complexity is equal to $O\left(n^{\log_2 7}\right) \approx O\left(n^{2.807}\right)$ (Cormen et al. 2001, Sect. 4.3).

An attentive reader will notice that the application of the recursive paradigm requires as the input data square matrices whose dimensions are the natural powers of two.

But this is not a problem; in the case of other matrix dimensions we can simply extend the matrix dimensions to the nearest natural power of 2, filling the new entries by zeros. Next, from the resulting matrix product, we can pick the upper left submatrix containing the desired matrix product.

In summary, the innovative idea of Strassen's algorithm consists of the following elements:

- calculate a certain number of auxiliary expressions from submatrices, in particular using matrix multiplication (but on half-sized matrices),
- construct a linear combination of the obtained sub-expression in such a way which ensures that the product entries fulfil the mathematical definition formula,
- apply this paradigm recursively.

Interestingly, it is not known which method Strassen used to obtain his famous formulae (3.3). One of the possible ways has been suggested by Gastinel (1971), pages 224–225. In general, we tend to aggregate additions/subtractions to a common product. Gastinel proposed to start from the aggregated product of main diagonal entries in one of the following forms

$$m_{1'} = (A_{11} - A_{22}\Theta') \cdot (B_{11} - \Theta B_{22}) \tag{3.5}$$

for certain square parameter matrices Θ, Θ'. We can observe, that in the target formula (3.2) the "mixed" products $A_{11}B_{22}$ and $A_{22}B_{11}$ do not exist, while the factors $A_{11}B_{11}$ and $A_{22}B_{22}$ are in expressions for diagonal elements of the AB product.

So now the idea is to use the product $m_1\prime$ in the construction of the linear combination for C_{11} and C_{22} to construct the non-diagonal elements C_{12}, C_{21} in the form of such an aggregated products, which will then allow us to cancel the mixed products $A_{11}B_{22}$, $A_{22}B_{11}$. Observe:

$$\begin{cases} C_{12} = A_{11}(B_{12} + \Theta B_{22}) + (A_{12} - A_{11}\Theta)B_{22} \\ C_{21} = A_{22}(B_{21} + \Theta' B_{11}) + (A_{21} - A_{22}\Theta')B_{11} \end{cases} \tag{3.6}$$

The respective products we denote as:

$$\begin{cases} m_3\prime = A_{11}(B_{12} + \Theta B_{22}) \\ m_5\prime = (A_{12} - A_{11}\Theta)B_{22} \\ m_4\prime = A_{22}(B_{21} + \Theta' B_{11}) \\ m_2\prime = (A_{21} - A_{22}\Theta')B_{11} \end{cases} \tag{3.7}$$

Constructing the formula for C_{11} entry we start by correcting the product $m_1\prime$ (3.5) gradually in a way, that will finally give the definitional $A_{11}B_{11} + A_{12}B_{21}$. The $m_1\prime$ product (3.5) contains four products of the general form $A_{ii} \cdot X \cdot B_{jj}$ from which we need $A_{11}B_{11}$ for C_{11}, and $A_{22}B_{22}$ for C_{22}, respectively. Thus, in C_{11} it is sufficient to add to $m_1\prime$ the missing term $A_{12}B_{21}$ and to subtract all the terms but $A_{11}B_{11}$. Introducing the additional condition $\Theta\Theta' = I$ into the parameter matrix, we obtain:

$$C_{11} = (m_1\prime + A_{12}B_{21}) + (-A_{22}B_{22} + \boxed{A_{22}\Theta' B_{11}} + \boxed{A_{11}\Theta B_{22}})$$

For clarity, we have highlighted the unnecessary "mixed" products. Similarly, we construct the second diagonal entry:

$$C_{22} = (m_1\prime + A_{21}B_{12}) + (-A_{11}B_{11} + \boxed{A_{22}\Theta' B_{11}} + \boxed{A_{11}\Theta B_{22}})$$

The next step is to cancel the terms $A_{22}\Theta' B_{11}$ and $A_{11}\Theta B_{22}$ by products $m_4\prime$, $m_3\prime$ (3.7), respectively. In parenthesis we have grouped those products that come directly from the previous form of C_{11} and C_{22}.

$$\begin{cases} C_{11} = m_1\prime + m_4\prime - A_{22}B_{21} + (A_{12}B_{21} - A_{22}B_{22} + \boxed{A_{11}\Theta B_{22}}) \\ C_{22} = m_1\prime + m_3\prime - A_{11}B_{12} + (A_{21}B_{12} - A_{11}B_{11} + \boxed{A_{22}\Theta' B_{11}}) \end{cases}$$

The last thing to do is to get rid of the last pair of "mixed" terms $A_{11}\Theta B_{22}$ and $A_{22}\Theta' B_{11}$. We do this with the use of $m_5\prime$, $m_2\prime$. Also in parentheses are the products of the previous form of C_{11} and C_{22}.

$$\begin{cases} C_{11} = m_1\prime + m_4\prime - m_5\prime + A_{12}B_{22} + (-A_{22}B_{21} + A_{12}B_{21} - A_{22}B_{22}) \\ C_{22} = m_1\prime + m_3\prime - m_2\prime + A_{21}B_{11} + (-A_{11}B_{12} + A_{21}B_{12} - A_{11}B_{11}) \end{cases}$$

We can see that there are now no unwanted mixed terms $A_{ii} \cdot X \cdot B_{jj}$ for $i \neq j$ in the last formula expressing C_{11} and C_{22}. Now we can see that each of the four products can be aggregated into a single product.

$$\begin{cases} C_{11} = m_{1'} + m_{4'} - m_{5'} + (A_{12} - A_{22})(B_{21} + B_{22}) \\ C_{22} = m_{1'} + m_{3'} - m_{2'} + (A_{21} - A_{11})(B_{11} + B_{12}) \end{cases}$$

The formula obtained together with (3.6) for C_{12} and C_{21} is equivalent to the Strassen formulae [(3.3) and (3.4)], if we choose for the parameter matrix $\Theta = -I$.

In the rest of the article Gastinel (1971) generalised Strassen's results to arbitrary size matrices. More on this we give in Sect. 3.3.

Another way to obtain Strassen formulae is proposed in the monograph Cormen et al. (2001) in Section 28.2.[3] Yet another ways of finding the Strassen formulae is suggested by Brent (1970), pages 6–9, Yuval (1978) and Paterson (1974) in Sect. 2, pages 484–485. Nevertheless, the Strassen formulae are still easy to check manually.

In summary, the Strassen algorithm has reduced the exponent to 2.807, but at the cost of increasing the number of matrix additions to 18. In the next section we will show how to reduce this number without affecting the exponent. Moreover, we now have access to a number of algorithms that solve this problem in different ways, but using higher base matrix dimensions. We could have chosen only a part of them to describe in this monograph.

As can be easily predicted, Strassen's algorithm also introduces some new problems in order to achieve its effectiveness. Most of them have been at least mitigated or solved by later research. We have listed them below, together with the works that deal with each problem.

- The constant coefficients hidden in the $O(\cdot)$ complexity is higher than in the definitional algorithm, causing that the Strassen's algorithm is usable only for a matrices larger than 2×2. Indeed, there have been works determining the boundary dimension when the Strassen's algorithms become faster:
  - Higham (1990) proposed to select $n = 8$. D'Alberto and Niculau (2007) proposed a software that adaptively determines the threshold for a given computer system.
  - Huang et al. (2016) dealt with most of—as they called—'conventional wisdoms' of the Strassen's algorithm and showed that it can be faster even for small matrices.
- The matrices to be multiplied should possibly be square in order to outperform the standard algorithm. Again, the work of Huang et al. (2016) broke this traditional claim.
- In the same work, Huang et al. (2016) showed that Strassen algorithm can be implemented in a way that does not require more work space than BLAS. Jianyu et al. (2020) successfully implemented Strassen's algorithm on GPU.

[3] In the 3rd edition of this book from 2009, this lecture is no longer available.

- Huang et al. (2016) also demonstrated speedup and practicality on:
  - multi-core architectures,
  - on distributed memory implementations.
- The numerical stability of Strassen is weaker than that of the standard algorithm.
  - An analysis by Higham (1990) showed that the loss of accuracy was exaggerated.
  - Another analysis of the numerical stability was carried out by Bini and Lotti (1980). It concluded, that the Strassen's construction is more stable than the Winograd variant of the algorithm (to be presented in the next section) and that the algorithms with coefficients in a set $\{-1, 0, 1\}$ are more stable than the others, e.g. Brockett and Dobkin (1976, 1978).

3.2 The Modified Strassen Algorithm

Article Winograd (1973) modified the Strassen algorithm (1969), reducing the number of additions significantly. The Strassen algorithm works in two steps: aggregate the multiplications into 7 auxiliary terms, then obtain a desired product by constructing a smart linear combination of them.

The Winograd (1973) algorithm works in four steps:

- First, we calculate 3 series of auxiliary terms:
  - the first series of auxiliary terms contains of 8 sums $S_.$, each of which is a single addition,
  - the second series calculates 7 aggregated products $M_.$, using already calculated sums $S_.$,
  - find the last series of 3 auxiliary terms $T_.$ with the use of $M_.$ and also a single addition.
- Finally, we linearly combine $T_.$ and $M_.$ term to obtain the expressions, that are equivalent to the definition of the matrix product.

The formulae for the auxiliary variables proposed by Winograd have the form:

$$\text{step 1}: \begin{cases} S_1 = A_{21} + A_{22} \\ S_2 = S_1 - A_{11} \\ S_3 = A_{11} - A_{21} \\ S_4 = A_{12} - S_2 \\ S_5 = B_{12} - B_{11} \\ S_6 = B_{22} - S_5 \\ S_7 = B_{22} - B_{12} \\ S_8 = S_6 - B_{21} \end{cases} \quad \text{step 2}: \begin{cases} M_1 = S_2 S_6 \\ M_2 = A_{11} B_{11} \\ M_3 = A_{12} B_{21} \\ M_4 = S_3 S_7 \\ M_5 = S_1 S_5 \\ M_6 = S_4 B_{22} \\ M_7 = A_{22} S_8 \end{cases} \quad \text{step 3}: \begin{cases} T_1 = M_1 + M_2 \\ T_2 = T_1 + M_4 \\ T_3 = T_1 + M_5 \end{cases} \tag{3.8}$$

The final product has the form:

$$\left[\begin{array}{c|c} M_2 + M_3 & T_3 + M_6 \\ \hline T_2 - M_7 & T_2 + M_5 \end{array}\right]$$

We see that the number of additions is 15, compared to Strassen (1969) 18, and the number of multiplications remains the same, i.e. 7. Thus the asymptotic complexity of the Winograd (1973) algorithm is the same as that of the Strassen (1969), but the constant factors hidden in the $O(\cdot)$ complexity are smaller.

A look at the formula (3.8) expressing the auxiliary terms reveals that, unlike the Strassen algorithm (3.3), the order in which the terms (3.8) are computed is important. This must be taken into account in parallel implementations of Winograd's algorithm.

- Yet another form of Strassen's algorithm was given by Paterson (1974), Sect. 2 page 484.
- Fischer (1974) proposed an optimal implementation of the Winograd (1973) with the smallest possible constant coefficient hidden in the "big O" notation.
- Probert (1976) and independently Bshouty (1995) showed that at least 15 additions are required for a 2×2 recursive scheme, which is the number of additions achieved by the Winograd (1973) algorithm.
- Yet another scheme for the 2×2 case is given by Waksman (1970), but in a commutative way and requiring additional divisions by 2 (already mentioned in Sect. 2.3).

One might ask whether there is any room for improvement in this case. The answer is given in depth by three articles:

- Hopcroft and Kerr (1969) proved that at least just 7 multiplications are required to multiply 2×2 matrices in any non-commutative algorithm.
- Winograd (1971) proved that even using the commutativity of the matrix entries, the minimum multiplication number is still 7.
- Landsberg (2005) proved that for 2×2 matrices it is not possible to construct so-called approximate algorithm with less non-scalar multiplications than the exact one algorithm, i.e. with less than 7. Another proof of this, but using different methods, is given in article Hauenstein, Ikenmeyer et al. (2013). More about approximate algorithms we give in a separate Chap. 5.

Thus, 2×2 dimensionality is the case when the minimum number of multiplications required for the commutative and non-commutative cases is the same. In the next section we present a generalisation of the Strassen algorithm.

3.3 The Generalization of Strassen Algorithm

After the appearance of Strassen's ingenious discovery, its generalisation seemed a natural way to further improve the efficiency of matrix multiplication, which was successfully realised by Gastinel (1971).

Gastinel proposed to calculate the following three series of auxiliary variables u_{ij}, v_{ij}, w_{ij}:

$$u_{ij} = (a_{ij} + a_{ii})b_{jj}, \quad v_{ij} = a_{ii}(b_{ij} - b_{jj}), \quad w_{ij} = (a_{ij} - a_{jj}) \sum_{s=1}^{n} b_{js}$$

for $i, j = 1, \ldots, n, \quad i \neq j$ and the last series $p_{ijl} = a_{il}b_{lj}$ for pairwise distinct i, j, l, complemented by a variable q aggregating all the diagonal elements of the matrices to be multiplied:

$$q = \left(\sum_{s=1}^{n} a_{ss}\right)\left(\sum_{s=1}^{n} b_{ss}\right) \tag{3.9}$$

We start by constructing the matrix product from the diagonal elements. The ith diagonal element should have the definitional form (3.10)

$$c_{ii} = a_{i1}b_{1i} + \cdots + (a_{ii}b_{ii}) + \cdots + a_{in}b_{ni}, \quad i = 1, \ldots \ n \tag{3.10}$$

We start the construction by expanding the q term (3.9). Observe:

$$\begin{aligned} q = & \boxed{a_{11}b_{11}} + \ldots + a_{11}b_{ii} + \ldots + a_{11}b_{nn} + \\ & \vdots \\ & + a_{ii}b_{11} + \ldots + \underline{|a_{ii}b_{ii}} + \ldots + a_{ii}b_{nn} + \\ & \vdots \\ & + a_{nn}b_{11} + \ldots + a_{nn}b_{ii} + \ldots + \boxed{a_{nn}b_{nn}} \end{aligned} \tag{3.11}$$

In q only the term $a_{ii}b_{ii}$ exists in the definition formula (3.10), so all the others must be corrected. We start it from properly chosen $w_{\cdot}$ terms. Observe

$$q + \sum_{\substack{s=1 \\ s \neq i}}^{n} w_{is} = q + \sum_{\substack{s=1 \\ s \neq i}}^{n} (\underline{|a_{is}} - \boxed{a_{ss}})(\underline{|b_{si}} + \boxed{b_{ss}} + \sum_{\substack{t=1 \\ t \neq i, t \neq s}}^{n} b_{st})$$

In the formula above, we can distinguish products made up of the following combinations of $a_{\cdot}$ and $b_{\cdot}$ terms:

- The pairs highlighted by $\underline{|a.}\cdot\underline{|b.}$ generate the desired value $\sum_{k=1,k\neq i}^{n} a_{ik}b_{ki}$, being the definition terms (3.10) that are missing in the formula (3.11) for q.
- Unnecessary pairs highlighted by $\underline{|a.}\cdot\boxed{b.}$ will be corrected by $-u.$ where:

$$u_{is} = (a_{ii} + a_{is})b_{ss}, \quad s = 1, \ldots, n, \quad s \neq i$$

By the way, using the above terms we can correct the unwanted products $a_{ii}b_{ss},\ i \neq s$ in q (3.11).

- The sum of the products $\underline{|a.}\cdot b.$ we correct directly by the auxiliary products p_{ijl}.
- Using the products $\boxed{a.}\cdot\boxed{b.}$ we correct the unwanted terms $a_{ss}b_{ss},\ s \neq i$ in q.
- The remaining pairs $-a_{ss}\sum_{t,\,t\neq s} b_{st},\quad s \neq i$ should be corrected, but now with a $+$ sign. We achieve this with $+v.$. Observe:

$$\begin{cases} v_{s1} = a_{ss}(b_{s1} - b_{11}) \\ \vdots \\ v_{s(s-1)} = a_{ss}\big(b_{s(s-1)} - b_{(s-1)(s-1)}\big) & s = 1, \ldots, n \\ v_{s(s+1)} = a_{ss}\big(b_{s(s+1)} - b_{(s+1)(s+1)}\big) & s \neq i \\ \vdots \\ v_{sn} = a_s(b_{sn} - b_{nn}) \end{cases}$$

At the same time we correct the last unwanted terms $a_{ss}b_{tt},\ s \neq i,\ t \neq s$ in q (3.11). Combining all the bullets together, we can obtain the desired formula for the i-th diagonal element of the matrix product:

$$\begin{aligned} c_{ii} &= q + \sum_{\substack{s=1\\ s\neq i}}^{n} w_{is} - \sum_{\substack{s=1\\ s\neq i}}^{n} u_{is} - \sum_{\substack{s=1\\ s\neq i}}^{n} \sum_{\substack{t=1\\ t\neq p,t\neq s}}^{n} p_{its} + \sum_{\substack{s=1\\ s\neq i}}^{n} \sum_{\substack{t=1\\ t\neq s}}^{n} v_{st} \\ &= q + \sum_{\substack{s=1\\ s\neq i}}^{n} \left(w_{is} - u_{is} - \sum_{\substack{t=1\\ t\neq i,t\neq s}}^{n} p_{its} + \sum_{\substack{t=1\\ t\neq s}}^{n} v_{st} \right) \end{aligned} \tag{3.12}$$

Let us now calculate the non-diagonal products. Although their number is approximately quadratic greater than the number of diagonal elements, the formulae expressing them are much less complex.

In the definitional form of the non-diagonal elements (1.1) for $i \neq j$ we can distinguish two groups of products:

- Terms with distinct all the $i,\ j,\ k$ indexes. There are exactly $n - 2$ such a terms.
- Terms, where $i = k$ or $k = j$.[4]

[4] The conjunction never happens, because for the non-diagonal elements $i \neq k$.

By grouping the definitional product terms according to the above rule we can rewrite (1.1) in the form:

$$c_{ij} = a_{ii}b_{ij} + a_{ij}b_{jj} + \sum_{\substack{k=1\\k\neq i,k\neq j}}^{n} a_{ik}b_{kj}, \quad i \neq j$$

The terms within the sum we can cover directly by the p_{ijk} product and the sum $a_{ii}b_{ij} + a_{ij}b_{jj}$ we can construct by the sum of properly chosen $u_{\cdot}$ and $v_{\cdot}$:

$$c_{ij} = u_{ij} + v_{ij} + \sum_{\substack{k=1\\k\neq i,k\neq j}}^{n} p_{ikj}, \quad i \neq j \tag{3.13}$$

Obviously, for $n = 2$ the method reduces to the Strassen's formulae (3.3).

All that remains is to count the number of multiplications of the non-commutative algorithm obtained. The number of variables u_{ij}, v_{ij}, w_{ij} is $3n(n-1)$, because the indices i, j are distinct. Similarly, the number of products $p_{ijl} = a_{il}b_{lj}$ is $n(n-1)(n-2)$ for distinct i, j, l. Now it is enough to add a single product q and the key issue in any fast matrix multiplication algorithm, i.e. the number of non-scalar[5] multiplications, is equal to:

$$3n(n-1) + n(n-1)(n-2) + 1 = n^3 - (n-1) \tag{3.14}$$

In summary, Gastinel reduced the number of necessary multiplications by a linear term $n - 1$. The advantage of the algorithm is undoubtedly its generality; Gastinel's formulae are applicable to any n.

3.3.1 Example—3 × 3 Case

In Sect. 2.2.1 we constructed a commutative algorithm for the 3×3 case. Now we want to use the Gastinel (1971) method to find a non-commutative algorithm for this size.

We construct the desired formula expressing the first diagonal element of the matrix product by using formula (3.12) with $i = 1$, $n = 3$:

$$\begin{aligned} c_{11} &= q + \sum_{\substack{s=1\\s\neq 1}}^{3} \left(w_{1s} - u_{1s} - \sum_{\substack{t=1\\t\neq 1,t\neq s}}^{3} p_{1ts} + \sum_{\substack{t=1\\t\neq s}}^{3} v_{st} \right) \\ &= q + (w_{12} - u_{12} - p_{132} + v_{21} + v_{23}) + (w_{13} - u_{13} - p_{123} + v_{31} + v_{32}) \end{aligned}$$

[5] The notion of non-scalar multiplications is described in Sect. 3.4.

$$= q + (w_{12} + w_{13}) - (u_{12} + u_{13}) - (p_{132} + p_{123}) + (v_{21} + v_{23} + v_{31} + v_{32})$$

Repeating the construction for the remaining two diagonal entries, we obtain:

$$\begin{cases} c_{22} = q + (w_{21} + w_{23}) - (u_{21} + u_{23}) - (p_{213} + p_{231}) + (v_{12} + v_{13} + v_{32} + v_{31}) \\ c_{33} = q + (w_{32} + w_{31}) - (u_{32} + u_{31}) - (p_{312} + p_{321}) + (v_{23} + v_{12} + v_{13} + v_{21}) \end{cases}$$

Let us now calculate the non-diagonal products. From (3.13) we obtain directly:

$$c_{12} = u_{12} + v_{12} + p_{132}$$

The remaining non-diagonal entries are constructed in the same way:

$$\begin{cases} c_{13} = u_{13} + v_{13} + p_{123} \\ c_{21} = u_{21} + v_{21} + p_{231} \\ c_{23} = u_{23} + v_{23} + p_{213} \\ c_{31} = u_{31} + v_{31} + p_{321} \\ c_{32} = u_{32} + v_{32} + p_{312} \end{cases}$$

We can see that for $n = 3$ number of multiplications is 25, which is worse than 24 obtained by the Winograd's method (Example 2.2.1), but in return we obtain the non-commutative formulae, thus applicable to matrices of arbitrary size by the recursion. Later, in Sect. 3.5, we show the best non-commutative result so far for 3×3 dimension with 23 multiplications.

3.4 The Brent Diophantine Equations Method

The Brent Diophantine equations method arose from the observation, that the Strassen's formulae (3.3) and (3.4) are essentially a linear combinations of aggregated products, with selected elements of the matrix multiplicand on the left, and of the multiplier on the right, respectively. Moreover, we can achieve this selection of elements simply by a linear combination. On this basis, Brent in his report (1970, Sect. 5) proposed to define r aggregated products in the general form:

$$m_p = \left(\sum_{i_1=1}^{m} \sum_{i_2=1}^{n} \alpha_{i_1 i_2}^{(p)} a_{i_1 i_2} \right) \left(\sum_{j_1=1}^{n} \sum_{j_2=1}^{l} \beta_{j_1 j_2}^{(p)} b_{j_1 j_2} \right), \quad p = 1, \ldots, r \tag{3.15}$$

where $a_{i_1 i_2}$ and $b_{j_1 j_2}$ are elements of the matrices A, B to be multiplied.

In general, we can allow the coefficients $\alpha_{i_1 i_2}^{(p)}$, $\beta_{j_1 j_2}^{(p)}$ to be real, rational or integer. The most desirable constructions are those that require all these coefficients to be from the set $\{-1, 0, 1\}$, such as the Strassen construction is. However, some valuable constructions, e.g. Brockett and Dobkin (1976, 1978), cited in 3.5, also require multiplication or division by two.

We state that the respective elements of the product $C = AB$ are the linear combinations of the aggregated products m_p (3.15):

$$c_{k_1k_2} = \sum_{p=1}^{r} \gamma_{k_1k_2}^{(p)} m_p \tag{3.16}$$

Probert (1974) gave a proof, that such a strategy is sufficient to obtain an optimal solution with respect to time.

Now is the right time to make a distinction between scalar multiplication and non-scalar multiplication in matrix algorithms. When a matrix variable is multiplied by another matrix variable, we speak of a non-scalar multiplication, and when a matrix variable is multiplied by a fixed constant, we speak of a scalar multiplication. In particular, in formula (3.15) the non-scalar multiplications are of the form $a_{i_1i_2} \cdot b_{j_1j_2}$ and the scalar multiplications are $\alpha_{i_1i_2}^{(p)} a_{i_1i_2}$ and $\beta_{j_1j_2}^{(p)} b_{j_1j_2}$. In the fast matrix multiplication algorithms we count only non-scalar multiplications, because only they impose they complexity exponent; when recursion is used, the matrix variables turn into matrices themselves and scalar multiplications are fast, performed in quadratic time, but the cost of non-scalar multiplications are of the $O(n^{2+\varepsilon})$ cost, for certain real positive ε. A more formal definition is given by Bini (1980) in Sect. 2.

Since the result of the sum (3.16) must comply with the definition $c_{k_1k_2} = \sum_k 1 \cdot (a_{k_1k} b_{kk_2})$ (where we have deliberately emphasised the "1" coefficient) in order to compare it, we must to regroup this sum with respect to the $a_{i_1i_2} b_{j_1j_2}$ terms. By inserting the aggregated products m_p (3.15) into (3.16) we obtain:

$$\begin{aligned} c_{k_1k_2} &= \sum_{p=1}^{r} \sum_{\substack{i_1=1,\ldots,m \\ i_2,j_1=1,\ldots,n \\ j_2=1,\ldots,l}} \gamma_{k_1k_2}^{(p)} \cdot \alpha_{i_1i_2}^{(p)} \beta_{j_1j_2}^{(p)} \cdot \left(a_{i_1i_2} b_{j_1j_2}\right) \\ &= \sum_{\substack{i_1=1,\ldots,m \\ i_2,j_1=1,\ldots,n \\ j_2=1,\ldots,l}} \left(a_{i_1i_2} b_{j_1j_2}\right) \sum_{p=1}^{r} \gamma_{k_1k_2}^{(p)} \cdot \alpha_{i_1i_2}^{(p)} \beta_{j_1j_2}^{(p)} \end{aligned} \tag{3.17}$$

From this form of the formula we can deduce that we must cancel all the $\sum_{p=1}^{r} \gamma_{k_1k_2}^{(p)} \cdot \alpha_{i_1i_2}^{(p)} \beta_{j_1j_2}^{(p)}$'s for $i_2 \neq j_1$, and otherwise make it equal to one. We can denote this rule consistently with the use of the Cronecker's symbol[6] as follows:

$$\sum_{p=1}^{r} \gamma_{k_1k_2}^{(p)} \cdot \alpha_{i_1i_2}^{(p)} \beta_{j_1j_2}^{(p)} = \delta_{i_2j_1} \cdot \delta_{k_1i_1} \delta_{k_2j_2} \tag{3.18}$$

[6] $\delta_{u,v} = 1$ for $u = v$ and $\delta_{u,v} = 0$ otherwise.

where $\delta_{k_1 i_1} \delta_{k_2 j_2}$ ensures the correct selection of those products, that correctly construct a given $c_{k_1 k_2}$ element of the resulting matrix.

3.4.1 Example—2 × 2 Case in the Strassen's Version

We illustrate the reasoning of Sect. 3.4 using the example of the 2×2 case in the Strassen version of the scheme. The p-th aggregated product (3.15) in the 2×2 case has the following form, for $p = 1, \ldots, 7$:

$$m_p = \left(\underline{\alpha_{11}^{(p)} a_{11}} + \overline{\alpha_{12}^{(p)} a_{12}} + \alpha_{21}^{(p)} a_{21} + \alpha_{22}^{(p)} a_{22} \right) \cdot \left(\underline{\beta_{11}^{(p)} b_{11}} + \beta_{12}^{(p)} b_{12} + \overline{\beta_{21}^{(p)} b_{21}} + \beta_{22}^{(p)} b_{22} \right) \tag{3.19}$$

In total, the product (3.19) gives 16 terms, but for each matrix element $c_{k_1 k_2}$ only 2 are desired. For example, we have highlighted the pairs corresponding to the c_{11} element, creating the definitional terms $a_{11} b_{11}$ and $a_{12} b_{21}$. The linear combination (3.16) for the c_{11} element, divided into two groups of elements, i.e. the definitional and the unwanted terms, has the form:

$$\begin{aligned} c_{11} &= \sum_{p=1}^{7} \left[\left(\gamma_{11}^{(p)} \alpha_{11}^{(p)} \beta_{11}^{(p)} \right) \cdot (a_{11} b_{11}) + \left(\gamma_{11}^{(p)} \alpha_{12}^{(p)} \beta_{21}^{(p)} \right) \cdot (a_{12} b_{21}) \right] \\ &+ \sum_{p=1}^{7} \sum_{\substack{i_1, i_2, j_1, j_2 \in \{1,2\} \\ i_2 \neq j_1 \\ i_1 \neq 1,\, j_2 \neq 1}} \gamma_{11}^{(p)} \alpha_{i_1 i_2}^{(p)} \beta_{j_1 j_2}^{(p)} \cdot \left(a_{i_1 i_2} b_{j_1 j_2} \right) \\ &= a_{11} b_{11} \sum_{p=1}^{7} \gamma_{11}^{(p)} \alpha_{11}^{(p)} \beta_{11}^{(p)} + a_{12} b_{21} \sum_{p=1}^{7} \gamma_{11}^{(p)} \alpha_{12}^{(p)} \beta_{21}^{(p)} \\ &+ \sum_{\substack{i_1, i_2, j_1, j_2 \in \{1,2\} \\ i_2 \neq j_1 \\ i_1 \neq 1,\, j_2 \neq 1}} \left(a_{i_1 i_2} b_{j_1 j_2} \right) \sum_{p=1}^{7} \gamma_{11}^{(p)} \alpha_{i_1 i_2}^{(p)} \beta_{j_1 j_2}^{(p)} \end{aligned} \tag{3.20}$$

Now it is quite clear that the, already rather complex, sum (3.20) gives the definition $c_{11} = a_{11} b_{11} + a_{12} b_{21}$ iff:

$$\sum_{p=1}^{7} \gamma_{11}^{(p)} \alpha_{11}^{(p)} \beta_{11}^{(p)} = \sum_{p=1}^{7} \gamma_{11}^{(p)} \alpha_{12}^{(p)} \beta_{21}^{(p)} = 1$$

and

$$\sum_{p=1}^{7} \gamma_{11}^{(p)} \alpha_{i_1 i_2}^{(p)} \beta_{j_1 j_2}^{(p)} = 0$$

for all the remaining 14 fours i_1, i_2, j_1, j_2. This result is simply equivalent to the Brent Diophantine Eq. (3.18) with $k_1 = k_2 = 1$. Analogous reasoning for the remaining elements of the matrix product leads to the rest of the $3 \cdot 16 = 48$ equations.

In summary, in the form of the aggregated products m_p (3.15) linearly combined in the sum (3.16) with $\gamma_{k_1 k_2}^{(p)}$ coefficients, the Strassen's algorithm can be represented as the following three series of seven 2×2 matrices:

$$A_{1 \le p \le 7}^{(p)} = \left(\begin{bmatrix} 1 & 0 \\ 0 & 1 \end{bmatrix}, \begin{bmatrix} 0 & 0 \\ 1 & 1 \end{bmatrix}, \begin{bmatrix} 1 & 0 \\ 0 & 0 \end{bmatrix}, \begin{bmatrix} 0 & 0 \\ 0 & 1 \end{bmatrix}, \begin{bmatrix} 1 & 1 \\ 0 & 0 \end{bmatrix}, \begin{bmatrix} -1 & 0 \\ 1 & 0 \end{bmatrix}, \begin{bmatrix} 0 & 1 \\ 0 & -1 \end{bmatrix} \right)$$

$$B_{1 \le p \le 7}^{(p)} = \left(\begin{bmatrix} 1 & 0 \\ 0 & 1 \end{bmatrix}, \begin{bmatrix} 1 & 0 \\ 0 & 0 \end{bmatrix}, \begin{bmatrix} 0 & 1 \\ 0 & -1 \end{bmatrix}, \begin{bmatrix} -1 & 0 \\ 1 & 0 \end{bmatrix}, \begin{bmatrix} 0 & 0 \\ 0 & 1 \end{bmatrix}, \begin{bmatrix} 1 & 1 \\ 0 & 0 \end{bmatrix}, \begin{bmatrix} 0 & 0 \\ 1 & 1 \end{bmatrix} \right)$$

$$\Gamma_{1 \le p \le 7}^{(p)} = \left(\begin{bmatrix} 1 & 0 \\ 0 & 1 \end{bmatrix}, \begin{bmatrix} 0 & 0 \\ 1 & -1 \end{bmatrix}, \begin{bmatrix} 0 & 1 \\ 0 & 1 \end{bmatrix}, \begin{bmatrix} 1 & 0 \\ 1 & 0 \end{bmatrix}, \begin{bmatrix} -1 & 1 \\ 0 & 0 \end{bmatrix}, \begin{bmatrix} 0 & 0 \\ 0 & 1 \end{bmatrix}, \begin{bmatrix} 1 & 0 \\ 0 & 0 \end{bmatrix} \right)$$

It is one thing to derive the Brent equations, but quite another to solve them. This is due to the galactic scale of the problem, which makes it practically impossible to solve in realistic time even for small sizes. In fact, in Brent's equation, the indices range over integers $1 \le i_1, k_1 \le m$, $1 \le i_2, j_1 \le n$ and $1 \le j_2, k_2 \le l$, so in total it consists of $(mnl)^2$ equations with $r(ml + mn + nl)$ unknowns, because we have r aggregated products m_p (3.15). So for square matrices of the $n \times n$ dimension the Brent's equation (3.18) consists of n^6 with $r \cdot 3n^2$ unknowns. In particular for $n = 2$ we have $2^6 = 64$ equations with $3 \cdot 2^2 \cdot 7 = 84$ unknowns, for scheme with $r = 7$ multiplications, and for $n = 3$ we already have as much as $3^6 = 729$ equations with $3 \cdot 3^2 \cdot r = 27r$ unknowns.

Heule et al. (2021) propose to solve the Brent's equation (3.18) using the so-called Gröbner bases (Buchberger 1965; Cox et al. 1992). Smirnov (2013) solved the Brent's equation numerically for some cases. In the next section we show the famous Laderman's result for the 3×3 case, which is claimed to be obtained just by solving the Brent's equation.

3.5 Non-commutative Results for the 3 × 3 Case Problem

For the 3×3 dimension case we have already presented in Sect. 2.2.1 the methods by 24 multiplications in commutative version by Winograd (1968) and by 25, but non-commutative in return, by Gastinel (1971) in Sect. 3.3.1. Then Hopcroft and Kerr (1971) described the method of obtaining a non-commutative algorithm for this case with 24 multiplications. On this basis Hopcroft and Musinski (1973) constructed such an algorithm. Brockett and Dobkin (1976, 1978) obtained a commutative algorithm with 23 multiplications, but their algorithm requires additional multiplications and divisions by 2.

In 1976, Laderman gave explicit formulae for the non-commutative algorithm for size, **which is the fastest known non-commutative result for this dimension**, without any hidden multiplications by constants, i.e. with all coefficients in the respective linear combinations taken from the set $\{-1, 0, 1\}$. In Laderman's construction, the 23 aggregated products have the following form:

$$
\begin{cases}
m_1 = (a_{11} + a_{12} + a_{13} - a_{21} - a_{22} - a_{32} - a_{33})b_{22} \\
m_2 = (a_{11} - a_{21})(-b_{12} + b_{22}) \\
m_3 = a_{22}(-b_{11} + b_{12} + b_{21} - b_{22} - b_{23} - b_{31} + b_{33}) \\
m_4 = (-a_{11} + a_{21} + a_{22})(b_{11} - b_{12} + b_{22}) \\
m_5 = (a_{21} + a_{22})(-b_{11} + b_{12}) \\
m_6 = a_{11}b_{11} \\
m_7 = (-a_{11} + a_{31} + a_{32})(b_{11} - b_{13} + b_{23}) \\
m_8 = (-a_{11} + a_{31})(b_{13} - b_{23}) \\
m_9 = (a_{31} + a_{32})(-b_{11} + b_{13}) \\
m_{10} = (a_{11} + a_{12} + a_{13} - a_{22} - a_{23} - a_{31} - a_{32})b_{23} \\
m_{11} = a_{32}(-b_{11} + b_{13} + b_{21} - b_{22} - b_{23} - b_{31} + b_{32}) \\
m_{12} = (-a_{13} + a_{32} + a_{33})(b_{22} + b_{31} - b_{32}) \\
m_{13} = (a_{13} - a_{33})(b_{22} - b_{32}) \\
m_{14} = a_{13}b_{31} \\
m_{15} = (a_{32} + a_{33})(-b_{31} + b_{32}) \\
m_{16} = (-a_{13} + a_{22} + a_{23})(b_{23} + b_{31} - b_{33}) \\
m_{17} = (a_{13} - a_{23})(b_{23} - b_{33}) \\
m_{18} = (a_{22} + a_{23})(-b_{31} + b_{33}) \\
m_{19} = a_{12}b_{21} \\
m_{20} = a_{23}b_{32} \\
m_{21} = a_{21}b_{13} \\
m_{22} = a_{31}b_{12} \\
m_{23} = a_{33}b_{33}
\end{cases}
\tag{3.21}
$$

Next, we combine them linearly in the following way:

$$
\begin{cases}
c_{11} = m_6 + m_{14} + m_{19} \\
c_{12} = m_1 + m_4 + m_5 + m_6 + m_{12} + m_{14} + m_{15} \\
c_{13} = m_6 + m_7 + m_9 + m_{10} + m_{14} + m_{16} + m_{18} \\
\\
c_{21} = m_2 + m_3 + m_4 + m_6 + m_{14} + m_{16} + m_{17} \\
c_{22} = m_2 + m_4 + m_5 + m_6 + m_{20} \\
c_{23} = m_{14} + m_{16} + m_{17} + m_{18} + m_{21} \\
\\
c_{31} = m_6 + m_7 + m_8 + m_{11} + m_{12} + m_{13} + m_{14} \\
c_{32} = m_{12} + m_{13} + m_{14} + m_{15} + m_{22} \\
c_{33} = m_6 + m_7 + m_8 + m_9 + m_{23}
\end{cases}
$$

Verification of the correctness of the result is straightforward, simply by substituting the relevant $c_{\cdot}$ products into the matrix product definition and comparing them.

To obtain the scheme from the Brent's equations, it is necessary to solve 729 equations with $27 \cdot 13 = 621$ unknowns. It is still not known how Laderman solved this set of 729 equations. In his paper from 1976, he claimed that he did not use computers, but instead combined the solution of four smaller systems into one. He also promised to present in detail his method for solving the Brent equations for the 3×3 case in another article, but he never fulfilled this promise.

Sedoglavic (2017) proposed how to obtain Laderman's scheme from the Strassen's scheme.

To improve the Strassen's result on the 3×3 "base" dimension, we need a non-commutative algorithm with at most 21 multiplications, because $\log_3 21 < \log_2 7 < \log_3 22$. Pan (1972) proved in Theorem 1, point 6, on page 249, that at least 18 multiplications are needed for the 3×3 non-commutative matrix multiplication problem. Later Blaser (2003) increased this lower bound to 19 multiplications. Unfortunately, no scheme with even 22 or—let alone—less multiplications is known, despite many efforts to find such a scheme.

Johnson and McLoughlin (1986) proposed two alternative families of non-commutative algorithms for 3×3 matrix multiplication. However, they depend on one or two parameters and contain rational coefficients. Heule and Kauers et al. (2021) tackled the 3×3 problem using so-called Boolean satisfiability solvers (SAT). Using severe (at the time of research) computer hardware resources (a few hundred of cores and over a terabyte of operating memory), they obtained over 17,000 alternative schemes with 23 multiplications, but no scheme with fewer multiplications was found. This shows how hard the problem of matrix multiplication really is.

In the next section we present a non-commutative algorithm formulated for an arbitrary size square matrices, for which the Winograd's and Laderman's formulae are special cases. We shall return to the 3×3 case in Chap. 5, in the context of the so-called arbitrary precision algorithms.

3.6 The Sykora's Algorithm

To formulate his algorithm in 1977, Sykora first defined the so-called modified Latin square as follows:

- Let $N = \{1, 2, \ldots, n\}$ and $i, j, s \in N$.
- A modified Latin square,[7] denoted by L_{nijs}, is a $n \times n$ matrix constructed according to the following rules:
  - in the ith row and jth column it contains only the number s,
  - every other row and column contains all the elements of N.

In addition, we define two sets: $S = N - \{i\}$, $T = N - \{j\}$.

Sykora divided the elements of the desired product into three groups according to the way in which the products were calculated.

I. The element c_{ij} is calculated from the definition, i.e. $c_{ij} = \sum_{k=1}^{n} P_k$ where $P_k = a_{ik} b_{kj}$.

II. The elements of the matrix $C = AB$, with the exception of the ith row and jth column, are calculated as follows:

- Denote:

$$\begin{cases} M_{qr1} = \left(-a_{i,m_{qr}} + a_{q,m_{qr}} + a_{qs}\right)\left(b_{m_{qr},j} + b_{sr} - b_{m_{qr},r}\right) \\ M_{qr2} = \left(a_{q,m_{qr}} + a_{qs}\right)\left(b_{m_{qr},r} - b_{m_{qr},j}\right) \\ M_{qr3} = \left(a_{i,m_{qr}} - a_{q,m_{qr}}\right)\left(b_{sr} - b_{m_{qr},r}\right) \end{cases} \tag{3.22}$$

where $m_{qr} = L_{nijs}[q; r]$ is the element lying on the intersection of the qth row and the rth column of the modified Latin square, $q \in S$, $r \in T$.

- Moreover:

$$H_{qrp} = a_{qp} b_{pr}, \quad \text{where } p \in N,\ p \neq m_{qr},\ p \neq s \tag{3.23}$$

Then:

$$c_{qr}\big|_{(q,r)\neq(i,j)} = P_{m_{qr}} + \sum_{k=1}^{3} M_{qrk} + \sum_{\substack{p=1 \\ p\neq s,\ p\neq m_{qr}}}^{n} H_{qrp} \tag{3.24}$$

III. The elements of the ith row and jth column of the product, except for the element c_{ij} (already calculated in the first group), are calculated in two steps:

- Let us denote for c_{ir}, $r \in T$ (for the ith row):

[7] More information about the Latin squares is given in the following part of the section.

$$R_r|_{r\neq j} = \left[\sum_{k=1}^{n} a_{ik} - \sum_{\substack{q=1\\q\neq i}}^{n} \left(a_{qs} + a_{q,m_{qr}}\right)\right] b_{sr}, \quad r \in T \tag{3.25}$$

then

$$c_{ir}|_{r\neq j} = R_r + \sum_{\substack{q=1\\q\neq i}}^{n} \left(M_{qr1} + M_{qr2} + P_{m_{qr}}\right) \tag{3.26}$$

- For c_{qj}, $q \in S$ (for the jth column):

$$Q_q\big|_{q\neq i} = a_{qs}\left[b_{sj} - \sum_{\substack{k=1\\k\neq s}}^{n} b_{kj} + \sum_{\substack{r=1\\r\neq j}}^{n} \left(b_{m_{qr},r} - b_{sr}\right)\right], \quad q \in S \tag{3.27}$$

then

$$c_{qj}\big|_{q\neq i} = Q_q + \sum_{\substack{r=1\\r\neq i}}^{n} \left(M_{qr1} + M_{qr3} + P_{m_{qr}}\right) \tag{3.28}$$

The structure of the P_k and H_{qrp} quantities is analogous, but we distinguish them, because the former refers to the indices on the ith row and jth column, and the latter to the rest.

The proof for the group I, i.e. for the element beyond the ith row and jth column, is straightforward; after expanding all the parenthesis and reducing the terms in formula (3.24), we obtain:

$$c_{qr}\big|_{\substack{q\neq i\\r\neq j}} = a_{qs}b_{sr} + a_{qm_{qr}}b_{m_{qr}r} + \sum_{\substack{p=1\\p\neq s,\ p\neq m_{qr}}}^{n} a_{qp}b_{pr}$$

thus the two terms $a_{qs}b_{sr}$ and $a_{qm_{qr}}b_{m_{qr}r}$ are exactly those that are missing in the sum of the H_{qrp} auxiliary terms (3.23) so the expression is equivalent to the element of the matrix multiplication.

Less straightforward is the proof for the product elements lying only on the ith row and jth column [with the exception of the (i,j)-th element]. Expanding and reducing the terms of the sub-expression $M_{qr1} + M_{qr2} + P_{m_{qr}}$ of formula (3.26) leads to:

$$M_{qr1} + M_{qr2} + P_{m_{qr}} = a_{im_{qr}}b_{m_{qr}r} + a_{qm_{qr}}b_{sr} + a_{qs}b_{sr} - a_{im_{qr}}b_{sr}$$

So from (3.25) and (3.26), after the reduction, we get:

$$c_{ir}|_{r\neq j} = \sum_{k=1}^{n} a_{ik}b_{sr} + \sum_{\substack{q=1\\q\neq i}}^{n} a_{im_{qr}}b_{m_{qr}r} - \sum_{\substack{q=1\\q\neq i}}^{n} a_{im_{qr}}b_{sr} \tag{3.29}$$

Observe:

$$\sum_{k=1}^{n} a_{ik}b_{sr} - \sum_{\substack{q=1\\q\neq i}}^{n} a_{im_{qr}}b_{sr} = \left[\sum_{k=1}^{n} a_{ik} - \sum_{q=1}^{n} a_{im_{qr}}\right] b_{sr} + a_{im_{ir}}b_{sr}$$

The first sum $\sum_{k=1}^{n} a_{ik}$ is simply the sum of the ith row elements of the multiplicand matrix. More complex is the analysis of the second sum, i.e. $\sum_{q=1}^{n} a_{im_{qr}}$. To do this, we need to go back to the properties of the modified Latin square:

- The index term m_{qr} means that under the sum we are choosing the rth column of the modified Latin square as the correct index for the $a_{.}$ elements.
- Importantly, $r \in T$ and $N - \{j\}$, so $r \neq j$ and the rth column of the modified Latin square contains all the elements from N, but in a permuted order.
- The above means, that the sum $\sum_{q=1}^{n} a_{im_{qr}}$ is just the permutation of the sum $\sum_{k=1}^{n} a_{ik}$ and the second sum contains all the elements of the same ith row.

We can summarise the above bullets by the following equality:

$$\sum_{k=1}^{n} a_{ik} = \sum_{q=1}^{n} a_{im_{qr}}$$

and we can rewrite expression (3.29), for the element of the ith row of the desired product, in the form:

$$c_{ir}|_{r\neq j} = a_{im_{ir}}b_{sr} + \sum_{\substack{q=1\\q\neq i}}^{n} a_{im_{qr}}b_{m_{qr}r} = a_{im_{ir}}b_{sr} + \sum_{q=1}^{n} a_{im_{qr}}b_{m_{qr}r} - a_{im_{ir}}b_{m_{ir}r}$$

Considering that the modified Latin square L_{nijs} contains s only on the ith row, $b_{m_{qr}r}$ for $q = i$ is equal to b_{sr}. Furthermore, $r \neq j$, so the sum $\sum_{q=1}^{n} a_{im_{qr}}b_{m_{qr}r}$ contains all the terms of the definition, but permuted.

Let us now prove the correctness of the formula expressing the jth column. Combining (3.22) and (3.27) with (3.28) by the reduction we obtain:

$$c_{qj}|_{q\neq i} = \left(a_{qs}b_{sj} + \sum_{\substack{r=1\\r\neq j}}^{n} a_{qm_{qr}}b_{m_{qr}j}\right) + \left(\sum_{\substack{r=1\\r\neq j}}^{n} a_{qs}b_{m_{qr}j} - \sum_{\substack{k=1\\k\neq s}}^{n} a_{qs}b_{kj}\right)$$

Using the same reasoning as for the ith line formula, we can deduce that the second parenthesis disappears:

$$a_{qs}\left[\sum_{\substack{r=1\\ r\neq j}}^{n} b_{m_{qr}j} - \sum_{\substack{k=1\\ k\neq s}}^{n} b_{kj}\right] = 0$$

and the terms in the first parentheses directly form the definitional product. In fact, from the definition of the modified Latin square it follows that the number s cannot be placed beyond the ith row and jth column of the L_{nijs} matrix, because each row and column must contain all the elements from 1 to n, including, which is already in the ith row and jth column. This means that we cannot find the s number twice or more times in any row or column other than the ith row and jth column. So in the first parentheses we sum all the definitional products except for $a_{qs}b_{sj}$, which is added separately next.

Now let us count the number of multiplications of the Sykora algorithm. The consecutive parts of the algorithm require the following number of multiplications:

- Part I: exactly n multiplications for P_k, because the indices i and j are fixed.
- Part II: for M_{qr1}, M_{qr2} and M_{qr3} [formula (3.22)] we need $3(n-1)^2$ multiplications, because q and r receives $n-1$ different values. For H_{qrp} quantity [formula (3.24)] we need $(n-1)^2(n-2)$, because p receives $n-2$ values and in total this part gives $(n-1)^2(n+1)$ multiplications to be performed.
- Part III: the only multiplications here are in the formulae (3.25) (expressing R_r) and (3.27) (expressing Q_q), each of which requires $n-1$ multiplication. Thus the part III takes $2(n-1)$ multiplications.

So the total number of multiplications required by the Sykora algorithm is:

$$n + (n-1)^2(n+1) + 2(n-1) = n^3 - (n-1)^2 \tag{3.30}$$

In summary, Sykora reduced the number of multiplications required by a quadratic term $(n-1)^2$. It is non-commutative, applicable to any matrix dimension and the multiplication number gain is by a linear better than the Gastinel's formula, which require $n^3 - (n-1)^1$ multiplications.

3.6.1 *Example—Sykora's Scheme in 3 × 3 Case*

The formula (3.30) for $n = 3$ Sykora algorithms leads to a $3^3 - (3-1)^2 = 23$ multiplication scheme. Let us construct a particular formula for the following modified Latin square:

$$L_{3,1,1,2} = \left[\begin{array}{c|cc} 2 & 2 & 2 \\ \hline 2 & 1 & 3 \\ 2 & 3 & 1 \end{array}\right] \tag{3.31}$$

thus $n = 3$, $i = j = 1$, $s = 2$ and the sets $S = T = \{2, 3\}$. We have highlighted the 2×2 submatrix, creating the classical Latin square. First we want to construct the $M_{qr1} \div M_{qr3}$ aggregated products, defined by the formula (3.22). To do this, we have to peek from the modified Latin square (3.31) value $m_{qr} = L_{nijs}[q\,; r]$ for each pair $(q, r) \in S \times T$ separately. The results are summarized in Table 3.1.

The elements of the matrix product $C = AB$ from the algorithm's part II, i.e. laying outside the "ij-cross", are calculated using the formula (3.24). For the c_{22} entry we have $q = r = 2$ and $m_{22} = L_{3,1,1,2}[2; 2] = 1$, thus $P_{m_{qr}} = P_1$:

$$\begin{aligned} c_{22} &= P_1 + \sum_{k=1}^{3} M_{22k} + \sum_{\substack{p=1 \\ p \neq 2,\ p \neq 1}}^{3} H_{22p} \\ &= a_{11}b_{11} + M_{22,1} + M_{22,2} + M_{22,3} + a_{23}b_{32} \end{aligned}$$

The remaining values c_{23}, c_{32}, c_{33} are obtained in the same way:

$$\begin{aligned} c_{23} &= a_{13}b_{31} + M_{23,1} + M_{23,2} + M_{23,3} + a_{21}b_{13} \\ c_{32} &= a_{13}b_{31} + M_{32,1} + M_{32,2} + M_{32,3} + a_{31}b_{12} \\ c_{33} &= a_{11}b_{11} + M_{33,1} + M_{33,2} + M_{33,3} + a_{33}b_{33} \end{aligned}$$

Table 3.1 Form of the aggregated products $M_{qr1} \div M_{qr3}$ for a modified Latin square (3.31)

	r	
q	2	3
2	$m_{22} = L_{3,1,1,2}[2; 2] = 1$	$m_{23} = L_{3,1,1,2}[2; 3] = 3$
	$M_{22,1} = (-a_{11} + a_{21} + a_{22}) \cdot (b_{11} + b_{22} - b_{12})$ $M_{22,2} = (a_{21} + a_{22})(b_{12} - b_{11})$ $M_{22,3} = (a_{11} - a_{21})(b_{22} - b_{12})$	$M_{23,1} = (-a_{13} + a_{23} + a_{22}) \cdot (b_{31} + b_{23} - b_{33})$ $M_{23,2} = (a_{23} + a_{22})(b_{33} - b_{31})$ $M_{23,3} = (a_{13} - a_{23})(b_{23} - b_{33})$
3	$m_{32} = L_{3,1,1,2}[3; 2] = 3$	$m_{33} = L_{3,1,1,2}[3; 3] = 1$
	$M_{32,1} = (-a_{13} + a_{33} + a_{32}) \cdot (b_{31} + b_{22} - b_{32})$ $M_{32,2} = (a_{33} + a_{32})(b_{32} - b_{31})$ $M_{32,3} = (a_{13} - a_{33})(b_{22} - b_{32})$	$M_{33,1} = (-a_{11} + a_{31} + a_{32}) \cdot (b_{11} + b_{23} - b_{13})$ $M_{33,2} = (a_{31} + a_{32})(b_{13} - b_{11})$ $M_{33,3} = (a_{11} - a_{31})(b_{23} - b_{13})$

Now we calculate the elements from part IIIa, i.e. from the 1st row. We use formulae (3.25) and (3.26). Observe:

$$R_2 = (a_{11} + a_{12} + a_{13} - a_{22} - a_{32} - a_{21} - a_{33})b_{22}$$
$$R_3 = (a_{11} + a_{12} + a_{13} - a_{22} - a_{32} - a_{23} - a_{31})b_{23}$$

then:

$$c_{12} = R_2 + M_{22,1} + M_{22,2} + M_{32,1} + M_{32,2} + a_{11}b_{11} + a_{13}b_{31}$$
$$c_{13} = R_3 + M_{23,1} + M_{23,2} + M_{33,1} + M_{33,2} + a_{11}b_{11} + a_{13}b_{31}$$

The elements from part IIIb are on the 1st column. Using formulae (3.27) and (3.28) we obtain:

$$Q_2 = a_{22}(b_{21} - b_{11} - b_{31} + b_{12} - b_{22} + b_{33} - b_{23})$$
$$Q_3 = a_{32}(b_{21} - b_{11} - b_{31} + b_{32} - b_{22} + b_{13} - b_{23})$$

then:

$$c_{21} = Q_2 + M_{22,1} + M_{22,3} + M_{23,1} + M_{23,3} + a_{11}b_{11} + a_{13}b_{31}$$
$$c_{31} = Q_3 + M_{32,1} + M_{32,3} + M_{33,1} + M_{33,3} + a_{11}b_{11} + a_{13}b_{31}$$

In summary, we can notice, that the Sykora scheme for $n = 3$, with the modified Latin square of the form (3.31), is equivalent to the Laderman's result, presented in Sect. 3.5. Table 3.2 shows the assignment of $M_{qr1} \div M_{qr3}$ to the products used by Laderman's scheme.

Moreover, $Q_2 = m_3$, $Q_3 = m_{11}$,$R_2 = m_1$, $R_3 = m_{10}$ and m_6, m_{14}, $m_{19} \div m_{23}$ contains the remaining, non-aggregated products.

A separate problem which is inherently embedded into the Sykora's algorithm are the Latin squares. An intelligible general introduction to the topic of Latin squares can be found in Schumer's monograph (2004), pages 76–81. A deeper insight into the

Table 3.2 Assignment of the aggregated products $M_{qr1} \div M_{qr3}$ to the aggregated products $m_{\cdot}$ (3.21) in the Laderman algorithm

	r	
q	2	3
2	$M_{22,1} = m_4$ $M_{22,2} = m_5$ $M_{22,3} = m_2$	$M_{23,1} = m_{16}$ $M_{23,2} = m_{18}$ $M_{23,3} = m_{17}$
3	$M_{32,1} = m_{12}$ $M_{32,2} = m_{15}$ $M_{32,3} = m_{13}$	$M_{33,1} = m_7$ $M_{33,2} = m_9$ $M_{33,3} = m_8$

topic is given by another monograph devoted only to this topic, also with applications, by Laywine and Mullen (1998).

The most important step regarding Latin squares in the Sykora algorithm is to obtain a Latin square of a given size. For small sizes, the simplest and most effective method may be to generate random permutations and check whether the Latin square condition is satisfied by trial and error. For larger sizes, the algorithm of Jacobson and Matthews (1996), which generates Latin squares of arbitrary size, may be helpful, giving as an example one of order 17.

Another, much more problematic issue is the number of different Latin squares of a given dimension $n \times n$. There is no formula that makes it easy to calculate the exact number of Latin squares. Instead, we have two possibilities:

- Find the approximate number of Latin squares. A monograph van Lint and Wilson (1992) pages 161–162 gives the following estimate:

$$\frac{(n!)^{2n}}{n^{n^2}} \leq L_n \leq \prod_{k=1}^{n} (k!)^{\frac{n}{k}}$$

where L_n in the $n \times n$ Latin square with symbols $1, \ldots, n$.

- Use the exact formula given by Shao and Wei (1992):

$$L_n = n! \sum_{A \in B_n} (-1)^{\sigma_0(A)} \binom{per(A)}{n} \tag{3.32}$$

where B_n is the set of $n \times n$ matrices with allowed values for each entry in the set $\{0, 1\}$, $\sigma_0(A)$ is the number of zero elements of the matrix A and $per(A)$ is the permanent of the matrix A.

Shao and Wei (1992) on page 296 says about their formula (3.32) that, unfortunately, "*this formula does not provide an efficient algorithm for computing the value of L_n, since the number of terms in the formula is exponential on n. Thus a different type of formula which is efficient in practical computation is still desirable.*" Indeed, in overall we have as much as 2^{n^2} matrices in the set B_n.

In addition, for each $A \in B_n$ we need to calculate the so-called permanent of the matrix. Its definition is similar to that of the determinant:

$$per(A) = \sum a_{1j_1} a_{1j_2} \cdot \cdots \cdot a_{1j_n}$$

where the sum is over all permutations $j_1 j_2 \cdots j_n$ of the set $\{1, 2, \ldots, n\}$. We see that the only difference from the determinant definition is that we sum all terms with positive sign, without checking the sign of the permutation. Although the permanent definition is simpler than the determinant definition, it is much harder to compute. There is no algorithm that calculates the permanent value in polynomial time. One of the known formulae is given by Ryser (1963) page 27:

$$per(A) = \sum_{(\varepsilon_1,\ldots,\varepsilon_n)\in\{0,1\}^n} (-1)^{n-\sum_{k=1}^{n}\varepsilon_k} \prod_{i=1}^{n}\sum_{j=1}^{n}\varepsilon_j a_{ij} \tag{3.33}$$

In other words, the outer sum ranges over all 2^n choices of ε_1 ..., ε_n equal to 0 or 1 independently. Yet another known formula is presented in article Glynn (2013), but no faster algorithm than of the exponential class is known. We can also use the formula similar to Laplace's expansion for the determinants along a row, a column or a diagonal, but only with + signs, but the time efficiency remains exponential (Percus 1971, p 12).

Let us return to the Sykora result. It uses the so-called modified Latin square L_{nijs}, which also contains all the elements of $\{1, \ldots, n\}$ in every row and column, but with an additional condition, that a pair of a given ith row and a given jth column contains a fixed number s. Considering that s can be chosen arbitrarily from the set $\{1, \ldots, n\}$, and that each row and column containing only the number s, can be placed independently in n positions, the number of possible modified Latin squares of size $n \times n$ is:

$$L_{nijs} = (n^2) \cdot n \cdot L_{n-1} = n^3 L_{n-1} \tag{3.34}$$

where L_{n-1} is the number of "classical" Latin squares of the size $(n-1) \times (n-1)$, which can be calculated, for example, using formula (3.32).

3.6.2 Example—Number of 3 × 3 Size Modified Latin Squares

In this section we show how to calculate the number of possible modified Latin squares of 3 × 3 size, using the results given in Sect. 3.6. We calculate it by the formula (3.34), so at first we need the number of "classical" Latin squares of the 2 × 2 size. We do this using the Shao, Wei formula (3.32), and the necessary permanents in this formula we calculate using the Ryser formula (3.33).

First of all, let us calculate the formula for calculating the permanent of the $(3-1) \cdot (3-1)$ size matrix.[8] For such a dimension equality (3.33) receives the following form:

$$per(A)|_{n=2} = \sum_{\varepsilon_1,\,\varepsilon_2=0}^{1} (-1)^{2-\varepsilon_1-\varepsilon_2} \prod_{i=1}^{2}\sum_{j=1}^{2}\varepsilon_j a_{ij}$$

[8] Of course, we can obtain the formula expressing the permanent for 2 × 2 matrix faster directly from its definition, by the sum of proper permutations, but our aim is to show by the example how the Rysers's formula (3.33) works.

$$= \sum_{\varepsilon_1, \varepsilon_2=0}^{1} (-1)^{2-\varepsilon_1-\varepsilon_2}(\varepsilon_1 a_{11} + \varepsilon_2 a_{12})(\varepsilon_1 a_{21} + \varepsilon_2 a_{22}) \tag{3.35}$$

So the sum in (3.35) has four summands:

$$\begin{cases} \text{for } \varepsilon_1 = \varepsilon_2 = 1: & [(a_{11} + a_{12})(a_{21} + a_{22})] \\ \text{for } \varepsilon_1 = \varepsilon_2 = 0: & [0] \\ \text{for } \varepsilon_1 = 1,\ \varepsilon_2 = 0: & [-a_{11}a_{21}] \\ \text{for } \varepsilon_1 = 0,\ \varepsilon_2 = 1: & [-a_{12}a_{22}] \end{cases}$$

Thus the permanent for 2×2 matrix size have the form:

$$per(A)|_{n=2} = (a_{11} + a_{12})(a_{21} + a_{22}) - a_{11}a_{21} - a_{12}a_{22} = a_{11}a_{22} + a_{12}a_{21} \tag{3.36}$$

Now let us return to the calculation of the number of Latin squares. We do this using the Shao and Wei (1992) formula (3.32). The Newton's binomial in this formula is nonzero only for those summands, for which $per(A) \geq n(= 2)$. Combining the fact that the matrix entries of the summand's $A \in B_n$ can only have values in the set 0 or 1 with the form of the permanent for $n = 2$, expressed by (3.36), the maximum value for $per(A)|_{n=2}$ is just 2 and this value is achieved only for a single four $a_{11} = a_{22} = a_{12} = a_{21} = 1$. So $\sigma_0(A) = 0$ and the sum (3.32) is equal to:

$$L_2 = 2! \sum_{A=\begin{bmatrix}1 & 1\\ 1 & 1\end{bmatrix}} (-1)^{\sigma_0(A)} \binom{2}{2} + 2! \sum_{A \in B_n - \begin{bmatrix}1 & 1\\ 1 & 1\end{bmatrix}} (-1)^{\sigma_0(A)} \binom{per(A)}{2} = 2 + 0 = 2$$

Let us return to the calculation of the number of Sykora's 3×3 *modified* Latin squares. By (3.34) it is:

$$L_{3ijs} = 3^3 L_2 = 3^3 \cdot 2 = 54$$

As a conclusion, it can be observed that Sykora's algorithm gives a number of equally efficient schemes to compute 3×3 matrix products, but it still remains only a subset of the possible schemes with the same efficiency. In fact, Heule and Kauers et al. (2021), using different methods, obtained more than a dozen of thousands of equivalent schemes with 23 multiplications, while Sykora's method, by changing the applied modified Latin square, gives only 54.

The monograph Ryser (1963) pages 36–37 gives the values of the number of Latin squares for $n = 1 \div 7$. Table 3.3 summarises these values, together with the number of the corresponding modified Latin squares.

Table 3.3 Number of the Latin squares and modified Latin squares with respect to the dimension

n	L_n	$L_{nijs} = n^3 L_{n-1}$
1	1	1
2	2	8
3	12	54
4	576	768
5	161, 280	72, 000
6	812, 851, 200	34, 836, 480
7	61, 479, 419, 904, 000	13, 279, 554, 699, 264, 000

3.7 Duality of Rectangular Matrices Multiplication

The results presented in the previous sections were for square matrices. By the Definition 1.1, the inner dimensions of the matrix factors must be equal, so all the matrix dimensions in the equation $C = AB$ for square matrices are equal. In general, however, we multiply matrices of dimensions $m \times n$ by $n \times p$. In the following, we will abbreviate this as (m, n, p) matrix multiplication.

The direct, definition-based algorithm for the (m, n, p)-size matrix product, requires $m \cdot n \cdot p$ scalar multiplication. As well as fast matrix multiplication for square matrices, the literature also provides results also for rectangular matrices. Some special but useful examples will be given later in this chapter.

The duality problem of matrix multiplication can be formulated as follows:

- We have a fast multiplication algorithm for the matrices with (m, n, p) dimensions in r multiplications, $r < mnp$.
- How to use this algorithm, to obtain the scheme to multiply all the remaining $(3! - 1) = 5$ permutations of this (m, n, p) dimensions also in r scalar multiplications.

The problem in question is straightforward in the case of (p, n, m)-size matrix multiplication. In fact, we can do it on the basis of a basic matrix algebra, i.e. the equality $AB = (B^T A^T)^T$ [property (1.3)].

The remaining four cases of the order of the dimensions to be multiplied require more advanced tools, which were provided in 1973 by an article by Hopcroft and Musinski. A slightly different approach to this problem is given in the article Probert (1974). Let A, B and C be $m \times n$, $n \times p$ and $m \times p$ matrices. We define so-called vector representation for these three matrices, e.g. we represent the matrix A as a vector in a row order, i.e. $A \triangleq a = [a_{11}, a_{12}, \ldots, a_{1n}, \ a_{21}, \ldots, a_{2n}, \cdots, \ a_{m1}, \ldots, a_{mn}]^T$. For the vectors we use what is known in the literature as the Hadamard product (Horn and Johnson 2012, Sect. 7.5). For the vector case it has the following form, denoted by a single dot:

$$\begin{bmatrix} x_1 \\ \vdots \\ x_n \end{bmatrix} \cdot \begin{bmatrix} y_1 \\ \vdots \\ y_n \end{bmatrix} = \begin{bmatrix} x_1 y_1 \\ \vdots \\ x_n y_n \end{bmatrix}$$

Thus, each term in the series of r aggregated products, representing a given fast matrix multiplication scheme for an AB product, can be divided into its left part Pa and its right part Rb, where the left part collects the linear combination of the A matrix elements, and the right part linearly collects the elements of the B matrix, respectively. Thus $\dim[P] = r \times mn$ and $\dim[R] = r \times np$. The aggregated products are given by $Pa \cdot Rb$.

The linear combination (3.16) can be constructed by another matrix M of the $mp \times r$ size—in a vector representation—as follows:

$$AB \triangleq M\,(Pa \cdot Rb) \tag{3.37}$$

linearly combining the respective aggregated products. We are now ready to present the duality theorem formulated in Hopcroft and Musinski (1973).

3.7.1 Matrix Multiplication Duality Theorem

If $M\,(Pa \cdot Rb)$ calculates AB in row order by r scalar multiplications for (m, n, p) dimensions, then:

(a) *$P^T(M^T c \cdot Rb)$ calculates CB^T for (m, p, n) dimensions,*
(b) *$R^T(M^T c \cdot Pa)$ calculates $(C^T A)^T$ for (p, m, n) dimensions,*
(c) *$M\,(Rb \cdot Pa)$ calculates $(B^T A^T)^T$ for (p, n, m) dimensions,*
(d) *$P^T(Rb \cdot M^T c)$ calculates $(BC^T)^T$ for (n, p, m) dimensions,*
(e) *$R^T(Pa \cdot M^T c)$ calculates $A^T C$ for (n, m, p) dimensions,*

in row order also by r scalar multiplications.

Proof

To prove the Theorem 3.7.1 we use the following auxiliary lemma.

- *Lemma: If the equality $x^T Y = x^T Z$ holds true for all x, where x is n-dimensional vector and Y, Z are matrices of matching dimensions, then $Y = Z$.*

Proof of the Lemma

From $x^T Y = x^T Z$ we immediately get $x^T(Y - Z) = 0$. Let us denote by x_j a vector with 1 on jth position, and zeros elsewhere. The equality to be proved claims to be true for any vector x, and therefore in particular also for the x_j, for each $j = 1, \ldots, n$.

Observe the expression:

$$x_j^T(Y - Z)$$

It selects (in other words, is equal to) the j-th row of $Y - Z$. So we can conclude that in order for the j-th row of $x^T(Y - Z)$ expression to contain only zeros for any x, the j-th row of $Y - Z$ must also be zero. Ranging over all $j = 1, \ldots, n$, we obtain the statement of the lemma.

■

Let us now present the proof of the Theorem 3.7.1 using case (a) as an example. First, we define a matrix D_B as a block diagonal matrix of B matrices, repeated as many times as the A matrix contains the rows, that is, of size $mn \times mp$:

$$D_B = \begin{bmatrix} B & & 0 \\ & \ddots & \\ 0 & & B \end{bmatrix}$$

It is easy to check that, using the vector representation of a matrix, the expression $a^T D_B$ gives the AB product just in a—here horizontal—vector representation. Thus, considering (3.37), we can claim that:

$$\left(a^T D_B\right)^T = M(Pa \cdot Rb)$$

Let us apply the transposition to both sides of the last equality. Observe:

$$a^T D_B = [M(Pa \cdot Rb)]^T = (Pa \cdot Rb)^T M^T \tag{3.38}$$

To transpose the $M(Pa \cdot Rb)$ term it is useful to recall how we multiply a horizontal vector by a diagonal matrix, i.e. $[\,x_1\ x_2\,]\,diag[y_1\ y_2] = [x_1 y_1 \quad x_2 y_2]$. So the right-hand side of (3.38) can be further modified in the following way:

$$(Pa \cdot Rb)^T M^T = (Pa)^T diag(Rb) M^T = a^T P^T diag(Rb) M^T$$

And so:

$$a^T D_B = a^T P^T diag(Rb) M^T \tag{3.39}$$

At this point, we can conclude the following equality:

$$D_B = P^T diag(Rb) M^T \tag{3.40}$$

because the lemma $\underset{x}{\forall}(x^T Y = x^T Z) \Rightarrow Y = Z$ applies to Eq. (3.39).

Now we can move on to the thesis of the case (a), which can be written in vector form as $CB^T \triangleq (c^T D_B^T)^T$. After left-side multiplication by transposition of (3.40) by c^T we obtain:

$$c^T D_B^T = c^T M\ diag(Rb) P = (M^T c)^T diag(Rb) P$$

Finally, we return to the element-by-element product form of the formula. After transposition, we obtain the thesis case (a):

$$\left[c^T D_B^T\right]^T = P^T \left[(M^T c)^T diag(Rb)\right]^T = P^T (M^T c \cdot Rb)$$

■

To illustrate the result of Hopcroft and Musinski (1973) (3.7.1), we have chosen another important result which provides an algorithm for the (m, n, p)-size case of fixed inner dimension $n = 2$. More details are given in the following section.

3.7.2 Example—3 × 2 × 4 Case

Article Hopcroft and Kerr (1968), together with the technical report by the same authors from 1969, gives non-commutative algorithm for multiplying rectangular matrices of dimensions $m \times 2$ by $2 \times p$ in $\lceil (3mp + \max(m, p))/2 \rceil$ multiplications.

Suppose we want to multiply $A = \left[a_{ij}\right]$ matrix of size 3×2 by $B = \left[b_{ij}\right]$ matrix of size 2×4. The series of the respective aggregated products for $m = 3,\ p = 4$, constructed on the basis of the works Hopcroft and Kerr (1968) and (1969), have the following form:

$$\left\{\begin{aligned}
m_1 &= a_{31}b_{13}\\
m_2 &= a_{32}b_{23}\\
m_3 &= a_{12}(b_{11}+b_{21})\\
m_4 &= a_{21}(b_{12}+b_{22})\\
m_5 &= (a_{11}-a_{12})b_{11}\\
m_6 &= (a_{21}-a_{22})b_{22}\\
m_7 &= (a_{11}+a_{21})(b_{11}+b_{12})\\
m_8 &= (a_{12}+a_{21})(b_{11}-b_{22})\\
m_9 &= (a_{12}+a_{22})(b_{21}+b_{22})\\
m_{10} &= (a_{11}+a_{31})(b_{11}+b_{13}+b_{21}+b_{23})\\
m_{11} &= (a_{11}+2a_{21}+a_{31})(b_{12}+b_{14})\\
m_{12} &= (a_{12}+a_{22}+a_{32})(b_{14}+b_{24})\\
m_{13} &= (a_{12}+2a_{22}+a_{32})(b_{22}+b_{24})\\
m_{14} &= (a_{11}-a_{12}+a_{31})(b_{11}+b_{21}+b_{23})\\
m_{15} &= (a_{11}-a_{12}+a_{31}-a_{32})(b_{21}+b_{23})\\
m_{16} &= (a_{12}+a_{21}+a_{22}+a_{32})(b_{14}-b_{22})\\
m_{17} &= (a_{11}+a_{21}+2a_{31})(b_{13}+b_{14}+b_{23}+b_{24})\\
m_{18} &= (a_{11}-a_{12}+a_{21}-a_{22}+a_{31}-a_{32})b_{14}\\
m_{19} &= (a_{11}-a_{12}+a_{21}-a_{22}+2a_{31}-2a_{32})(b_{23}+b_{24})\\
m_{20} &= (a_{11}-a_{12}+a_{21}-a_{22}+2a_{31}-a_{32})(b_{14}+b_{23}+b_{24})
\end{aligned}\right. \tag{3.41}$$

So the gain is 4 products over the definitional $3 \cdot 2 \cdot 4 = 24$ multiplications. We construct the desired matrix product $Y = AB$ as follows:

$$\left\{\begin{aligned}
y_{11} &= m_3+m_5\\
y_{12} &= -m_4-m_5+m_7-m_8\\
y_{13} &= -m_1-m_3+m_{10}-m_{14}\\
y_{14} &= m_2-m_6+2m_{12}-m_{13}-m_{16}+2m_{18}+m_{19}-m_{20}\\
\\
y_{21} &= -m_3+m_6+m_8+m_9\\
y_{22} &= m_4-m_6\\
y_{23} &= -m_1-m_2+m_3-m_{10}-m_{12}+m_{14}+m_{17}-m_{20}\\
y_{24} &= m_6-m_{12}+m_{13}+m_{16}\\
\\
y_{31} &= -m_2-m_5+m_{14}-m_{15}\\
y_{32} &= -m_4+m_5+m_6-m_7+m_8+m_{11}-m_{16}-m_{18}\\
y_{33} &= m_1+m_2\\
y_{34} &= -m_2-m_{18}-m_{19}+m_{20}
\end{aligned}\right. \tag{3.42}$$

Now we need to identify the matrices M, P and R that satisfy the formula $AB \overset{\wedge}{=} M\,(Pa{\cdot}Rb)$. By analysing the form of the aggregated products (3.41), we can construct P and R matrices corresponding to the multipliers and multiplicands of the aggregated products (3.41).

$$
P = \begin{bmatrix}
0 & 0 & 0 & 0 & 1 & 0 \\
0 & 0 & 0 & 0 & 0 & 1 \\
0 & 1 & 0 & 0 & 0 & 0 \\
0 & 0 & 1 & 0 & 0 & 0 \\
1 & -1 & 0 & 0 & 0 & 0 \\
0 & 0 & 1 & -1 & 0 & 0 \\
1 & 0 & 1 & 0 & 0 & 0 \\
0 & 1 & 1 & 0 & 0 & 0 \\
0 & 1 & 0 & 1 & 0 & 0 \\
1 & 0 & 0 & 0 & 1 & 0 \\
1 & 0 & 2 & 0 & 1 & 0 \\
0 & 1 & 0 & 1 & 0 & 1 \\
0 & 1 & 0 & 2 & 0 & 1 \\
1 & -1 & 0 & 0 & 1 & 0 \\
1 & -1 & 0 & 0 & 1 & -1 \\
0 & 1 & 1 & 1 & 0 & 1 \\
1 & 0 & 1 & 0 & 2 & 0 \\
1 & -1 & 1 & -1 & 1 & -1 \\
1 & -1 & 1 & -1 & 2 & -2 \\
1 & -1 & 1 & -1 & 2 & -1
\end{bmatrix}, \quad
R = \begin{bmatrix}
0 & 0 & 1 & 0 & 0 & 0 & 0 & 0 \\
0 & 0 & 0 & 0 & 0 & 0 & 1 & 0 \\
1 & 0 & 0 & 0 & 1 & 0 & 0 & 0 \\
0 & 1 & 0 & 0 & 0 & 1 & 0 & 0 \\
1 & 0 & 0 & 0 & 0 & 0 & 0 & 0 \\
0 & 0 & 0 & 0 & 0 & 1 & 0 & 0 \\
1 & 1 & 0 & 0 & 0 & 0 & 0 & 0 \\
1 & 0 & 0 & 0 & 0 & -1 & 0 & 0 \\
0 & 0 & 0 & 0 & 1 & 1 & 0 & 0 \\
1 & 0 & 1 & 0 & 1 & 0 & 1 & 0 \\
0 & 1 & 0 & 1 & 0 & 0 & 0 & 0 \\
0 & 0 & 0 & 1 & 0 & 0 & 0 & 1 \\
0 & 0 & 0 & 0 & 0 & 1 & 0 & 1 \\
1 & 0 & 0 & 0 & 1 & 0 & 1 & 0 \\
0 & 0 & 0 & 0 & 1 & 0 & 1 & 0 \\
0 & 0 & 0 & 1 & 0 & -1 & 0 & 0 \\
0 & 0 & 1 & 1 & 0 & 0 & 1 & 1 \\
0 & 0 & 0 & 1 & 0 & 0 & 0 & 0 \\
0 & 0 & 0 & 0 & 0 & 0 & 1 & 1 \\
0 & 0 & 0 & 1 & 0 & 0 & 1 & 1
\end{bmatrix} \tag{3.43}
$$

To construct the matrix M, we should analyse the linear combinations that express the matrix product $Y = AB$ (3.42):

$$
M = \begin{bmatrix}
0 & 0 & 1 & 0 & 1 & 0 & 0 & 0 & 0 & 0 & 0 & 0 & 0 & 0 & 0 & 0 & 0 & 0 & 0 & 0 \\
0 & 0 & 0 & -1 & -1 & 0 & 1 & -1 & 0 & 0 & 0 & 0 & 0 & 0 & 0 & 0 & 0 & 0 & 0 & 0 \\
-1 & 0 & -1 & 0 & 0 & 0 & 0 & 0 & 0 & 1 & 0 & 0 & 0 & -1 & 0 & 0 & 0 & 0 & 0 & 0 \\
0 & 1 & 0 & 0 & 0 & -1 & 0 & 0 & 0 & 0 & 0 & 2 & -1 & 0 & 0 & -1 & 0 & 2 & 1 & -1 \\
0 & 0 & -1 & 0 & 0 & 1 & 0 & 1 & 1 & 0 & 0 & 0 & 0 & 0 & 0 & 0 & 0 & 0 & 0 & 0 \\
0 & 0 & 0 & 1 & 0 & -1 & 0 & 0 & 0 & 0 & 0 & 0 & 0 & 0 & 0 & 0 & 0 & 0 & 0 & 0 \\
-1 & -1 & 1 & 0 & 0 & 0 & 0 & 0 & 0 & -1 & 0 & -1 & 0 & 1 & 0 & 0 & 1 & 0 & 0 & -1 \\
0 & 0 & 0 & 0 & 0 & 1 & 0 & 0 & 0 & 0 & 0 & -1 & 1 & 0 & 0 & 1 & 0 & 0 & 0 & 0 \\
0 & -1 & 0 & 0 & -1 & 0 & 0 & 0 & 0 & 0 & 0 & 0 & 0 & 1 & -1 & 0 & 0 & 0 & 0 & 0 \\
0 & 0 & 0 & -1 & 1 & 1 & -1 & 1 & 0 & 0 & 1 & 0 & 0 & 0 & 0 & -1 & 0 & -1 & 0 & 0 \\
1 & 1 & 0 & 0 & 0 & 0 & 0 & 0 & 0 & 0 & 0 & 0 & 0 & 0 & 0 & 0 & 0 & 0 & 0 & 0 \\
0 & -1 & 0 & 0 & 0 & 0 & 0 & 0 & 0 & 0 & 0 & 0 & 0 & 0 & 0 & 0 & 0 & -1 & -1 & 1
\end{bmatrix} \tag{3.44}
$$

Let us illustrate how the Theorem 3.7.1 works by the example of the case (a), which produces an algorithm for the case of size $(3, 4, 2)$. For clarity, we will denote

B^T matrix directly by X. So we want to multiply matrices of the following form (including the result Z):

$$C = \begin{bmatrix} c_{11} & c_{12} & c_{13} & c_{14} \\ c_{21} & c_{22} & c_{23} & c_{24} \\ c_{31} & c_{32} & c_{33} & c_{34} \end{bmatrix}, \quad B^T = \begin{bmatrix} b_{11} & b_{21} \\ b_{12} & b_{22} \\ b_{13} & b_{23} \\ b_{14} & b_{24} \end{bmatrix}, \quad X = \begin{bmatrix} x_{11} & x_{12} \\ x_{21} & x_{22} \\ x_{31} & x_{32} \\ x_{41} & x_{42} \end{bmatrix}, \quad Z = \begin{bmatrix} z_{11} & z_{12} \\ z_{21} & z_{22} \\ z_{31} & z_{32} \end{bmatrix} \tag{3.45}$$

to find the product $Z = CX$ of the size 3×2. These matrices (3.45) in a row order have the following vector representations:

$$\begin{cases} C \triangleq c = \left[c_{11}\ c_{12}\ c_{13}\ c_{14} \,\middle|\, c_{21}\ c_{22}\ c_{23}\ c_{24} \,\middle|\, c_{31}\ c_{32}\ c_{33}\ c_{34} \right] \\ X \triangleq x = \left[x_{11}\ x_{12} \,\middle|\, x_{21}\ x_{22} \,\middle|\, x_{31}\ x_{32} \,\middle|\, x_{41}\ x_{42} \right] \\ Z \triangleq z = \left[z_{11}\ z_{12} \,\middle|\, z_{21}\ z_{22} \,\middle|\, z_{31}\ z_{32} \right] \end{cases} \tag{3.46}$$

Using (3.46), it is now easy to find the aggregated products $m_.$ for the matrix product $Z = CB^T = CX$. We construct it as the internal product $(M^T c \cdot bx)$ of the expression $P^T(M^T c \cdot bx)$ from case (a) of Theorem 3.7.1:

$$\begin{cases} m_1 = (-c_{13} - c_{23} + c_{33})x_{31} \\ m_2 = (c_{14} - c_{23} - c_{31} + c_{33} - c_{34})x_{32} \\ m_3 = (c_{11} - c_{13} - c_{21} + c_{23})(x_{11} + x_{12}) \\ m_4 = (-c_{12} + c_{22} - c_{31})(x_{21} + x_{22}) \\ m_5 = (c_{11} - c_{12})x_{11} \\ m_6 = (-c_{14} + c_{21} - c_{22} + c_{24} + c_{31})x_{22} \\ m_7 = (c_{12} - c_{31})(x_{11} + x_{21}) \\ m_8 = (-c_{12} + c_{21} + c_{31})(x_{11} - x_{22}) \\ m_9 = c_{21}(x_{12} + x_{22}) \\ m_{10} = (c_{13} - c_{23})(x_{11} + x_{12} + x_{31} + x_{32}) \\ m_{11} = c_{31}(x_{21} + x_{41}) \\ m_{12} = (2c_{14} - c_{23} - c_{24})(x_{41} + x_{42}) \\ m_{13} = (-c_{14} + c_{24})(x_{22} + x_{42}) \\ m_{14} = (-c_{13} + c_{23} + c_{31})(x_{11} + x_{12} + x_{32}) \\ m_{15} = c_{31}(-x_{12} - x_{32}) \\ m_{16} = (-c_{14} + c_{24} - c_{31})(-x_{22} + x_{41}) \\ m_{17} = c_{23}(x_{31} + x_{32} + x_{41} + x_{42}) \\ m_{18} = (2c_{14} - c_{31} - c_{34})x_{41} \\ m_{19} = (c_{14} - c_{34})(x_{32} + x_{42}) \\ m_{20} = (-c_{14} - c_{23} + c_{34})(x_{32} + x_{41} + x_{42}) \end{cases} \tag{3.47}$$

Finally, we construct the result $Z = CX$ as a linear combination of aggregated products (3.47), directly by the formula $P^T(M^T c \cdot bx)$:

$$\begin{cases} z_{11} = m_5 + m_7 + m_{10} + m_{11} + m_{14} + m_{15} + m_{17} + m_{18} + m_{19} + m_{20} \\ z_{12} = m_3 - m_5 + m_8 + m_9 + m_{12} + m_{13} - m_{14} - m_{15} + m_{16} - m_{18} - m_{19} - m_{20} \\ \\ z_{21} = m_4 + m_6 + m_7 + m_8 + 2m_{11} + m_{16} + m_{17} + m_{18} + m_{19} + m_{20} \\ z_{22} = -m_6 + m_9 + m_{12} + 2m_{13} + m_{16} - m_{18} - m_{19} - m_{20} \\ \\ z_{31} = m_1 + m_{10} + m_{11} + m_{14} + m_{15} + 2m_{17} + m_{18} + 2m_{19} + 2m_{20} \\ z_{32} = m_2 + m_{12} + m_{13} - m_{15} + m_{16} - m_{18} - 2m_{19} - m_{20} \end{cases} \tag{3.48}$$

In summary, Theorem 3.7.1 of Hopcroft and Musinski (1973) allows us construct the algorithm for rectangular matrix multiplication of the (m, n, p)-size for all the possible six orders of the dimensions in a given number of multiplications, without having to reconstruct the algorithm from the very beginning each time it is used. Instead, we can start with a given algorithm for a given, arbitrary order of dimensions and derive the remaining five algorithms simply from the Hopcroft and Musinski result.

In an illustration of the Hopcroft and Musinski (1973), Theorem 3.7.1, we also mentioned the article Hopcroft and Kerr (1968), which shows how to construct fast algorithm for the multiplication of rectangular matrices of dimensions $m \times 2$ by $2 \times p$, for any m and p.

In the Example 3.7.2 we deal with dimensions 3×2 by 2×4. Below we have shown ready to use formulae for, in practice often occurring, case 3×2 by 2×3 in 15 multiplications. The formulae below was derived from the article Hopcroft and Kerr (1968) in the same article Hopcroft and Musinski (1973), pages 162–163, where the duality theorem in question was published:

$$A|_{3\times 2} \cdot B|_{2\times 3} = \begin{bmatrix} m_1 + m_2 & -m_2 - m_3 + m_7 - m_8 & -m_1 - m_5 - m_{13} + m_{15} \\ -m_1 - m_4 + m_8 - m_9 & m_3 + m_4 & -m_3 - m_6 + m_{11} - m_{12} \\ -m_2 - m_6 + m_{13} - m_{14} & -m_4 - m_5 + m_{10} - m_{11} & m_5 + m_6 \end{bmatrix} \tag{3.49}$$

where:

$$\begin{cases} m_1 = (a_{11} - a_{12})b_{11} \\ m_2 = a_{12}(b_{11} + b_{21}) \\ m_3 = a_{21}b_{12}, \quad m_4 = a_{22}b_{22} \\ m_5 = a_{31}(b_{13} + b_{23}) \\ m_6 = (-a_{31} + a_{32})b_{23} \\ m_7 = (a_{11} + a_{21})(b_{11} + b_{12} + b_{21} + b_{22}) \\ m_8 = (a_{11} - a_{12} + a_{21})(b_{11} + b_{21} + b_{22}) \\ m_9 = (a_{11} - a_{12} + a_{21} - a_{22})(b_{21} + b_{22}) \\ m_{10} = (a_{22} + a_{32})(b_{12} + b_{13} + b_{22} + b_{23}) \\ m_{11} = (a_{22} - a_{31} + a_{32})(b_{12} + b_{13} + b_{23}) \\ m_{12} = (-a_{21} + a_{22} - a_{31} + a_{32})(b_{12} + b_{13}) \\ m_{13} = (a_{12} + a_{31})(b_{11} - b_{23}) \\ m_{14} = (-a_{12} - a_{32})(b_{21} + b_{23}) \\ m_{15} = (a_{11} + a_{31})(b_{11} + b_{13}) \end{cases} \tag{3.50}$$

Ready-to-use formulae for 4×2 by 2×4 case in 26 multiplications, instead of 32 by the definition gives, are given in work Probert (1974). The aggregated products are enumerated on page 79, and the matrix showing how to combine them linearly is G on the page 81 (this result is arranged in a column order).

We shall refer to the Hopcroft, Musinski duality result in the section devoted to the Bini's result in Sect. 5.4 on so-called arbitrary precision approximating algorithms.

3.8 Methods of Constructing Matrix Multiplication Algorithms

In this section we have presented general methods for constructing a new algorithm for fast matrix multiplication by combining or modifying existing ones. They are summarised in the article Probert and Fischer (1980).

- **Method I: Permutation rule**

The method includes the duality results that we presented in detail in Sect. 3.7. They can be briefly summarised as follows: We have a (m, n, p)-size problem realized by r non-scalar multiplications algorithm. By the Theorem 3.7.1 we can construct an algorithm for each of the six permutations of the given problem, realised in the same cost of r non-scalar multiplications.

- **Method II: Additive rule**

Suppose we have algorithms for (m, k_1, p) and (m, k_2, p) problems. We can replace them by the (m, n, p) problem with $n = k_1 + k_2$ number of multiplications equal to the sum of the two sub-problems multiplications. The rule follows directly from the formula for block matrix multiplication. We give a practical example at the end of this section.

- **Method III: Multiplicative rule**

This rule is realised by the recursion, i.e. the matrices of one subproblem act as submatrices of the second subproblem. Suppose we have algorithms for (i_1, j_1, k_1) and (i_2, j_2, k_2) problems. We can replace them by the (m, n, p) problem with $m = i_1 i_2,\ n = j_1 j_2,\ p = k_1 k_2$ dimensionality obtaining the algorithm with the number of multiplications equal to the product of the number of multiplications of the two sub-problems.

- **Method IV: Zero-padding rule**

Suppose we have algorithms for (m, n, p). We replace it with an $(m, n + 1, p)$ and the cost remains the same.

The zero-padding rule can be treated as a special case of additive rule, where the second subproduct with dimensionality $(m, 1, p)$ is filled with zeros.

3.8.1 Example—Additive Rule Applied to Obtain 3 x 3 Algorithm

Before the result Laderman (1976), the best non-commutative algorithm to multiply the 3×3 matrices required 24 multiplications, which is obtained by the following decomposition:

$$\left[\begin{array}{cc|c} a_{11} & a_{12} & a_{13} \\ a_{21} & a_{22} & a_{23} \\ a_{31} & a_{32} & a_{33} \end{array}\right] \cdot \left[\begin{array}{ccc} b_{11} & b_{12} & b_{13} \\ b_{21} & b_{22} & b_{23} \\ \hline b_{31} & b_{32} & b_{33} \end{array}\right]$$

$$= \begin{bmatrix} a_{11} & a_{12} \\ a_{21} & a_{22} \\ a_{31} & a_{32} \end{bmatrix} \cdot \begin{bmatrix} b_{11} & b_{12} & b_{13} \\ b_{21} & b_{22} & b_{23} \end{bmatrix} + \begin{bmatrix} a_{13} \\ a_{23} \\ a_{33} \end{bmatrix} \cdot \begin{bmatrix} b_{31} & b_{32} & b_{33} \end{bmatrix}$$

$$= \left[A_1 \,\middle|\, A_2 \right] \left[\begin{array}{c} B_1 \\ \hline B_2 \end{array}\right] = A_1 B_1 + A_2 B_2$$

We can distinguish two sub-products here:

- (3,2,3) which can be realised by the results from Hopcroft and Kerr (1968) in 15 multiplications, according to formulas (3.49) and (3.50).
- (3,1,3) implemented directly by the definition of matrix multiplication (1.1) in 9 operations.
- Using method II (additive rule) we get the algorithm for (3,2,3) + (3,1,3) = (3,2 + 1,3) = (3,3,3) problem.

So in total we need $15 + 9 = 24$ multiplications to multiply 3×3 matrices using this method.

3.9 Trilinear Identity Algorithm for Matrices with Even Dimensions

First, we introduce the so-called bilinear and corresponding trilinear representation of—non-commutative—matrix multiplication algorithms. In general they are defined for rectangular matrices e.g. in the article Pan (1984) in Sect. 2 (bilinear) and Sect. 9 (trilinear). Let us consider matrices $X = [x_{ij}]$, $Y = [Y_{jk}]$ with dimensions $m \times n$ and $n \times p$, respectively. To compute the XY product in the bilinear representation, we first evaluate the linear combinations of x and y variables:

$$L_q = \sum_i \sum_j f(i, j, q) x_{ij}, \quad L_q^* = \sum_j \sum_k f^*(j, k, q) y_{jk} \quad for\ q = 1, \ldots, r$$

where variables $i = 1, \ldots, m; j = 1, .., n; k = 1, .., p$. Next, we calculate the desired product linearly combining the series of products $L_q L_q^*$:

$$(XY)_{ik} = \sum_j x_{ij} y_{jk} = \sum_{q=1}^{r} f^{**}(k, i, q) L_q L_q^* \tag{3.51}$$

The number r of multiplications L_q by L_q^* is called the rank of the algorithm. The constants f, f^*, f^{**} are chosen in such a way which assures that (3.51) is an identity in the indeterminates x_{ij}, y_{jk}. In particular for Strassen's algorithm $r = 7$ and f, f^*, $f^{**} \in \{-1, 0, 1\}$.

When it comes to the space complexity of the bilinear algorithm, Pan (1982) proved in remark 2.2 on page 347 that it is quadratic, i.e. $O(n^2)$. The article Abrahamson (1991) in Sect. 7, pages 280–283, gives a deep study of the relation between the time efficiency of fast matrix multiplication and its memory requirements.

Now we move to the trilinear representation of the matrix multiplication algorithm. We multiply both sides of (3.51) by an additional z_{ki} variable and sum both sides on k, i to obtain (Pan 1984, Sect. 2):

$$\sum_{i,j,k} x_{ij} y_{jk} z_{ki} = Trace(XYZ) = \sum_{q=1}^{r} L_q L_q^* L_q^{**} \tag{3.52}$$

where:

$$L_q^{**} = \sum_{k,i} f^{**}(k, i, q) z_{ki}$$

So the z_{ki} variable stands for the $(XY)_{ik}$ element of the XY product.

Both the bilinear and the trilinear representations are equivalent. To return from the trilinear to the bilinear representation, we need to make the coefficients of the z_{ki} variable in Eq. (3.52) equal. For example, the Strassen's algorithm (3.3) and (3.4) in the trilinear representation consist of a single, but trilinear equality:

$$\begin{aligned}
&(x_{11}y_{11} + x_{12}y_{21})z_{11} + (x_{11}y_{12} + x_{12}y_{22})z_{21} \\
&\quad + (x_{21}y_{11} + x_{22}y_{21})z_{12} + (x_{21}y_{12} + x_{22}y_{22})z_{22} \\
&= (x_{11} + x_{22})(y_{11} + y_{22})(z_{11} + z_{22}) + (x_{21} + x_{22})y_{11}(z_{12} - z_{22}) \\
&\quad + x_{11}(y_{12} - y_{22})(z_{21} + z_{22}) + x_{22}(y_{21} - y_{11})(z_{11} + z_{12}) \\
&\quad + (x_{11} + x_{12})y_{22}(-z_{11} + z_{21}) + (x_{21} - x_{11})(y_{11} + y_{12})z_{22} \\
&\quad + (x_{12} - x_{22})(y_{21} + y_{22})z_{11}
\end{aligned}$$

or, equivalently, noticing that $L_q L_q^* = m_q$:

$$\begin{aligned}
\sum_{i,j,k=1}^{2} x_{ij}y_{jk}z_{ki} &= m_1(z_{11} + z_{22}) + m_2(z_{12} - z_{22}) + m_3(z_{21} + z_{22}) \\
&\quad + m_4(z_{11} + z_{12}) + m_5(-z_{11} + z_{21}) + m_6 z_{22} + m_7 z_{11}
\end{aligned}$$

Pan, in his article of 1972, proposed the following trilinear identity, valid for positive, even n:

$$\begin{aligned}
\sum_{i,j,k=0}^{n-1} x_{ij} \cdot y_{jk} \cdot z_{ki} \underset{x..y..z..}{\equiv} &\sum_{\substack{i,j,\,k=0 \\ i+j+k \text{ is even}}}^{n-1} \left(x_{ij} + x_{k+1,i+1}\right)\left(y_{jk} + y_{i+1,j+1}\right)\left(z_{ki} + z_{j+1,k+1}\right) + \\
&- \sum_{i,k=0}^{n-1} x_{k+1,i+1} \sum_{\substack{j=0 \\ j:\, i+j+k \text{ is even}}}^{n-1} \left(y_{jk} + y_{i+1,j+1}\right) z_{ki} + \\
&- \sum_{i,j=0}^{n-1} x_{ij} \cdot y_{i+1,j+1} \sum_{\substack{k=0 \\ k:\, i+j+k \text{ is even}}}^{n-1} \left(z_{ki} + z_{j+1,k+1}\right) + \\
&- \sum_{k,j=0}^{n-1} \sum_{\substack{i=0 \\ i:\, i+j+k \text{ is even}}}^{n-1} \left(x_{ij} + x_{k+1,i+1}\right) y_{jk} \cdot z_{j+1,k+1}
\end{aligned} \tag{3.53}$$

with "border" conditions:

$$f_{qn} = f_{q0};\ f_{nq} = f_{0q}, \quad q = 0, 1, \ldots, n, \quad \mathit{for}\ f = x;\ y;\ z \tag{3.54}$$

In particular, $f_{nn} = f_{00}$. It should be noted that in the matrix multiplication algorithm defined by the trilinear identity (3.53), the numbering of the matrix entries starts from zero.

In order to prove the trilinear identity (3.53), let us first smartly expand the trilinear aggregate and group it into pairs:

$$\begin{aligned}
&\left(x_{ij} + x_{k+1,i+1}\right)\left(y_{jk} + y_{i+1,j+1}\right)\left(z_{ki} + z_{j+1,k+1}\right) \\
&= \left[x_{ij} \cdot y_{jk} \cdot z_{ki} + x_{k+1,i+1} \cdot y_{i+1,j+1} \cdot z_{j+1,k+1}\right] \\
&\quad + \left(x_{k+1,i+1} \cdot y_{i+1,j+1} \cdot z_{ki} + x_{k+1,i+1} \cdot y_{jk} \cdot z_{ki}\right) \\
&\quad + \left(x_{ij} \cdot y_{i+1,j+1} \cdot z_{j+1,k+1} + x_{ij} \cdot y_{i+1,j+1} \cdot z_{ki}\right) \\
&\quad + \left(x_{k+1,i+1} \cdot y_{jk} \cdot z_{j+1,k+1} + x_{ij} \cdot y_{jk} \cdot z_{j+1,k+1}\right)
\end{aligned} \tag{3.55}$$

The core of the proof is to notice, that the condition that $i + j + k$ is even ensures that the sum (3.53) contains all the definitional matrix multiplication terms. Indeed, if $i + j + k$ is even, then $(i + 1) + (j + 1) + (k + 1)$ is odd. So every triple (i, j, k) with even $i + j + k$ value [with conditions (3.54)] has a corresponding one triple $(i + 1, j + 1, k + 1)$, and vice versa. Furthermore, each triple $(i + 1, j + 1, k + 1)$ has a corresponding one permutation $(k + 1,\ i + 1,\ j + 1)$. Thus, for even n, the triples (i, j, k) and $(k + 1,\ i + 1,\ j + 1)$ create separate, equinumerous sets with sum covering the whole Cartesian product $(0,\ 1, \ldots,\ n - 1)^3$. So:

$$\sum_{\substack{i,j,k=0 \\ i+j+k \text{ is even}}}^{n-1} \left[x_{ij} \cdot y_{jk} \cdot z_{ki} + x_{k+1,i+1} \cdot y_{i+1,j+1} \cdot z_{j+1,k+1}\right] = \sum_{i,j,k=0}^{n-1} x_{ij} \cdot y_{jk} \cdot z_{ki} \tag{3.56}$$

It remains to find a way of calculating the correction terms that does inflict the complexity of the trilinear matrix multiplication algorithm in the form (3.53). For the remaining three pairs in (3.55) we can find the following simple identities:

$$\begin{cases}
x_{k+1,i+1} \cdot y_{i+1,j+1} \cdot z_{ki} + x_{k+1,i+1} \cdot y_{jk} \cdot z_{ki} = x_{k+1,i+1}\left(y_{i+1,j+1} + y_{jk}\right)z_{ki} \\
x_{ij} \cdot y_{i+1,j+1} \cdot z_{j+1,k+1} + x_{ij} \cdot y_{i+1,j+1} \cdot z_{ki} = x_{ij} \cdot y_{i+1,j+1}\left(z_{j+1,k+1} + z_{ki}\right) \\
x_{k+1,i+1} \cdot y_{jk} \cdot z_{j+1,k+1} + x_{ij} \cdot y_{jk} \cdot z_{j+1,k+1} = \left(x_{k+1,i+1} + x_{ij}\right)y_{jk} \cdot z_{j+1,k+1}
\end{cases}$$

Now we need to sum the above identities in a time-optimised way. To design an efficient scheme, it is sufficient to use the standard summation methods:

$$\begin{aligned}
\sum_{i,j,k} x_{k+1,i+1}\left(y_{i+1,j+1} + y_{jk}\right)z_{ki} &= \sum_{i,k}\sum_{j} x_{k+1,i+1}\left(y_{i+1,j+1} + y_{jk}\right)z_{ki} \\
&= \sum_{i,k} x_{k+1,i+1}\left[\sum_{j}\left(y_{i+1,j+1} + y_{jk}\right)z_{ki}\right]
\end{aligned}$$

$$= \sum_{i,k} x_{k+1,i+1} \left[\sum_{j} \left(y_{i+1,j+1} + y_{jk} \right) \right] z_{ki}$$

Only in the parenthesis we have mixed i, j and k indices, so we should sum them in the most internal loop. We do the same for the second of the three identities:

$$\begin{aligned}
&\sum_{i,j,k} x_{ij} \cdot y_{i+1,j+1} \left(z_{j+1,k+1} + z_{ki} \right) \\
&= \sum_{i,j} \sum_{k} x_{ij} \cdot y_{i+1,j+1} \left(z_{j+1,k+1} + z_{ki} \right) \\
&= \sum_{i,j} x_{ij} \cdot y_{i+1,j+1} \sum_{k} \left(z_{j+1,k+1} + z_{ki} \right)
\end{aligned}$$

And for the last one:

$$\begin{aligned}
&\sum_{i,j,k} \left(x_{k+1,i+1} + x_{ij} \right) y_{jk} \cdot z_{j+1,k+1} \\
&= \sum_{j,k} \sum_{i} \left(x_{k+1,i+1} + x_{ij} \right) y_{jk} \cdot z_{j+1,k+1} \\
&= \sum_{j,k} \left[\sum_{i} \left(x_{k+1,i+1} + x_{ij} \right) \right] y_{jk} \cdot z_{j+1,k+1}
\end{aligned}$$

In this way we have proved the famous trilinear identity (3.53). To transform the matrix multiplication algorithm expressed by the trilinear identity in question into the more convenient in practice bilinear form (3.51), we define four series of aggregated products:

$$\left\{ \begin{aligned}
P_{ijk}^{(0)} &= \left(x_{ij} + x_{k+1,i+1} \right) \left(y_{jk} + y_{i+1,j+1} \right) \\
P_{ki}^{(1)} &= x_{k+1,i+1} \left[\sum_{j: i+j+k \text{ is even}} \left(y_{i+1,j+1} + y_{jk} \right) \right] \\
P_{ij}^{(2)} &= x_{ij} \cdot y_{i+1,j+1} \\
P_{jk}^{(3)} &= \left[\sum_{i: i+j+k \text{ is even}} \left(x_{k+1,i+1} + x_{ij} \right) \right] y_{jk}
\end{aligned} \right. \tag{3.57}$$

for $i, j, k = 0, \ldots, n-1$. The trilinear identity can be rewritten as:

$$T = \sum_{i,j,k=0}^{n-1} x_{ij} \cdot y_{jk} \cdot z_{ki} \underset{x..y..z..}{\equiv} \sum_{\substack{i,j,k=0 \\ i+j+k \text{ is even}}}^{n-1} P_{ijk}^{(0)} \left(z_{ki} + z_{j+1,k+1} \right) - \sum_{i,k=0}^{n-1} P_{ki}^{(1)} z_{ki} +$$

$$
- \sum_{i,j=0}^{n-1} P_{ij}^{(2)} \sum_{\substack{k=0 \\ k:i+j+k \text{ is even}}}^{n-1} \left(z_{ki} + z_{j+1,k+1}\right) - \sum_{k,j=0}^{n-1} P_{jk}^{(3)} z_{j+1,k+1}
$$

$$
= T_0 - T_1 - T_2 - T_3
$$

with consecutive assignment of the $T_\cdot$ terms to the corresponding sum. To make the coefficients of the $z_\cdot$ variable equal we regroup the T_0, T_2 terms:

$$
T_0 = \sum_{\substack{i,j,k=0 \\ i+j+k \text{ is even}}}^{n-1} P_{ijk}^{(0)} \left(z_{ki} + z_{j+1,k+1}\right)
$$

$$
= \sum_{\substack{i,j,k=0 \\ i+j+k \text{ is even}}}^{n-1} P_{ijk}^{(0)} z_{ki} + \sum_{\substack{i,j,k=0 \\ i+j+k \text{ is even}}}^{n-1} P_{ijk}^{(0)} z_{j+1,k+1}
$$

$$
= \sum_{k,i=0}^{n-1} \left[\sum_{j:i+j+k \text{ is even}} P_{ijk}^{(0)} \right] z_{ki} + \sum_{j,k=0}^{n-1} \left[\sum_{i:i+j+k \text{ is even}} P_{ijk}^{(0)} \right] z_{j+1,k+1}
$$

$$
T_2 = \sum_{i,j=0}^{n-1} P_{ij}^{(2)} \sum_{\substack{k=0 \\ k:i+j+k \text{ is even}}}^{n-1} \left(z_{ki} + z_{j+1,k+1}\right)
$$

$$
= \sum_{\substack{i,j,k=0 \\ i+j+k \text{ is even}}}^{n-1} P_{ij}^{(2)} \left(z_{ki} + z_{j+1,k+1}\right)
$$

$$
= \sum_{k,i=0}^{n-1} \left[\sum_{j:i+j+k \text{ is even}} P_{ij}^{(2)} \right] z_{ki} + \sum_{j,k=0}^{n-1} \left[\sum_{i:i+j+k \text{ is even}} P_{ij}^{(2)} \right] z_{j+1,k+1}
$$

Finally, we regroup the terms with respect to the coefficients z_{ki} and $z_{j+1,k+1}$:

$$
T = \sum_{k,i=0}^{n-1} \left(-P_{ki}^{(1)} + \sum_{j:i+j+k \text{ is even}} \left[P_{ijk}^{(0)} - P_{ij}^{(2)} \right] \right) z_{ki}
$$

$$
+ \sum_{j,k=0}^{n-1} \left(-P_{jk}^{(3)} + \sum_{i:i+j+k \text{ is even}} \left[P_{ijk}^{(0)} - P_{ij}^{(2)} \right] \right) z_{j+1,k+1} \tag{3.58}
$$

Now we associate the coefficients z_{ki} with $(XY)^{(1)}$ matrix, and $z_{j+1,k+1}$ with $(XY)^{(2)}$. We linearly combine the products (3.57) to obtain these matrices:

$$\begin{cases} (XY)^{(1)}_{ik} = -P^{(1)}_{ki} + \sum_{j:i+j+k \text{ is even}} \left[P^{(0)}_{ijk} - P^{(2)}_{ij}\right] \\ (XY)^{(2)}_{k+1,j+1} = -P^{(3)}_{jk} + \sum_{i:i+j+k \text{ is even}} \left[P^{(0)}_{ijk} - P^{(2)}_{ij}\right] \end{cases} \tag{3.59}$$

The desired product XY is obtained as a sum $(XY)^{(1)} + (XY)^{(2)}$. We compute each of the correction terms series $P^{(1)}_{ki}$, $P^{(2)}_{ij}$, $P^{(3)}_{jk}$ in n^2 non-scalar multiplications, also called as bilinear steps. The trilinear aggregate $P^{(0)}_{ijk}$, due to the condition that the sum of the (i, j, k) variables is even, we compute $n^3/2$ times, so the Pan's trilinear equality (3.53) defines the bilinear (non-commutative) matrix multiplication algorithm with $r = n^3/2+3n^2$ non-scalar multiplications. The function $\log_n(r)$ reaches the minimum for $n = 34$, giving the complexity exponent $\omega \leq \log_{34} 23120 = 2.849\,53$.

3.9.1 *Example—4 × 4 Algorithm Obtained from the Trilinear Identity*

We derive a matrix multiplication algorithm for the 4×4 matrix size from the trilinear identity (3.53) in three steps:

- find the particular form of the aggregated products $P^{(0)}_{ijk}$, $P^{(1)}_{ki}$, $P^{(2)}_{ij}$, $P^{(3)}_{jk}$—formula (3.57),
- linearly combine the aggregated product according to formula (3.59), obtaining two auxiliary matrices $(XY)^{(1)}$ and $(XY)^{(2)}$,
- The algorithm is given by sum $XY = (XY)^{(1)} + (XY)^{(2)}$.

In Fig. 3.1 we have illustrated the distribution of the parity in the $i + j + k$ sum. Each entry contains the triple (ijk). We have highlighted in grey those triples, whose sum $i + j + k$ is even.

We can observe—literally—that each "even" triple (i, j, k) has its exclusive corresponding "odd" triple $(k+1,\ i+1,\ j+1)$. For example, for $k = 3$ we have numbered each even triple in the parenthesis. They lay on the fourth board. The corresponding triple, which we have marked with a "A" symbol, is in the zero row of the boards, covering all its odd places. Similarly, for even triples with $k = 2$ the corresponding odd triples are in the second row, and analogously for $k = 0,\ 1$. In total, the sets of odd and even triples have the same cardinality and have the amount $n^3/2$ and their union is the total Cartesian product $(0,\ 1,\ 2,\ 3)^3$.

Series of products $P^{(0)}_{ijk}$, of amount $n^3/2 = 4^3/2 = 32$, and series $P^{(2)}_{ij}$, consisting of $n^2 = 4^2 = 16$ products, have the (general) form independent of the index parity.

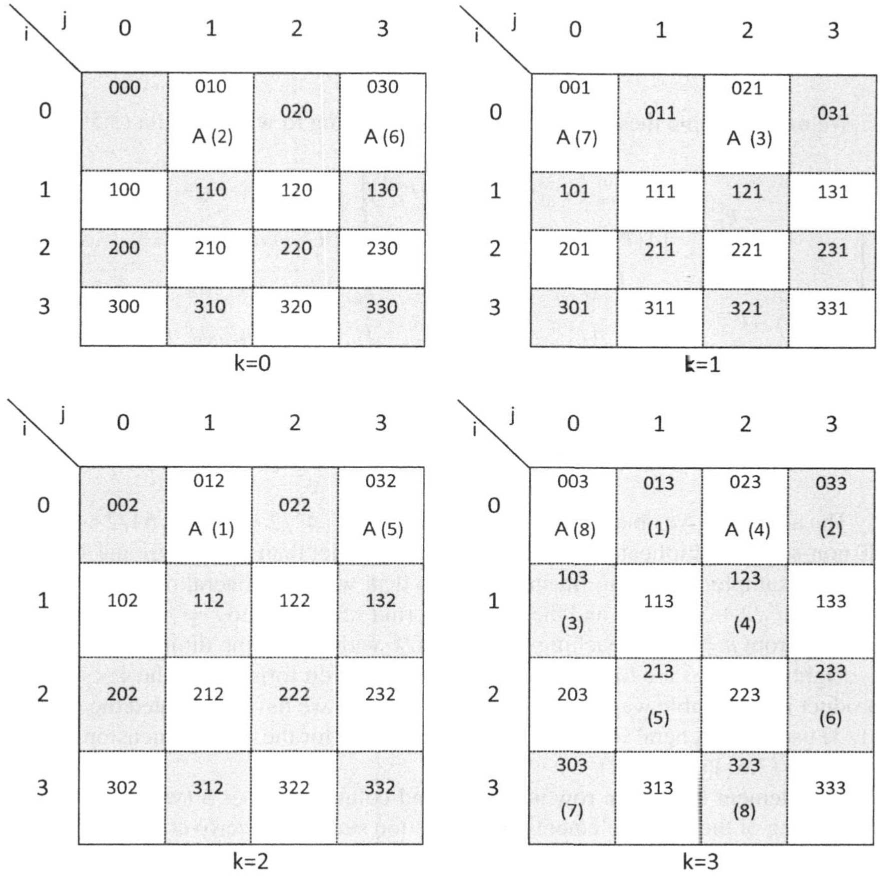

Fig. 3.1 Parity diagram of the $i + j + k$ sum

The remaining 16 products $P_{ki}^{(1)}$ for the matrix dimension $n = 4$ have the following form:

$$P_{ki}^{(1)} = \begin{cases} x_{k+1,i+1}\big[\big(y_{i+1,1} + y_{0k}\big) + \big(y_{i+1,3} + y_{2k}\big)\big], & (k+i)\ even \\ x_{k+1,i+1}\big[\big(y_{i+1,2} + y_{1k}\big) + \big(y_{i+1,0} + y_{3k}\big)\big], & (k+i)\ odd \end{cases} \tag{3.60}$$

because for even $k + i$ to make the whole sum $i + j + k$ even we must take also j even, and for $k + i$ odd to make $i + j + k$ even, we must take odd the j parameter. Moreover, by the virtue (3.54) $y_{i+1,4} = y_{i+1,0}$. Similarly $x_{k+1,4} = x_{k+1,0}$ and for $P_{jk}^{(3)}$ we have 16 products:

$$P_{jk}^{(3)} = \begin{cases} \left[\left(x_{k+1,1} + x_{0j}\right) + \left(x_{k+1,3} + x_{2j}\right)\right] y_{jk}, & (j+k) \ even \\ \left[\left(x_{k+1,2} + x_{1j}\right) + \left(x_{k+1,0} + x_{3j}\right)\right] y_{jk}, & (j+k) \ odd \end{cases} \tag{3.61}$$

We now combine these products linearly according to with formula (3.59):

$$\begin{cases} (XY)_{ik}^{(1)} = -P_{ki}^{(1)} + \begin{cases} \left[\left(P_{i0k}^{(0)} - P_{i0}^{(2)}\right) + \left(P_{i2k}^{(0)} - P_{i2}^{(2)}\right)\right], & (i+k) \ even; \ (j=0,\ 2) \\ \left[\left(P_{i1k}^{(0)} - P_{i1}^{(2)}\right) + \left(P_{i3k}^{(0)} - P_{i3}^{(2)}\right)\right], & (i+k) \ odd; \ (j=1,\ 3) \end{cases} \\ (XY)_{k+1,j+1}^{(2)} = -P_{jk}^{(3)} + \begin{cases} \left[\left(P_{0jk}^{(0)} - P_{0j}^{(2)}\right) + \left(P_{2jk}^{(0)} - P_{2j}^{(2)}\right)\right], & (j+k) \ even; \ (i=0,\ 2) \\ \left[\left(P_{1jk}^{(0)} - P_{1j}^{(2)}\right) + \left(P_{3jk}^{(0)} - P_{3j}^{(2)}\right)\right], & (j+k) \ odd; \ (i=1,\ 3) \end{cases} \end{cases} \tag{3.62}$$

And finally:

$$XY = (XY)^{(1)} + (XY)^{(2)} \tag{3.63}$$

The algorithm we obtained uses $r = n^3/2 + 3n^2 = 4^3/2 + 3 \cdot 4^2 = 64/2 + 3 \cdot 16 = 80$ non-scalar multiplications which is obviously larger than the definitional $4^3 = 64$, but the example is given to illustrate how to deal with the general result. For $n = 6$ we have $n^3/2 + 3n^2 = n^3$ and the trilinear form (3.53) leads to $r = n^3/2 + 3n^2 < n^3$ starting from $n = 8$ approaching the ratio 50% with increasing dimension.

Eighty products are too many to present the explicit formula for the 4×4 matrix product in a readable way, but as an example below we have calculated the element $(1, 3)$ using the general formula (3.59), exemplified for the 4×4 dimension case by formulae (3.61) and (3.62).

The element located in row number 1 and column number 3 (we recall that the numbering of the matrix elements in this section starts from zero) corresponds to z_{31} in the general trilinear formula (3.52). So for $(XY)^{(1)}$ we have $i = 1,\ k = 3$ and for $(XY)^{(2)}$ $j+1 = 3,\ k+1 = 1$, so $j = 2,\ k = 0$. The corresponding entries of $(XY)^{(1)}$ and $(XY)^{(2)}$ we obtain from formula (3.62). For $(XY)^{(1)}$ the sum $i + k = 1 + 3 = 4$ is even and for $(XY)^{(2)}$ the parity of the sum $j + k = 2 + 0 = 2$ is also even, so:

$$\begin{cases} (XY)_{ik=1,3}^{(1)} = -P_{31}^{(1)} + \left[\left(P_{103}^{(0)} - P_{10}^{(2)}\right) + \left(P_{123}^{(0)} - P_{12}^{(2)}\right)\right] \triangleq z_{3,1} \\ (XY)_{k+1,j+1=0+1,2+1}^{(2)} = -P_{20}^{(3)} + \left[\left(P_{020}^{(0)} - P_{02}^{(2)}\right) + \left(P_{220}^{(0)} - P_{22}^{(2)}\right)\right] \triangleq z_{3,1} \end{cases} \tag{3.64}$$

Now all we have to do is to find the non-scalar products in (3.64). $P_{ijk}^{(0)}$ and $P_{ij}^{(2)}$ are calculated directly from the general formula (3.57):

$$\begin{cases} \underset{(ijk)}{P^{(0)}_{103}} = (x_{10} + \underset{=0,2}{x_{3+1,2}})(y_{03} + y_{21}) \\ \underset{(ijk)}{P^{(0)}_{123}} = (x_{12} + \underset{=0,2}{x_{3+1,2}})(y_{23} + y_{23}) \\ \underset{(ijk)}{P^{(0)}_{020}} = (x_{02} + x_{11})(y_{20} + y_{13}) \\ \underset{(ijk)}{P^{(0)}_{220}} = (x_{22} + x_{13})(y_{20} + y_{33}) \end{cases} \quad \begin{cases} \underset{(ij)}{P^{(2)}_{10}} = x_{10}y_{21} \\ \underset{(ij)}{P^{(2)}_{12}} = x_{12}y_{23} \\ \underset{(ij)}{P^{(2)}_{02}} = x_{02}y_{13} \\ \underset{(ij)}{P^{(2)}_{22}} = x_{22}y_{33} \end{cases}$$

The products $P^{(1)}_{ki}$ and $P^{(3)}_{jk}$ are calculated from formulae (3.60) and (3.61):

$$\begin{cases} \underset{(ki)}{P^{(1)}_{31}} = \underset{=0,2}{x_{3+1,2}}\big[(y_{21} + y_{03}) + (y_{23} + y_{23})\big] \\ \underset{(jk)}{P^{(3)}_{20}} = [(x_{11} + x_{02}) + (x_{13} + x_{22})]y_{20} \end{cases}$$

The matrix entry (1,3) we are looking for is given by the following sum, in accordance with (3.63) and then (3.64):

$$(XY)_{13} = (XY)^{(1)}_{13} + (XY)^{(2)}_{13} \tag{3.65}$$

In his article from 1980, Pan presented a similar but more sophisticated algorithm that computes the matrix product in $(n^3 - 4n)/3 + 6n^2$ bilinear steps, i.e. non-scalar multiplications. For $n = 70$ it gives an exponent $\omega_{70} \le \log_{70} 143\,640 = 2.795\ 12$. We present it in detail in Sect. 4.2.

3.10 Double Disjoint Matrix Multiplication

One of the most important techniques for speeding up matrix multiplication is called disjoint matrix multiplication. Its general working principle is to compute more than one matrix product at a time, with less total computational cost than computing each of them separately. Let us start with the algorithm presented in Pan (1984) for computing two independent matrix products, namely

$$\begin{cases} XY \\ UV \end{cases} \tag{3.66}$$

To find an algorithm for calculating both the products XY and UV at once, let us start from the following elementary algebraic identity valid for any $x,\ y,\ z$ and $u,\ v,\ w$ real variables:

$$(x + u)(y + v)(z + w) = xyz + uvw + x(y + v)w + uy(z + w) + (x + u)vz \tag{3.67}$$

Importantly, the above identity (3.67) holds the true without the use of commutativity. Since identity (3.67) holds for all real values, we can supplement the variables by smartly chosen indexes. After assignment $x := x_{ij},\ y := y_{jk},\ z := z_{ki}$ and $u := u_{jk},\ v := v_{ki},\ w := w_{ij}$ identity (3.67) takes the form:

$$\begin{aligned}
&(x_{ij} + u_{jk})(y_{jk} + v_{ki})(z_{ki} + w_{ij}) \\
&= x_{ij}y_{jk}z_{ki} + u_{jk}v_{ki}w_{ij} \\
&\quad + x_{ij}(y_{jk} + v_{ki})w_{ij} \\
&\quad + u_{jk}y_{jk}(z_{ki} + w_{ij}) \\
&\quad + (x_{ij} + u_{jk})v_{ki}z_{ki}
\end{aligned} \tag{3.68}$$

where by $x_{ij},\ y_{jk}$ we mean the respective entries of $X,\ Y$ matrices and by $u_{jk},\ v_{ki}$ entries of $U,\ V$ matrices in products (3.66):

$$\begin{cases} X = \left[x_{ij}\right], & Y = \left[y_{jk}\right] \\ U = \left[u_{jk}\right], & V = \left[v_{ki}\right] \end{cases}$$

with indices that vary in the ranges:

$$i = 0, 1, \ldots, m-1;\ j = 0, 1, \ldots, n-1;\ k = 0, 1, \ldots, p-1 \tag{3.69}$$

Thus the dimensions of matrices $X,\ Y$ are $m \times n,\ \ n \times p$ and of matrices $U,\ V$ are $n \times p,\ \ p \times m$ respectively. The relationship between z_{ki} and w_{ij} and the respective elements of matrices Z and W was clarified in Sect. 3.9 in the part presenting the trilinear representation of the matrix multiplication algorithm.

Let us now sum the identity (3.68) with respect to the three variables $i,\ j,\ k$. We obtain:

$$\begin{aligned}
&\sum_{i,j,k=0}^{m-1,n-1,p-1} (x_{ij} + u_{jk})(y_{jk} + v_{ki})(z_{ki} + w_{ij}) \\
&= \sum_{i,j,k=0}^{m-1,n-1,p-1} \left(x_{ij}y_{jk}z_{ki} + u_{jk}v_{ki}w_{ij}\right) \\
&\quad + \sum_{i,j,k=0}^{m-1,n-1,p-1} \left[x_{ij}(y_{jk} + v_{ki})w_{ij}\right] \\
&\quad + \sum_{i,j,k=0}^{m-1,n-1,p-1} \left[u_{jk}y_{jk}(z_{ki} + w_{ij})\right] \\
&\quad + \sum_{i,j,k=0}^{m-1,n-1,p-1} \left[(x_{ij} + u_{jk})v_{ki}z_{ki}\right]
\end{aligned} \tag{3.70}$$

where the notation $\sum_{i,j,k=0}^{m-1,n-1,p-1}$ means that the upper limit of a given variable $i,\ j,\ k$ can be read from the order of the list of upper limits. In the rest of the chapter, we will omit an explicit indication of the limits of $i,\ j,\ k$, assuming that they are given by the formula (3.69).

In (3.70) we refer to the aggregate:

$$\sum_{i,j,k} (x_{ij} + u_{jk})(y_{jk} + v_{ki})(z_{ki} + w_{ij})$$

denoted as T. Furthermore, we denote by $T_1,\ T_2,\ T_3$ the following correction terms in Eq. (3.70), after their optimisation with respect to the number of necessary non-scalar multiplications:

$$\begin{cases} T_1 = \sum_{i,j,k} \left[x_{ij}(y_{jk} + v_{ki})w_{ij}\right] = \sum_{i,j} \left[x_{ij} \sum_{k} (y_{jk} + v_{ki})\right] w_{ij} \\ T_2 = \sum_{i,j,k} \left[u_{jk} y_{jk}(z_{ki} + w_{ij})\right] = \sum_{j,k} \left[u_{jk} y_{jk}\right] \sum_{i} (z_{ki} - w_{ij}) \\ T_3 = \sum_{i,j,k} \left[(x_{ij} + u_{jk}) v_{ki} z_{ki}\right] = \sum_{k,i} \left[\sum_{j} (x_{ij} + u_{jk}) v_{ki}\right] z_{ki} \end{cases} \tag{3.71}$$

In the trilinear representation described in Sect. 3.9, the two matrix products (3.66) can be represented by the following trilinear form, which appears in (3.70):

$$Trace(XYZ + UVW) = \sum_{i,j,k} (x_{ij} y_{jk} z_{ki} + u_{jk} v_{ki} w_{ij})$$

Now, on the basis of (3.70), the desired trilinear form, representing the disjoint matrix multiplications $XY,\ UV$, can be represented by the aggregate T corrected by $T_1,\ T_2,\ T_3$:

$$Trace(XYZ + UVW) = T - T_1 - T_2 - T_3 \tag{3.72}$$

In order to transform the algorithm (3.72) into a bilinear form, that is closer to implementation, we define the following four series of products:

$$\begin{cases} P_{ijk}^{(0)} = (x_{ij} + u_{jk})(y_{jk} + v_{ki}) \\ P_{ij}^{(1)} = x_{ij} \sum_{k} (y_{jk} + v_{ki}) \\ P_{jk}^{(2)} = u_{jk} y_{jk} \\ P_{ki}^{(3)} = \left[\sum_{j} (x_{ij} + u_{jk})\right] v_{ki} \end{cases} \tag{3.73}$$

Using the aggregated products $P_{ijk}^{(0)}$, $P_{ij}^{(1)}$, $P_{jk}^{(2)}$, $P_{ki}^{(3)}$, we can rewrite the trilinear form (3.72) representing the algorithm as follows (Pan 1984, page 411):

$$\sum_{i,j,k}\left(x_{ij}y_{jk}z_{ki} + u_{jk}v_{ki}w_{ij}\right) = \sum_{i,j,k} P_{ijk}^{(0)}\left(z_{ki} + w_{ij}\right) - \sum_{i,j} P_{ij}^{(1)} w_{ij}$$
$$- \sum_{j,k} P_{jk}^{(2)} \sum_{i}\left(z_{ki} + w_{ij}\right) - \sum_{i,k} P_{ki}^{(3)} z_{ki} \qquad (3.74)$$

In order to obtain the formulae for the first matrix product of a pair XY, UV, we regroup the terms of (3.74) in such a way that the coefficients corresponding to the z_{ki} variable can be equated. The trilinear representation of a product XY, i.e. the sum $\sum_{i,j,k}\left(x_{ij}y_{jk}z_{ki}\right)$, can be rewritten equivalently as:

$$\sum_{i,j,k} x_{ij}y_{jk}z_{ki} = \sum_{i,k}\left[\sum_{j} x_{ij}y_{jk}\right] z_{ki}$$

Now [definition formula (1.1)] it is clear, that the inner sum represents a single entry of the XY product, in its ith row and sth column. In the trilinear representation we multiply it by z_{ki} and then we sum it over all entries i.e. i, k variables. In other words, an (i, k)th entry of the product XY is associated with z_{ki} variable. It is worth noting that here we have a kind of transposition, i.e. $z_{ki} \triangleq (XY)_{ik}$.

To present the formula expressing the XY product in a bilinear form let us return to the part of the trilinear equality (3.74) with z_{ki} variable. Observe:

$$\begin{aligned}
\sum_{i,j,k} x_{ij}y_{jk}z_{ki} &= \sum_{i,j,k} P_{ijk}^{(0)} z_{ki} - \sum_{j,k} P_{jk}^{(2)} \sum_{i} z_{ki} - \sum_{i,k} P_{ki}^{(3)} z_{ki} \\
&= \sum_{i,j,k} P_{ijk}^{(0)} z_{ki} - \sum_{i,j,k} P_{jk}^{(2)} z_{ki} - \sum_{i,k} P_{ki}^{(3)} z_{ki} \\
&= \sum_{i,k}\left[\sum_{j} P_{ijk}^{(0)}\right] z_{ki} - \sum_{i,k}\left[\sum_{j} P_{jk}^{(2)}\right] z_{ki} - \sum_{i,k} P_{ki}^{(3)} z_{ki} \\
&= \sum_{i,k}\left[-P_{ki}^{(3)} + \sum_{j}\left(P_{ijk}^{(0)} - P_{jk}^{(2)}\right)\right] z_{ki}
\end{aligned}$$

For the variable w_{ij}, the procedure is the same and the formulas expressing the desired two products in bilinear form are as follows:

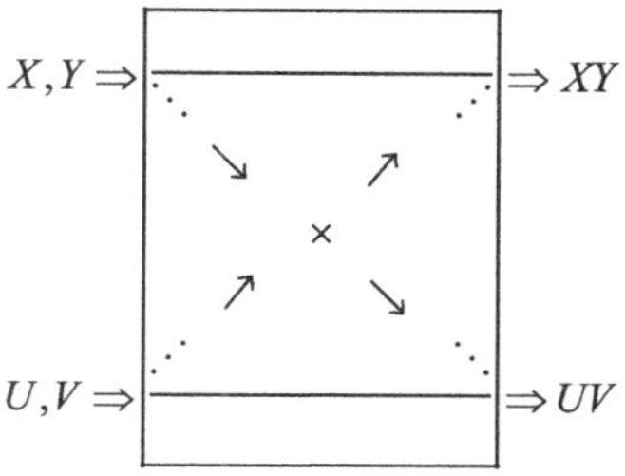

Fig. 3.2 Double disjoint matrix multiplication

$$\begin{cases} (XY)_{ik} = -P^{(3)}_{ki} + \sum_j \left(P^{(0)}_{ijk} - P^{(2)}_{jk}\right) \\ (UV)_{ji} = -P^{(1)}_{ij} + \sum_k \left(P^{(0)}_{ijk} - P^{(2)}_{jk}\right) \end{cases} \tag{3.75}$$

To sum up, considering the i, j, k variables ranges (3.69) and form of the products (3.73), the above formula (3.75) allows us to calculate at once two matrix products, of (m, n, p) and (n, p, m) dimensionality, in:

$$M_{m,n,p} = mnp + mn + np + pm$$

non-scalar products, instead of $2mnp$ non-scalar products by calculating it separately by the very matrix product definition (Pan 1984, page 398).

In particular, for $m = p = n$ we can apply the algorithm to multiply at once two pairs of square matrices with the same dimensions in $n^3 + 3n^2$ non-scalar multiplications, instead of $2n^3$ by the classical definition. The inequality $n^3 + 3n^2 < 2n^3$ shows that such a double disjoint approach requires fewer non-scalar multiplications for $n \geq 4$. The efficiency exponent in the square case is given by the inequality:

$$\omega \leq \log_n\left[M_{n,n,n}/2\right] = \log_n\left[(n^3 + 3n^2)/2\right]$$

which achieves minimum for $n = 13$ equal to $\omega \leq 2.810714$. In Chap. 5 we show how to use the disjoint matrix multiplication algorithm in case of single matrix product. Figure 3.2 graphically illustrates the mechanism of double disjoint matrix multiplication.

3.10.1 Example—Rectangular Double Disjoint Matrix Multiplication

Let us obtain the explicit formulae for double disjoint matrix multiplication in the case $m = 2,\ n = 2,\ p = 3$. Thus the dimensions of the matrices $X,\ Y$ are $2 \times 2,\ 2 \times 3$ and of the matrices $U,\ V$ are $2 \times 3,\ 3 \times 2$ respectively [formula (3.69)]. The indexes i, j, k change in the following ranges:

$$i = 0, 1; \quad j = 0, 1; \quad k = 0, 1, 2$$

Now we want to find all the products. For $P_{ijk}^{(0)}$ series of product we have elements of general formula (3.73) $2 \cdot 2 \cdot 3 = 12$:

$$\begin{cases} P_{000}^{(0)} = (x_{0,0} + u_{0,0})(y_{0,0} + v_{0,0}), \ P_{100}^{(0)} = (x_{1,0} + u_{0,0})(y_{0,0} + v_{0,1}), \ P_{010}^{(0)} = (x_{0,1} + u_{1,0})(y_{1,0} + v_{0,0}) \\ P_{110}^{(0)} = (x_{1,1} + u_{1,0})(y_{1,0} + v_{0,1}), \ P_{001}^{(0)} = (x_{0,0} + u_{0,1})(y_{0,1} + v_{1,0}), \ P_{101}^{(0)} = (x_{1,0} + u_{0,1})(y_{0,1} + v_{1,1}) \\ P_{011}^{(0)} = (x_{0,1} + u_{1,1})(y_{1,1} + v_{1,0}), \ P_{111}^{(0)} = (x_{1,1} + u_{1,1})(y_{1,1} + v_{1,1}), \ P_{002}^{(0)} = (x_{0,0} + u_{0,2})(y_{0,2} + v_{2,0}) \\ P_{102}^{(0)} = (x_{1,0} + u_{0,2})(y_{0,2} + v_{2,1}), \ P_{012}^{(0)} = (x_{0,1} + u_{1,2})(y_{1,2} + v_{2,0}), \ P_{112}^{(0)} = (x_{1,1} + u_{1,2})(y_{1,2} + v_{2,1}) \end{cases}$$

Similarly, for $P_{jk}^{(2)}$ series of products, we have $2 \cdot 3 = 6$ entries:

$$\begin{cases} P_{00}^{(2)} = u_{0,0} y_{0,0}, \ P_{10}^{(2)} = u_{1,0} y_{1,0}, \ P_{01}^{(2)} = u_{0,1} y_{0,1} \\ P_{11}^{(2)} = u_{1,1} y_{1,1}, \ P_{02}^{(2)} = u_{0,2} y_{0,2}, \ P_{12}^{(2)} = u_{1,2} y_{1,2} \end{cases}$$

And finally the series $P_{ij}^{(1)}$ and $P_{ki}^{(3)}$ we can calculate in accordance with (3.73):

$$P_{ij}^{(1)} = x_{ij} \sum_{k=0}^{2} \left(y_{jk} + v_{ki} \right) = x_{ij} \left[\left(y_{j0} + v_{0i} \right) + \left(y_{j1} + v_{1i} \right) + \left(y_{j2} + v_{12} \right) \right]$$

$$P_{ki}^{(3)} = \left[\sum_{j=0}^{1} \left(x_{ij} + u_{jk} \right) \right] v_{ki} = [(x_{i0} + u_{0k}) + (x_{i1} + u_{1k})] v_{ki}$$

So the series $P_{ij}^{(1)}$ have the following $2 \cdot 2 = 4$ entries:

$$\begin{cases} P_{00}^{(1)} = x_{0,0} \left[(y_{0,0} + v_{0,0}) + (y_{0,1} + v_{1,0}) + (y_{0,2} + v_{2,0}) \right] \\ P_{10}^{(1)} = x_{1,0} \left[(y_{0,0} + v_{0,1}) + (y_{0,1} + v_{1,1}) + (y_{0,2} + v_{2,1}) \right] \\ P_{01}^{(1)} = x_{0,1} \left[(y_{1,0} + v_{0,0}) + (y_{1,1} + v_{1,0}) + (y_{1,2} + v_{2,0}) \right] \\ P_{11}^{(1)} = x_{1,1} \left[(y_{1,0} + v_{0,1}) + (y_{1,1} + v_{1,1}) + (y_{1,2} + v_{2,1}) \right] \end{cases}$$

and the series $P_{ki}^{(3)}$ have $3 \cdot 2 = 6$ entries:

$$\begin{cases} P_{00}^{(3)} = \left[(x_{0,0} + u_{0,0}) + (x_{0,1} + u_{1,0}) \right] v_{0,0}, \ P_{01}^{(3)} = \left[(x_{1,0} + u_{0,0}) + (x_{1,1} + u_{1,0}) \right] v_{0,1} \\ P_{10}^{(3)} = \left[(x_{0,0} + u_{0,1}) + (x_{0,1} + u_{1,1}) \right] v_{1,0}, \ P_{11}^{(3)} = \left[(x_{1,0} + u_{0,1}) + (x_{1,1} + u_{1,1}) \right] v_{1,1} \\ P_{20}^{(3)} = \left[(x_{0,0} + u_{0,2}) + (x_{0,1} + u_{1,2}) \right] v_{2,0}, \ P_{21}^{(3)} = \left[(x_{1,0} + u_{0,2}) + (x_{1,1} + u_{1,2}) \right] v_{2,1} \end{cases}$$

To obtain the result i.e. two matrix products XY and UV we combine the products $P_{ijk}^{(0)}$, $P_{ij}^{(1)}$, $P_{jk}^{(2)}$, $P_{ki}^{(3)}$ linearly according to formula (3.75). Observe:

$$XY = \left[\begin{array}{c|c|c} -P^{(3)}_{00} + (P^{(0)}_{000} - P^{(2)}_{00}) + (P^{(0)}_{010} - P^{(2)}_{10}) & -P^{(3)}_{10} + (P^{(0)}_{001} - P^{(2)}_{01}) + (P^{(0)}_{011} - P^{(2)}_{11}) & -P^{(3)}_{20} + (P^{(0)}_{002} - P^{(2)}_{02}) + (P^{(0)}_{012} - P^{(2)}_{12}) \\ \hline -P^{(3)}_{01} + (P^{(0)}_{100} - P^{(2)}_{00}) + (P^{(0)}_{110} - P^{(2)}_{10}) & -P^{(3)}_{11} + (P^{(0)}_{101} - P^{(2)}_{01}) + (P^{(0)}_{111} - P^{(2)}_{11}) & -P^{(3)}_{21} + (P^{(0)}_{102} - P^{(2)}_{02}) + (P^{(0)}_{112} - P^{(2)}_{12}) \end{array}\right]$$

$$UV = \left[\begin{array}{c|c} -P^{(1)}_{00} + (P^{(0)}_{000} - P^{(2)}_{00}) + (P^{(0)}_{001} - P^{(2)}_{01}) + (P^{(0)}_{002} - P^{(2)}_{02}) & -P^{(1)}_{10} + (P^{(0)}_{100} - P^{(2)}_{00}) + (P^{(0)}_{101} - P^{(2)}_{01}) + (P^{(0)}_{102} - P^{(2)}_{02}) \\ \hline -P^{(1)}_{01} + (P^{(0)}_{010} - P^{(2)}_{10}) + (P^{(0)}_{011} - P^{(2)}_{11}) + (P^{(0)}_{012} - P^{(2)}_{12}) & -P^{(1)}_{11} + (P^{(0)}_{110} - P^{(2)}_{10}) + (P^{(0)}_{111} - P^{(2)}_{11}) + (P^{(0)}_{112} - P^{(2)}_{12}) \end{array}\right]$$

The total number of products is $M_{2,2,3} = 2 \cdot 2 \cdot 3 + 2 \cdot 2 + 2 \cdot 3 + 3 \cdot 2 = 28$.

References

Abrahamson K (1991) Time-space tradeoffs for algebraic problems on general sequential machines. J Comput Syst Sci 43(2):269–289

Bini D (1980) Relations between exact and approximate bilinear algorithms. Appl Calcolo 17:87–97

Bini D, Lotti G (1980) Stability of fast algorithms for matrix multiplication. Numerische Mathematik 36(1):63–72

Blaser M (2003) On the complexity of the multiplication of matrices of small formats. J Complex 19(1):43–60

Brent R (1970) Algorithms for matrix multiplication. Technical Report STAN-CS-70-157. Department of Computer Science, Stanford University

Brockett R, Dobkin D (1976) On the number of multiplications required for matrix multiplication. SIAM J Comput 5(4):624–628

Brockett R, Dobkin D (1978) Errata: On the number of multiplications required for matrix multiplication. SIAM J Comput 7(2):238

Bshouty NH (1995) On the additive complexity of 2 × 2 matrix multiplication. Inf Process Lett 56(6):329–335

Buchberger B (1965) Ein Algorithmus zum Auffinden der Basiselemente des Restklassenrings nach einem nulldimensionalen Polynomideal. Ph.D. thesis. Universität Innsbruck

Cormen TH, Leiserson ChE, Rivest RL, Stein C (2001) Introduction to algorithms 2. The MIT Press, Cambridge

Cox D, Little J, OShea D (1992) Ideals, varieties, and algorithms. In: Undergraduate Texts in Mathematics. Springer, Berlin

D'Alberto P, Niculau A (2007) Adaptive Strassen's matrix multiplications. In: Proceedings of the 21st annual international conference on supercomputing, pp 284–292

Egervary E (1954) On hypermatrices whose blocks are commutable in pairs and their application in lattice-dynamics. Acta Sci Math 15:211–222

Fischer PC (1974) Further schemes for combining matrix algorithms. In: Loeck J (ed) LNCS, vol 14. Springer, Berlin, pp 428–436

Friedman B (1961) Eigenvalues of composite matrices. Math Proc Cambridge Philos Society 57(1):37–49

Gantmacher FR (1960) The theory of matrices. Chelsea Publishing Company, New York, NY

Gastinel N (1971) Sur le calcul des produits de matrices. Numer Math 17:222–229 (in French)

Glynn DG (2013) Permanent formulae from the Veronesean. Des Codes Crypt 68(1–3):39–47
Hauenstein JD, Ikenmeyer C, Landsberg JM (2013) Equations for lower bounds on border rank. Exp Math 22(4):372–383
Heule MJH, Kauers M, Seidl M (2021) New ways to multiply 3 × 3-matrices. J Symbolic Comput 104:899–916
Higham NJ (1990) Exploiting fast matrix multiplication within the level 3 BLAS. ACM Trans Math Soft 16(4):352–368
Hopcroft JE, Kerr LR (1968) Some techniques for proving certain simple programs optimal. In: Proceedings on Tenth Annual Symposium on Switching and Automata Theory, pp 36–45
Hopcroft JE, Kerr LR (1969) On minimizing the number of multiplications necessary for matrix multiplication. Technical Report 69-44. Department of Computer Science, Cornell University, Ithaca, NY
Hopcroft JE, Kerr LR (1971) On minimizing the number of multiplications necessary for matrix multiplication. SIAM J Appl Math 20(1):30–36
Hopcroft J, Musinski J (1973) Duality applied to the complexity of matrix multiplication and other bilinear forms. SIAM J Comput 2:159–173
Horn RA, Johnson CR (2012) Matrix analysis. Cambridge University Press, Cambridge
Huang J, Smith TM, Henry GM, Geijn RA (2016) Strassen's algorithm reloaded. In: Proceedings of the International Conference for High Performance Computing, Networking, Storage and Analysis, Salt Lake City, USA
Jacobson MT, Matthews P (1996) Generating uniformly distributed random Latin squares. J Comb des 4(6):405–437
Jianyu H, Yu CD, Geijn RA (2020) Strassen's algorithm reloaded on GPUs. ACM Trans Math Softw (TOMS) 46(1):1–22
Johnson RW, McLoughlin AM (1986) Noncommutative bilinear algorithms for 3 × 3 matrix multiplication. SIAM J Comput 15(2):595–603
Laderman JD (1976) A noncommutative algorithm for multiplying 3x3 matrices using 23 multiplications. Bull Am Math Soc 82(1):126–128
Landsberg JM (2005) The border rank of the multiplication of 2 × 2 matrices is seven. J Am Math Soc 19(2):447–459
Laywine CF, Mullen GL (1998) Discrete mathematics using Latin squares. In: Series in Discrete Mathematics and Optimization. Wiley, New York, NY
Pan V (1972) On schemes for the evaluation of products and inverses of matrices. Uspekhi Matematicheskikh Nauk 27(5):167 (in Russian)
Pan V (1982) Fast matrix multiplication without APA-algorithms. Comput Math Appl 8(5):343–366
Pan V (1984) How can we speed up matrix multiplication? SIAM Rev 26(3):393–415
Paterson MS (1974) Complexity of product and closure algorithms for matrices. In: Proceedings of the International Congress of Mathematicians, Vancouver, pp 483–489
Pease MC (1965) Methods of matrix algebra. Academic Press, New York, London, Vol, p 16
Percus JK (1971) Combinatorial methods, vol 4. Springer-Verlag, New York, NY
Probert RL (1974) On the complexity of symmetric computations. INFOR Inf Syst Oper Res 12:71–86
Probert RL (1976) On the additive complexity of matrix multiplication. SIAM J Comput 5(2):187–203
Probert RL, Fischer PC (1980) Decomposition techniques for matrix multiplication problems. Utilitas Math 18:257–267
Ryser HJ (1963) Combinatorial mathematics. The Carus Mathematical Monographs 14, The Mathematical Association of America
Schumer PD (2004) Mathematical journeys. John Wiley & Sons, Inc., Hoboken, NJ
Scott RF (1880) Theory of determinants. Cambridge University Press, Cambridge
Sedoglavic A (2017) Laderman matrix multiplication algorithm can be constructed using Strassen algorithm and related tensor's isotropies. arXiv:1703.08298
Shao J, Wei W (1992) A formula for the number of Latin squares. Discret Math 110:293–296

Silvester JR (2000) Determinants of block matrices. Math Gaz 84(501):460–467
Smirnov AV (2013) The bilinear complexity and practical algorithms for matrix multiplication. Comput Math Math Phys 53(12):1781–1795
Strassen V (1969) Gaussian elimination is not optimal. Numer Math 13:354–356
van Lint JH, Wilson RM (1992) A course in combinatorics. Cambridge University Press, New York, NY
Waksman A (1970) On Winograd's algorithm for inner products. IEEE Trans Comput 19:360–361
Winograd S (1968) A new algorithm for inner product. IEEE Trans Comput 7:693–694
Winograd S (1973) Some remarks on fast multiplication of polynomials. In: Traub JF (ed) Complexity of sequential and parallel numerical algorithms. Academic Press, New York, pp 189–208
Winograd S (1971) On multiplication of 2 × 2 matrices. In: Linear algebra and its applications, vol 4, pp 381–388
Yuval G (1978) A simple proof of Strassen's result. Inf Process Lett 6(7):285–286

Chapter 4
Triple Disjoint Algorithms with Application to Feasible Matrix Multiplication

The algorithms presented so far have not overcome the complexity of Strassen's algorithm, which is equal to $\omega \approx 2.807\,354$. Moreover, the algorithm for disjoint matrix multiplication from Sect. 3.10, as its name suggests, is aimed at a different problem, but a single matrix product calculation.

The algorithms presented later in the monograph in Chap. 5, in order to achieve their favourable complexity exponents, require the use of recursive methods to transform a given disjoint matrix multiplication algorithm into an algorithm that computes a single matrix product (for these methods we devote a separate Sect. 6.1) and/or to use the Schonhage tau-Theorem (Sect. 6.2.2). This inevitable leads to a large constant hidden in the "big O" notation. Worse, the subclass of arbitrary precision algorithms, described later in Chap. 5, requires yet another series of such calls to work in an exact way.

On the contrary, Victor Pan proposed a branch of feasible fast matrix multiplication algorithms. They achieve a weaker complexity exponent (around 2.77), but in return what makes them stand out is that they do not require a long series of recursion calls to achieve this efficiency. More precisely, they do not need to use the tau-theorem to be efficient, nor the approximation methods (described later in Sect. 6.2).

In this chapter we have described matrix multiplication algorithms constructed on the basis of methods presented in the works of Pan (1978, 1980), Respondek (2023) and Laderman et al. (1992).

In general, the algorithms we present in this chapter are aimed at square matrices and they are constructed in the following order:

- Sections 4.1 and 4.3: Construct the triple disjoint matrix multiplication algorithm, for the problems of equal size. Section 4.3 optimises the basic algorithm from Sect. 4.1.
- Sect. 4.2: Adapt the obtained triple disjoint algorithm to a single disjoint algorithm.

J. S. Respondek, *Fast Matrix Multiplication with Applications*, Studies in Big Data 166,
https://doi.org/10.1007/978-3-031-76930-6_4

4.1 First Step to a Fast Triple Disjoint Matrix Multiplication Algorithm

In this section we present the algorithm for triple disjoint matrix multiplication, which was constructed on the basis of results from Sect. 5 of the paper Pan (1978), Sects. 3–4 of the paper Laderman et al. (1992) and the paper Respondek (2023). Let us consider three pairs of square matrices of the same dimensionality $n \times n$, $A = [a_{ij}]$, $B = [b_{jk}]$, $U = [u_{jk}]$, $V = [v_{ki}]$ and $X = [x_{ki}]$, $Y = [y_{ij}]$ with indices from 0 to $n-1$. Our goal is to obtain the following three matrix products at once, with significantly fewer non-commutative multiplications than the definitional three times to the third power of their dimension:

$$\begin{cases} AB \\ UV \\ XY \end{cases} \tag{4.1}$$

In general, the algorithm we are looking for can be illustrated by the diagram in Fig. 4.1.

This means that each element of a given matrix product participates in the calculation the other two products, but does not affect their resulting value. To design the triple disjoint algorithm, we start from the following elementary algebraic identity:

$$\begin{aligned} abc + uvw + xyz = {} & (a+u+x)(b+v+y)(c+w+z) \\ & - \left[ay(c+w+z) + ub(w+z+c) + xv(z+c+w)\right] \\ & - \left[a(b+v)w + u(v+y)z + x(y+b)c \right. \\ & \left. + (a+u)vc + (u+x)yw + (x+a)bz\right] \\ & - \left[avz + uyc + xbw\right] \end{aligned}$$

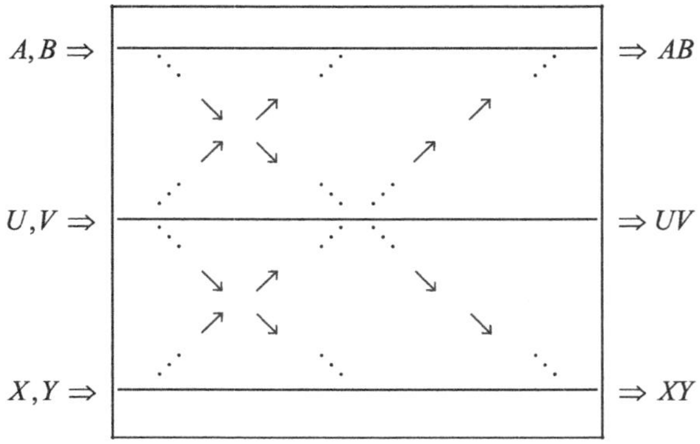

Fig. 4.1 Triple disjoint matrix multiplication

is valid for any real variable without using commutativity. Let us assign $a := a_{ij}$, $b := b_{jk}$, $c := c_{ki}$, $u := u_{jk}$, $v := v_{ki}$, $w := w_{ij}$ and $x := x_{ki}$, $y := y_{ij}$, $z := z_{jk}$:

$$\begin{aligned}
& f(a_{ij}, b_{jk}, c_{ki}, u_{jk}, v_{ki}, w_{ij}, x_{ki}, y_{ij}, z_{jk}) \\
& \quad = a_{ij}b_{jk}c_{ki} + u_{jk}v_{ki}w_{ij} + x_{ki}y_{ij}z_{jk} \\
& \quad = (a_{ij} + u_{jk} + x_{ki})(b_{jk} + v_{ki} + y_{ij})(c_{ki} + w_{ij} + z_{jk}) \\
& \quad\quad \left.\begin{aligned}
& -\big[a_{ij}y_{ij}(c_{ki} + w_{ij} + z_{jk}) + u_{jk}b_{jk}(w_{ij} + z_{jk} + c_{ki}) + x_{ki}v_{ki}(z_{jk} + c_{ki} + w_{ij})\big] \\
& -\big[a_{ij}\big(b_{jk} + v_{ki}\big)w_{ij} + u_{jk}\big(v_{ki} + y_{ij}\big)z_{jk} + x_{ki}\big(y_{ij} + b_{jk}\big)c_{ki} \\
& +\big(a_{ij} + u_{jk}\big)v_{ki}c_{ki} + \big(u_{jk} + x_{ki}\big)y_{ij}w_{ij} + \big(x_{ki} + a_{ij}\big)b_{jk}z_{jk}\big]
\end{aligned}\right\} \tilde{f} \\
& \quad -\big[a_{ij}v_{ki}z_{jk} + u_{jk}y_{ij}c_{ki} + x_{ki}b_{jk}w_{ij}\big] \;\big\}\; \hat{f}
\end{aligned} \tag{4.2}$$

In Eq. (4.2) we can distinguish the following groups of terms:

- The terms directly corresponding to the definition terms of the matrix multiplication, i.e. $a_{ij}b_{jk}c_{ki}$, $u_{jk}v_{ki}w_{ij}$ and $x_{ki}y_{ij}z_{jk}$. They are called *desired terms*, because they represent the matrix multiplication terms in their trilinear representation.
- Single aggregate term $(a_{ij} + u_{jk} + x_{ki})(b_{jk} + v_{ki} + y_{ij})(c_{ki} + w_{ij} + z_{jk})$.
- *Acceptable correction terms*, grouped by $\tilde{f}$ function. They are not desired, because they do not represent matrix multiplication terms, but they are acceptable, because when all the terms are summed at indices i, j and k, they create six series of n^2 number each.
- *Unacceptable correction terms*, grouped by $\hat{f}$ function. They are unacceptable, because they do not represent a matrix multiplication term and, perhaps even worse, each of them, when added together, requires n^3 non-scalar multiplications.

Now it is clear that in order to obtain an efficient triple disjoin matrix multiplication algorithm based on equality (4.2) we have to get rid of the unacceptable correction terms $a_{ij}v_{ki}z_{jk}$, $u_{jk}y_{ij}c_{ki}$ and $x_{ki}b_{jk}w_{ij}$. To do this, Laderman et al. (1992) proposed an ingenious idea in Sect. 4, which we describe below. They proposed to consider each of the triple disjoint products AB, UV and XY as a 2×2 block matrix with $n \times n$ blocks embedded in it. For example, for the AB product we have constructed the matrices as follows:

$$A = \left[\begin{array}{c|c} (a_{ij}) & (a_{ij^+}) \\ \hline (a_{i^+j}) & (a_{i^+j^+}) \end{array}\right], \quad B = \left[\begin{array}{c|c} (b_{jk}) & (b_{jk^+}) \\ \hline (b_{j^+k}) & (b_{j^+k^+}) \end{array}\right], \quad AB = \left[\begin{array}{c|c} (c_{ki}) & (c_{k^+i}) \\ \hline (c_{ki^+}) & (c_{k^+i^+}) \end{array}\right] \tag{4.3}$$

where the indices i, j, k vary in the ranges from 0 to $n - 1$ and the indices $i^+ = i + n$, $j^+ = j + n$ and $k^+ = k + n$. The remaining two products UV and XY we represent analogously:

$$U = \left[\begin{array}{c|c} (u_{jk}) & (u_{jk^+}) \\ \hline (u_{j^+k}) & (u_{j^+k^+}) \end{array}\right], \quad V = \left[\begin{array}{c|c} (v_{ki}) & (v_{ki^+}) \\ \hline (v_{k^+i}) & (v_{k^+i^+}) \end{array}\right], \quad UV = \left[\begin{array}{c|c} (w_{ij}) & (w_{i^+j}) \\ \hline (w_{ij^+}) & (w_{i^+j^+}) \end{array}\right]$$

$$X = \left[\begin{array}{c|c} (x_{ki}) & (x_{ki^+}) \\ \hline (x_{k^+i}) & (x_{k^+i^+}) \end{array}\right], \quad Y = \left[\begin{array}{c|c} (y_{ij}) & (y_{ij^+}) \\ \hline (y_{i^+j}) & (y_{i^+j^+}) \end{array}\right], \quad XY = \left[\begin{array}{c|c} (z_{jk}) & (z_{j^+k}) \\ \hline (z_{jk^+}) & (z_{j^+k^+}) \end{array}\right] \tag{4.4}$$

Now let us sum the trilinear identity (4.2) 8 times, each time with properly chosen series of $a_{ij}, b_{jk}, c_{ki}, u_{jk}, v_{ki}, w_{ij}, x_{ki}, y_{ij}, z_{jk}$ indeterminates. We hope the reader will appreciate the extra spaces we have inserted between the indeterminates belonging to the respective matrix products to make the lecture more readable. Observe:

$$\begin{array}{r|llll}
(1): & f(a_{ij}, b_{jk}, c_{ki}, & u_{jk}, v_{ki}, w_{ij}, & x_{ki}, y_{ij}, z_{jk}) \\
(2): & +f(-a_{ij}, b_{jk^+}, -c_{k^+i}, & u_{j^+k}, v_{ki}, w_{ij^+}, & x_{ki^+}, y_{i^+j}, z_{jk}) \\
(3): & +f(a_{ij^+}, b_{j^+k}, c_{ki}, & -u_{jk}, v_{ki^+}, -w_{i^+j}, & x_{k^+i}, y_{ij}, z_{jk^+}) \\
(4): & +f(a_{i^+j}, b_{jk}, c_{ki^+}, & u_{jk^+}, v_{k^+i}, w_{ij}, & -x_{ki}, y_{ij^+}, -z_{j^+k}) \\
(5): & +f(-a_{i^+j^+}, b_{j^+k}, -c_{ki^+}, & u_{jk^+}, v_{k^+i^+}, w_{i^+j}, & x_{k^+i}, y_{ij^+}, z_{j^+k^+}) \\
(6): & +f(a_{i^+j}, b_{jk^+}, c_{k^+i^+}, & -u_{j^+k^+}, v_{k^+i}, -w_{ij^+}, & x_{ki^+}, y_{i^+j^+}, z_{j^+k}) \\
(7): & +f(a_{ij^+}, b_{j^+k^+}, c_{k^+i}, & u_{j^+k}, v_{ki^+}, w_{i^+j^+}, & -x_{k^+i^+}, y_{i^+j}, -z_{jk^+}) \\
(8): & +f(a_{i^+j^+}, b_{j^+k^+}, c_{k^+i^+}, & u_{j^+k^+}, v_{k^+i^+}, w_{i^+j^+}, & x_{k^+i^+}, y_{i^+j^+}, z_{j^+k^+})
\end{array} \tag{4.5}$$

The above scheme was presented in the form of Tables 2–9 in Sect. 4 of work Laderman et al. (1992). Its analogous version is also presented in articles by Pan (1978, 1980). In practice, this means that in order to obtain the three matrix products AB, UV and XY we have to use—let us call it 'base'—the trilinear equality (4.2) 8 times. In the sum (4.5), each of the unwanted terms appears in some sense in four pairs and, most importantly, in each of the pairs with opposite signs. For example, the general form unwanted term $a_{ij}v_{ki}z_{jk}$ appears in the following four pairs of $f(\cdot)$ calls:

- $a_{ij}v_{ki}z_{jk}$: (call no. 1, with + sign) and (call no. 2, with − sign).
- $a_{i^+j}v_{k^+i}z_{j^+k}$: (call no. 4, with − sign) and (call no. 6, with + sign).
- $a_{ij^+}v_{ki^+}z_{jk^+}$: (call no. 3, with + sign) and (call no. 7, with − sign).
- $a_{i^+j^+}v_{k^+i^+}z_{j^+k^+}$: (call no. 5, with − sign) and (call no. 8, with + sign).

Now it is clear that in the total sum of 8 copies of the trilinear identity (4.2) none of the four forms of the undesired terms $a_{ij}v_{ki}z_{jk}$, i.e. $a_{ij}v_{ki}z_{jk}$, $a_{i^+j}v_{k^+i}z_{j^+k}$, $a_{ij^+}v_{ki^+}z_{jk^+}$ and $a_{i^+j^+}v_{k^+i^+}z_{j^+k^+}$ no more appear.

However, it should be noted that the distribution of the undesired terms pairs of opposite signs among the 8 summands in (4.5) is different for each of the 3 series of unwanted terms $a_{ij}v_{ki}z_{jk}$, $u_{jk}y_{ij}c_{ki}$ and $x_{ki}b_{jk}w_{ij}$. For example, calls number 4 and 6 of the trilinear form $f(\cdot)$, diagram (4.5), return $-(a_{i^+j}v_{k^+i}z_{j^+k})$ and $(a_{i^+j}v_{k^+i}z_{j^+k})$ values for the general $a_{ij}v_{ki}z_{jk}$ term, respectively. So the cancellation works for this unwanted term. But for the remaining two unwanted terms, these calls return values with different undefined variables and/or with the signs that do not cancel each other, i.e. return values $u_{jk^+}y_{ij^+}c_{ki^+}$, $-u_{j^+k^+}y_{i^+j^+}c_{k^+i^+}$ and $-x_{ki}b_{jk}w_{ij}$, $-x_{ki^+}b_{jk^+}w_{ij^+}$, respectively. In Table 4.1 we have summarized the distribution of the undesired pairs of terms with opposite signs among the respective function calls, according to diagram (4.5).

Let us show how to obtain the matrix products AB, UV and XY in the explicit, bilinear form, using the example of the (c_{ki^+}) block of the product AB. We have chosen this element because performing the reasoning for a block, for example (c_{ki}), would hide some essential aspects of the construction.

From the very definition of matrix multiplication (1.1) in its block version (3.2) the c_{ki^+} element in the notation of diagram (4.3) is equal to:

$$c_{ki^+} = \sum_j \left(a_{i^+j}b_{jk} + a_{i^+j^+}b_{j^+k}\right)$$

From the total trilinear sum (4.5), we now select those terms that contain c_{ki^+} terms. They appear in the calls no. 4 and 5, numbered according to the diagram in question (4.5). Observe:

$$\underbrace{a_{i^+j}b_{jk}c_{ki^+}}_{\substack{f(\cdot)\text{ call no. (4)}^{(*)}\\ ^{(*)}\text{(left part)}}} + \underbrace{\left(-a_{i^+j^+}\right)b_{j^+k}\left(-c_{ki^+}\right)}_{\substack{f(\cdot)\text{ call no. (5)}^{(*)}\\ ^{(*)}\text{(left part)}}} = \left(a_{i^+j}b_{jk} + a_{i^+j^+}b_{j^+k}\right)c_{ki^+}$$

$$=$$

$$\left.\begin{aligned}
&= (a_{i^+j} + u_{jk^+} - x_{ki})(b_{jk} + v_{k^+i} + y_{ij^+})c_{ki^+} + \\
&-\left[a_{i^+j}y_{ij^+}c_{ki^+} + u_{jk^+}b_{jk}c_{ki^+} + (-x_{ki})v_{k^+i}c_{ki^+}\right] + \\
&-\left[(-x_{ki})\left(y_{ij^+} + b_{jk}\right)c_{ki^+} + \left(a_{i^+j} + u_{jk^+}\right)v_{k^+i}c_{ki^+}\right]
\end{aligned}\right\} \begin{gathered} f(\cdot)\text{ call no. (4)}^{(*)} \\ ^{(*)}\text{(right part)}\end{gathered}$$

$$+$$

$$\left.\begin{aligned}
&+(-a_{i^+j^+} + u_{jk^+} + x_{k^+i})(b_{j^+k} + v_{k^+i^+} + y_{ij^+})(-c_{ki^+}) + \\
&-\left[(-a_{i^+j^+})y_{ij^+}(-c_{ki^+}) + u_{jk^+}b_{j^+k}(-c_{ki^+}) + x_{k^+i}v_{k^+i^+}(-c_{ki^+})\right] + \\
&-\left[x_{k^+i}\left(y_{ij^+} + b_{j^+k}\right)(-c_{ki^+}) + \left(-a_{i^+j^+} + u_{jk^+}\right)v_{k^+i^+}(-c_{ki^+})\right]
\end{aligned}\right\} \begin{gathered} f(\cdot)\text{ call no. (5)}^{(*)} \\ ^{(*)}\text{(right part)}\end{gathered}$$

Table 4.1 Distribution of the corresponding pairs of unwanted terms with opposite signs

$a_{ij}v_{ki}z_{jk}$			$u_{jk}y_{ij}c_{ki}$			$x_{ki}b_{jk}w_{ij}$		
$a_{ij}v_{ki}z_{jk}$	(1) +	(2) −	$u_{jk}y_{ij}c_{ki}$	(1) +	(3) −	$x_{ki}b_{jk}w_{ij}$	(1) +	(4) −
$a_{i^+j}v_{k^+i}z_{j^+k}$	(4) −	(6) +	$u_{j^+k}y_{i^+j}c_{k^+i}$	(2) −	(7) +	$x_{k^+i}b_{j^+k}w_{i^+j}$	(3) −	(5) +
$a_{ij^+}v_{ki^+}z_{jk^+}$	(3) +	(7) −	$u_{jk^+}y_{ij^+}c_{ki^+}$	(4) +	(5) −	$x_{ki^+}b_{jk^+}w_{ij^+}$	(2) +	(6) −
$a_{i^+j^+}v_{k^+i^+}z_{j^+k^+}$	(5) −	(8) +	$u_{j^+k^+}y_{i^+j^+}c_{k^+i^+}$	(6) −	(8) +	$x_{k^+i^+}b_{j^+k^+}w_{i^+j^+}$	(7) −	(8) +

On the right-hand side, we can further rearrange with respect to c_{ki+} terms, while retaining the allocation to separate $f(\cdot)$ calls. Notice how the 'main' summation operator changes its sign:

$$\begin{aligned}
&\left(a_{i+j}b_{jk} + a_{i+j+}b_{j+k}\right)c_{ki+} \\
&= \Big\{(a_{i+j} + u_{jk+} - x_{ki})(b_{jk} + v_{k+i} + y_{ij+}) - \left[a_{i+j}y_{ij+} + u_{jk+}b_{jk} + (-x_{ki})v_{k+i}\right] \\
&-\left[(-x_{ki})\left(y_{ij+} + b_{jk}\right) + \left(a_{i+j} + u_{jk+}\right)v_{k+i}\right]\Big\}c_{ki+} \\
&- \Big\{(-a_{i+j+} + u_{jk+} + x_{k+i})(b_{j+k} + v_{k+i+} + y_{ij+}) - \left[(-a_{i+j+})y_{ij+} + u_{jk+}b_{j+k} + x_{k+i}v_{k+i+}\right] \\
&-\left[x_{k+i}\left(y_{ij+} + b_{j+k}\right) + \left(-a_{i+j+} + u_{jk+}\right)v_{k+i+}\right]\Big\}c_{ki+}
\end{aligned} \tag{4.6}$$

We can see that in the above trilinear equality (4.6) we can isolate two parts with the following general structure (we deliberately return to the form without grouping it according to the c_{ki+} terms).

$$\begin{aligned}
&(a_{ij} + u_{jk} + x_{ki})(b_{jk} + v_{ki} + y_{ij})c_{ki} - \left[a_{ij}y_{ij}c_{ki} + u_{jk}b_{jk}c_{ki} + x_{ki}v_{ki}c_{ki}\right] \\
&\quad - \left[x_{ki}\left(y_{ij} + b_{jk}\right)c_{ki} + \left(a_{ij} + u_{jk}\right)v_{ki}c_{ki}\right]
\end{aligned} \tag{4.7}$$

Now it is time to sum up the trilinear relation (4.7) with respect to the i, j and k indices. After optimising with respect to the number of terms, analogous to the expressions used e.g. in Eq. (3.71) in the section on double disjoint matrix multiplication 3.10, we obtain:

$$\begin{aligned}
&\sum_{i,j,k} (a_{ij} + u_{jk} + x_{ki})(b_{jk} + v_{ki} + y_{ij})c_{ki} \\
&\quad - \sum_{i,j} a_{ij}y_{ij} \sum_{k} c_{ki} - \sum_{j,k} u_{jk}b_{jk} \sum_{i} c_{ki} - \sum_{k,i} x_{ki}v_{ki} \sum_{j} c_{ki} \\
&\quad - \sum_{k,i} x_{ki}\left[\sum_{j} \left(y_{ij} + b_{jk}\right)\right]c_{ki} - \sum_{k,i}\left[\sum_{j} \left(a_{ij} + u_{jk}\right)v_{ki}\right]c_{ki}
\end{aligned} \tag{4.8}$$

In order to present the result in the bilinear form—as in previous sections, especially those devoted to double disjoint matrix multiplication—let us define some auxiliary bilinear product series. The first product is the main aggregate, which depends on three indices:

$$P_{ijk}\left(a_{ij}, u_{jk}, x_{ki}, b_{jk}, v_{ki}, y_{ij}\right) = (a_{ij} + u_{jk} + x_{ki})(b_{jk} + v_{ki} + y_{ij})$$

There are now two groups of acceptable correction terms, depending on two indices. The first group is as follows:

$$\begin{cases} P_{ij}^{(0)}\left(a_{ij}, y_{ij}\right) = a_{ij}y_{ij} \\ P_{jk}^{(1)}\left(u_{jk}, b_{jk}\right) = u_{jk}b_{jk} \\ P_{ki}^{(2)}\left(x_{ki}, v_{ki}\right) = x_{ki}v_{ki} \end{cases}$$

and the last one group of terms is:

$$\begin{cases} Q_{ij}^{(0)}\left(a_{ij}, b_{jk}, v_{ki}\right) = a_{ij}\left[\sum_{k}\left(b_{jk} + v_{ki}\right)\right] \\ Q_{jk}^{(0)}\left(u_{jk}, v_{ki}, y_{ij}\right) = u_{jk}\left[\sum_{i}\left(v_{ki} + y_{ij}\right)\right] \\ Q_{ki}^{(0)}\left(x_{ki}, y_{ij}, b_{jk}\right) = x_{ki}\left[\sum_{j}\left(y_{ij} + b_{jk}\right)\right] \end{cases}, \quad \begin{cases} Q_{ij}^{(1)}\left(u_{jk}, x_{ki}, y_{ij}\right) = \left[\sum_{k}\left(u_{jk} + x_{ki}\right)\right]y_{ij} \\ Q_{jk}^{(1)}\left(x_{ki}, a_{ij}, b_{jk}\right) = \left[\sum_{i}\left(x_{ki} + a_{ij}\right)\right]b_{jk} \\ Q_{ki}^{(1)}\left(a_{ij}, u_{jk}, v_{ki}\right) = \left[\sum_{j}\left(a_{ij} + u_{jk}\right)\right]v_{ki} \end{cases} \tag{4.9}$$

Let us rewrite the trilinear relation (4.8) using the above bilinear products P_{ijk}, $P_{..}^{(\cdot)}$ and $Q_{..}^{(\cdot)}$. It is equal to:

$$\sum_{i,j,k} P_{ijk}c_{ki} - \left[\sum_{i,j} P_{ij}^{(0)} \sum_{k} c_{ki} + \sum_{j,k} P_{jk}^{(1)} \sum_{i} c_{ki} + \sum_{k,i} P_{ki}^{(2)} \sum_{j} c_{ki}\right] - \left[\sum_{k,i} Q_{ki}^{(0)} c_{ki} + \sum_{k,i} Q_{ki}^{(1)} c_{ki}\right] \tag{4.10}$$

To obtain the c_{ki} value that we are looking for, it is sufficient to rewrite the respective sums in the trilinear expression (4.10) in the form with indexes i and k shifted to the outer sum. Observe:

$$\begin{aligned} &\sum_{i,j,k} P_{ijk}c_{ki} - \sum_{i,j,k} P_{ij}^{(0)}c_{ki} - \sum_{i,j,k} P_{jk}^{(1)}c_{ki} - \sum_{i,j,k} P_{ki}^{(2)}c_{ki} - \sum_{k,i} Q_{ki}^{(0)}c_{ki} - \sum_{k,i} Q_{ki}^{(1)}c_{ki} \\ &= \sum_{i,k}\left[\sum_{j}\left(P_{ijk} - P_{ij}^{(0)} - P_{jk}^{(1)} - P_{ki}^{(2)}\right) - Q_{ki}^{(0)} - Q_{ki}^{(1)}\right]c_{ki} \end{aligned} \tag{4.11}$$

The equality (4.11) can be optimised,[1] because the $P_{ki}^{(2)}$ term does not depend on j, i.e. the $P_{ki}^{(2)}$ term is simply added n times. Finally, we can write the general sum of

[1] In the next section, where we will apply the considered triple disjoint matrix multiplication algorithm to a single matrix product, such an optimisation is impossible, because we no longer use the set sum $\bigcup_{i,j,k}(i, j, k)$ indexes that create a cube $(0, 1, ..., n-1)^3$. Thus the inner sum index j range will depend on the current values of the remaining indices i, k.

the expression (4.7) over all i, j and k indices in the following form:

$$\sum_{i,j,k}\left\{(a_{ij}+u_{jk}+x_{ki})(b_{jk}+v_{ki}+y_{ij})c_{ki}-\left[a_{ij}y_{ij}c_{ki}+u_{jk}b_{jk}c_{ki}+x_{ki}v_{ki}c_{ki}\right]\right.$$
$$\left.-\left[x_{ki}\left(y_{ij}+b_{jk}\right)c_{ki}+\left(a_{ij}+u_{jk}\right)v_{ki}c_{ki}\right]\right\}=\sum_{i,k}\left(\widehat{AB}\right)_{ik}c_{ki} \tag{4.12}$$

where (denoted $\left(\widehat{AB}\right)_{ik}\left(a_{ij}, b_{jk}, u_{jk}, v_{ki}, x_{ki}, y_{ij}\right)$ as $\left(\widehat{AB}\right)_{ik}$, $P_{ijk}\left(a_{ij}, u_{jk}, x_{ki}, b_{jk}, v_{ki}, y_{ij}\right)$ as P_{ijk} etc. for brevity):

$$\left(\widehat{AB}\right)_{ik}=\sum_{j}\left(P_{ijk}-P_{ij}^{(0)}-P_{jk}^{(1)}\right)-nP_{ki}^{(2)}-Q_{ki}^{(0)}-Q_{ki}^{(1)} \tag{4.13}$$

We have marked it with a caret because it does not directly express the matrix product. It does not contain all the terms of the equality (4.2) on which it is based. The unacceptable correction terms, denoted by $\hat{f}$, are not included. Nevertheless, such an expression $\left(\widehat{AB}\right)_{ik}$ is useful because after applying the 2×2 block scheme, these unacceptable terms disappear when they are summed up, as shown in diagram (4.5) and Table 4.1.

Finally, we can write down the trilinear expression for this block (c_{ki^+}) of the AB matrix product using the expression $\left(\widehat{AB}\right)_{ik}$. To do this, we sum up the equality (4.6) with respect to the i, j and k indices, twice using the general trilinear equality (4.12):

$$\sum_{i,j,k}\left(a_{i^+j}b_{jk}+a_{i^+j^+}b_{j^+k}\right)c_{ki^+}=\sum_{i,k}\left[\sum_{j}\left(a_{i^+j}b_{jk}+a_{i^+j^+}b_{j^+k}\right)\right]c_{ki^+}$$
$$=\sum_{i,k}\left[\left(\widehat{AB}\right)_{ik}(a_{i^+j}, b_{jk}, u_{jk^+}, v_{k^+i}, -x_{ki}, y_{ij^+})\right.$$
$$\left.-\left(\widehat{AB}\right)_{ik}(-a_{i^+j^+}, b_{j^+k}, u_{jk^+}, v_{k^+i^+}, x_{k^+i}, y_{ij^+})\right]c_{ki^+}$$

And the explicit formula for the elements of the (c_{ki^+}) block is:

$$\begin{cases} c_{ki^+}=\sum_{j}\left(a_{i^+j}b_{jk}+a_{i^+j^+}b_{j^+k}\right) \\ \quad=\left(\widehat{AB}\right)_{ik}(a_{i^+j}, b_{jk}, u_{jk^+}, v_{k^+i}, -x_{ki}, y_{ij^+}) \\ \quad-\left(\widehat{AB}\right)_{ik}(-a_{i^+j^+}, b_{j^+k}, u_{jk^+}, v_{k^+i^+}, x_{k^+i}, y_{ij^+}) \end{cases}$$

where $\left(\widehat{AB}\right)_{ik}$ is given by the linear combination of the bilinear products (4.13). By analogy, we obtain the entries of the remaining blocks of the C product:

$$\begin{cases} c_{ki} = \left(\widehat{AB}\right)_{ik}(a_{ij}, b_{jk}, u_{jk}, v_{ki}, x_{ki}, y_{ij}) \\ \quad +\left(\widehat{AB}\right)_{ik}(a_{ij^+}, b_{j^+k}, -u_{jk}, v_{ki^+}x_{k^+i}, y_{ij}) \\ c_{k^+i} = \left(\widehat{AB}\right)_{ik}(a_{ij^+}, b_{j^+k^+}, u_{j^+k}, v_{ki^+}, -x_{k^+i^+}, y_{i^+j}) \\ \quad -\left(\widehat{AB}\right)_{ik}(-a_{ij}, b_{jk^+}, u_{j^+k}, v_{ki}, x_{ki^+}, y_{i^+j}) \\ c_{k^+i^+} = \left(\widehat{AB}\right)_{ik}(a_{i^+j}, b_{jk^+}, -u_{j^+k^+}, v_{k^+i}, x_{ki^+}, y_{i^+j^+}) \\ \quad +\left(\widehat{AB}\right)_{ik}(a_{i^+j^+}, b_{j^+k^+}, u_{j^+k^+}, v_{k^+i^+}x_{k^+i^+}, y_{i^+j^+}) \end{cases}$$

Using the same construction, we obtain the bilinear expressions for the remaining two matrix products. Namely, for the UV product we obtain:

$$\begin{cases} w_{ij} = \left(\widehat{UV}\right)_{ji}(a_{ij}, b_{jk}, u_{jk}, v_{ki}, x_{ki}, y_{ij}) \\ \quad +\left(\widehat{UV}\right)_{ji}(a_{i^+j}, b_{jk}, u_{jk^+}, v_{k^+i}, -x_{ki}, y_{ij^+}) \\ w_{ij^+} = \left(\widehat{UV}\right)_{ji}(-a_{ij}, b_{jk^+}, u_{j^+k}, v_{ki}, x_{ki^+}, y_{i^+j}) \\ \quad -\left(\widehat{UV}\right)_{ji}(a_{i^+j}, b_{jk^+}, -u_{j^+k^+}, v_{k^+i}, x_{ki^+}, y_{i^+j^+}) \\ w_{i^+j} = -\left(\widehat{UV}\right)_{ji}(a_{ij^+}, b_{j^+k}, -u_{jk}, v_{ki^+}, x_{k^+i}, y_{ij}) \\ \quad +\left(\widehat{UV}\right)_{ji}(-a_{i^+j^+}, b_{j^+k}, u_{jk^+}, v_{k^+i^+}, x_{k^+i}, y_{ij^+}) \\ w_{i^+j^+} = \left(\widehat{UV}\right)_{ji}(a_{ij^+}, b_{j^+k^+}u_{j^+k}, v_{ki^+} - x_{k^+i^+}, y_{i^+j}) \\ \quad +\left(\widehat{UV}\right)_{ji}(a_{i^+j^+}, b_{j^+k^+}u_{j^+k^+}, v_{k^+i^+}, x_{k^+i^+}, y_{i^+j^+}) \end{cases}$$

where $\left(\widehat{UV}\right)_{ji}$ is the following linear combination:

$$\left(\widehat{UV}\right)_{ji} = \sum_k \left(P_{ijk} - P_{jk}^{(1)} - P_{ki}^{(2)}\right) - nP_{ij}^{(0)} - Q_{ij}^{(0)} - Q_{ij}^{(1)}$$

And for the third product XY:

$$\begin{cases} z_{jk} = \left(\widehat{XY}\right)_{kj}(a_{ij}, b_{jk}, u_{jk}, v_{ki}, x_{ki}, y_{ij}) \\ \quad +\left(\widehat{XY}\right)_{kj}(-a_{ij}, b_{jk^+}, u_{j^+k}, v_{ki}, x_{ki^+}, y_{i^+j}) \\ z_{jk^+} = \left(\widehat{XY}\right)_{kj}(a_{ij^+}, b_{j^+k}, -u_{jk}, v_{ki^+}, x_{k^+i}, y_{ij}) \\ \quad -\left(\widehat{XY}\right)_{kj}(a_{ij^+}, b_{j^+k^+}, u_{j^+k}, v_{ki^+}, -x_{k^+i^+}, y_{i^+j}) \\ z_{j^+k} = -\left(\widehat{XY}\right)_{kj}(a_{i^+j}, b_{jk}, u_{jk^+}, v_{k^+i}, -x_{ki}, y_{ij^+}) \\ \quad +\left(\widehat{XY}\right)_{kj}(a_{i^+j}, b_{jk^+} - u_{j^+k^+}, v_{k^+i}, x_{ki^+}, y_{i^+j^+}) \\ z_{j^+k^+} = \left(\widehat{XY}\right)_{kj}(-a_{i^+j^+}, b_{j^+k}, u_{jk^+}, v_{k^+i^+}, x_{k^+i}, y_{ij^+}) \\ \quad +\left(\widehat{XY}\right)_{kj}(a_{i^+j^+}, b_{j^+k^+}, u_{j^+k^+}, v_{k^+i^+}, x_{k^+i^+}, y_{i^+j^+}) \end{cases} \tag{4.14}$$

where, as before, $\left(\widehat{XY}\right)_{kj}$ is a linear combination of bilinear products:

$$\left(\widehat{XY}\right)_{kj} = \sum_i \left(P_{ijk} - P_{ij}^{(0)} - P_{ki}^{(2)}\right) - nP_{jk}^{(1)} - Q_{jk}^{(0)} - Q_{jk}^{(1)} \tag{4.15}$$

To summarise, in this sub-section we have obtained a triple disjoint matrix multiplication algorithm, which allows us to compute three matrix products at once. Each of the products is made with the same, even square dimensionality $(2n, 2n, 2n)$ which requires:

$$M^{(1)}_{\text{disjoint}}(2n) = 8\left(n^3 + 9n^2\right) = (2n)^3 + 18(2n)^2 \tag{4.16}$$

non-commutative products, with the exponent equation:

$$\omega \le \log_{2n}\left[M^{(1)}_{\text{disjoint}}(2n)/3\right] = \log_{2n}\left[\frac{(2n)^3}{3} + 6(2n)^2\right]$$

with minimum $\omega \le 2.795\,292$ obtained for $2n = 72$. Moreover, information when it gives less products than carrying out these three matrix multiplications from the definition we obtain from the inequality:

$$(2n)^3 + 18(2n)^2 < 3 \cdot (2n)^3$$

which is fulfilled from $2n = 10$.

4.1.1 Example—First Step to Triple Disjoint Matrix Multiplication

Giving a readable example is not an obvious task, because the number of products $M(2n)$ increases rapidly (equality (4.16)). Even with n equal to 2, the expression (4.16) produces as many products as $(2 \cdot 2)^3 + 18(2 \cdot 2)^2 = 4^3 + 18 \cdot 4^2 = 352$. However, in order to make the lecture as clear as possible, we have decided to illustrate how the algorithm in question works in practice, using the example of part of one of the products AB, XY and UV. Namely, let us present in detail the construction of the z_{j^+k} block of the XY product for the matrices dimensions equal $2n \times 2n = 4 \times 4$, thus the z_{j^+k} block dimensionality is $n \times n = 2 \times 2$. Its entries are given by the following sum (Eq. (4.14)):

$$\begin{aligned} z_{j^+k} = &-\left(\widehat{XY}\right)_{kj}(a_{i^+j}, b_{jk}, u_{jk^+}, v_{k^+i}, -x_{ki}, y_{ij^+}) \\ &+\left(\widehat{XY}\right)_{kj}(a_{i^+j}, b_{jk^+} - u_{j^+k^+}, v_{k^+i}, x_{ki^+}, y_{i^+j^+}) \end{aligned} \tag{4.17}$$

where $\left(\widehat{XY}\right)_{kj}$ is the linear combination given by equality (4.15). Let us expand this linear combination. If we assume that all the products work on the series of indeterminates $(a_{ij}, b_{jk}, u_{jk}, v_{ki}, x_{ki}, y_{ij})$, we obtain it in a general form:

$$\begin{aligned} \left(\widehat{XY}\right)_{kj} &= \left(P_{0jk} - P_{0j}^{(0)} - P_{k0}^{(2)}\right) + \left(P_{1jk} - P_{1j}^{(0)} - P_{k1}^{(2)}\right) - 2P_{jk}^{(1)} - Q_{jk}^{(0)} - Q_{jk}^{(1)} \\ &= \left[(a_{0j} + u_{jk} + x_{k0})(b_{jk} + v_{k0} + y_{0j}) - a_{0j}y_{0j} - x_{k0}v_{k0}\right] \\ &\quad + \left[(a_{1j} + u_{jk} + x_{k1})(b_{jk} + v_{k1} + y_{1j}) - a_{1j}y_{1j} - x_{k1}v_{k1}\right] \\ &\quad - 2u_{jk}b_{jk} - u_{jk}\left[\left(v_{k0} + y_{0j}\right) + \left(v_{k1} + y_{1j}\right)\right] \\ &\quad - \left[\left(x_{k0} + a_{0j}\right) + \left(x_{k1} + a_{1j}\right)\right]b_{jk} \end{aligned} \tag{4.18}$$

To obtain the single block z_{j^+k} we must use the linear combination (4.18) twice for different indeterminate series, in accordance with equality (4.17). Considering that $i^+ = i + n$ etc., with $n = 2$ we obtain (now the signs and indices of indeterminates play a crucial role, so they must be explicitly stated):

$$\begin{aligned} &\left(\widehat{XY}\right)_{kj}(a_{i^+j}, b_{jk}, u_{jk^+}, v_{k^+i}, -x_{ki}, y_{ij^+}) \\ &= \left[(a_{2j} + u_{j(k+2)} - x_{k0})(b_{jk} + v_{(k+2)0} + y_{0(j+2)}) - a_{2j}y_{0(j+2)} + x_{k0}v_{(k+2)0}\right] \\ &\quad + \left[(a_{3j} + u_{j(k+2)} - x_{k1})(b_{jk} + v_{(k+2)1} + y_{1(j+2)}) - a_{3j}y_{1(j+2)} + x_{k1}v_{(k+2)1}\right] \\ &\quad - 2u_{j(k+2)}b_{jk} - u_{j(k+2)} \\ &\quad \times \left[\left(v_{(k+2)0} + y_{0(j+2)}\right) + \left(v_{(k+2)1} + y_{1(j+2)}\right)\right] - \left[\left(-x_{k0} + a_{2j}\right) + \left(-x_{k1} + a_{3j}\right)\right]b_{jk} \\ &= \left(a_{2j}v_{(k+2)0} + a_{3j}v_{(k+2)1}\right) - \left(x_{k0}y_{0(j+2)} + x_{k1}y_{1(j+2)}\right) \end{aligned}$$

and:

$$\begin{aligned}
&\left(\widehat{XY}\right)_{kj}(a_{i^+j}, b_{jk^+} - u_{j^+k^+}, v_{k^+i}, x_{ki^+}, y_{i^+j^+}) \\
&= \left[(a_{2j} - u_{(j+2)(k+2)} + x_{k2})(b_{j(k+2)} + v_{(k+2)0} + y_{2(j+2)}) - a_{2j}y_{2(j+2)} - x_{k2}v_{(k+2)0}\right] \\
&\quad + \left[(a_{3j} - u_{(j+2)(k+2)} + x_{k3})(b_{j(k+2)} + v_{(k+2)1} + y_{3(j+2)}) - a_{3j}y_{3(j+2)} - x_{k3}v_{(k+2)1}\right] \\
&\quad + 2u_{(j+2)(k+2)}b_{j(k+2)} + u_{(j+2)(k+2)}\left[\left(v_{(k+2)0} + y_{2(j+2)}\right) + \left(v_{(k+2)1} + y_{3(j+2)}\right)\right] \\
&\quad - \left[\left(x_{k2} + a_{2j}\right) + \left(x_{k3} + a_{3j}\right)\right]b_{j(k+2)} \\
&= \left(a_{2j}v_{(k+2)0} + a_{3j}v_{(k+2)1}\right) + \left(x_{k2}y_{2(j+2)} + x_{k3}y_{3(j+2)}\right)
\end{aligned} \tag{4.19}$$

Finally, we obtain the z_{j^+k} block elements as a sum of $\left(\widehat{XY}\right)_{kj}(a_{i^+j}, b_{jk}, u_{jk^+}, v_{k^+i}, -x_{ki}, y_{ij^+})$ with negative sign and $\left(\widehat{XY}\right)_{kj}(a_{i^+j}, b_{jk^+} - u_{j^+k^+}, v_{k^+i}, x_{ki^+}, y_{i^+j^+})$ (equality (4.17)):

$$\begin{aligned}
z_{(j+2)k} &= -\left(a_{2j}v_{(k+2)0} + a_{3j}v_{(k+2)1} - x_{k0}y_{0(j+2)} - x_{k1}y_{1(j+2)}\right) \\
&\quad + \left(a_{2j}v_{(k+2)0} + a_{3j}v_{(k+2)1} + x_{k2}y_{2(j+2)} + x_{k3}y_{3(j+2)}\right) \\
&= -\underbrace{\left(a_{2j}v_{(k+2)0} + a_{3j}v_{(k+2)1}\right)}_{=\rightarrow} + \underbrace{\left(a_{2j}v_{(k+2)0} + a_{3j}v_{(k+2)1}\right)}_{=\leftarrow} \\
&\quad - \left(-x_{k0}y_{0(j+2)} - x_{k1}y_{1(j+2)}\right) + \left(x_{k2}y_{2(j+2)} + x_{k3}y_{3(j+2)}\right) \\
&= 0 + x_{k0}y_{0(j+2)} + x_{k1}y_{1(j+2)} + x_{k2}y_{2(j+2)} + x_{k3}y_{3(j+2)} \\
&= \sum_{i=0}^{3} x_{ki}y_{i(j+2)}
\end{aligned}$$

The sum obtained is nothing more than the upper right 2×2 block of the desired XY matrix product; see the diagram (4.4).

4.2 Fast Single Matrix Multiplication Algorithm

In this section we show how to adapt the triple disjoint matrix multiplication algorithm from previous Sect. 4.1 to construct a fast single matrix multiplication algorithm. In general, any triple (or even s-disjoint, for any positive integer s) disjoint algorithm working on square matrices of the same size can be used to obtain an algorithm for single matrix multiplication by the methods developed in Sect. 5 of the article Pan (1984), which we presented in Sect. 6.1 of this monograph. Nevertheless, on the basis of the results from Sect. 4.1, the article Laderman et al. (1992) in Sect. 6 has built a specialised construction. Let us start with the basic trilinear representation of a AB matrix product, where A and B are $n \times n$ matrices:

$$\sum_{i,j,k=0}^{n-1} a_{ij} b_{jk} c_{ki}$$

We can see that the set sum indices $\bigcup_{i,j,k=0,\ldots,n-1} (i, j, k)$ form a cube $(0, 1, \ldots, n-1)^3$. Let us rewrite the trilinear representation of a AB product as follows:

$$\sum_{i,j,k=0}^{n-1} a_{ij} b_{jk} c_{ki} = \sum_{i=0}^{n-1} a_{ii} b_{ii} c_{ii} + \sum_{(i,j,k)\in S} \left(a_{ij} b_{jk} c_{ki} + a_{jk} b_{ki} c_{ij} + a_{ki} b_{ij} c_{jk} \right) \tag{4.20}$$

To ensure that the left side of Eq. (4.20) is equal to its right counterpart, we need to find such a set S of triples (i, j, k) which, together with their cyclic shifts (j, k, i) and (k, i, j), will form the whole cube in question except its main diagonal. More precisely, the obtained cube must fulfil an additional condition, i.e. it must be constructed as a sum of sets without common parts. Such a set was constructed, for example, in Sect. 6 of the article Laderman et al. (1992):

$$S = \{ (i, j, k) : \ 0 \le i \le j < k \le n-1 \quad \text{or} \quad 0 \le k < j \le i \le n-1 \} \tag{4.21}$$

4.2.1 Lemma

The sets of triples (i, j, k) S *defined by* (4.21) *together with the sets containing their corresponding cyclic shifts* (j, k, i) and (k, i, j):

$$S^{(2)} = \bigcup_{(i,j,k)\in S} (j, k, i), \qquad S^{(3)} = \bigcup_{(i,j,k)\in S} (k, i, j)$$

have the following properties:

- *Make a cube* $(0, 1, \ldots, n-1)^3$ *except for its main diagonal, i.e.:*

$$S \cup S^{(2)} \cup S^{(3)} = (0, 1, \ldots, n-1)^3 \setminus \{ (0, 0, 0), \ (1, 1, 1), \ldots, (n-1, n-1, n-1) \}$$

- *The sets* S, $S^{(2)}$ *and* $S^{(3)}$ *are disjoint in pairs.*

To prove the lemma at first let us divide the set S into two disjoint parts:

$$S = S_1 \cup S_2$$

where

$$S_1 = \{\,(i,j,k):\ 0 \le i < j < k \le n-1 \quad \text{or} \quad 0 \le k < j < i \le n-1\,\}$$
$$S_2 = \{\,(i,j,k):\ 0 \le i = j < k \le n-1 \quad \text{or} \quad 0 \le k < j = i \le n-1\,\}$$

Let us start with the subset S_1 that includes the case of sharp inequalities between the indices i, j and k. The subset of a cube $(0, 1, ..., n-1)^3$ with these sharp inequalities consists of $3! = 6$ orders of the (i, j, k) triples. The subset S_1 contains two of them i.e. $i < j < k$ and $k < j < i$. Each of them, after cyclic shifts, induces another two orders of the i, j, k indices. In addition, each of the four arrangements obtained is different. Thus the subset S_1, together with the cyclic shifts of the indices, covers all six possible orders of the (i, j, k) triples with sharp order, and the six groups of indices obtained are also pairwise disjoint.

The subset S_2 contains those (i, j, k) triples, where $i = j$. In this case, the third index k is either lower or higher than $i(= j)$. After two cyclic shifts we have two remaining cases, where $j = k$ or $k = i$. For each of these two cases the construction of S_2 ensures, that the third variable can be smaller or larger than either of the two. In this way we obtain all six orders of (i, j, k) triples for the case of a non-sharp inequality and they are pairwise distinct. ■

Now let us return to the expression (4.20), which represents a single matrix multiplication AB. To adapt to it the triple disjoint matrix multiplication algorithm from the previous section Laderman et al. (1992) devised an ingenious method of "dispatching" the indeterminate variables of a single matrix multiplication (AB) between the variables AB, UV and XY of the triple disjoint multiplication algorithm. Namely, we assign:

$$\begin{cases} u_{ij} := a_{ij} \\ x_{ij} := a_{ij} \end{cases}, \quad \begin{cases} v_{jk} := b_{jk} \\ y_{jk} := b_{jk} \end{cases}, \quad \begin{cases} w_{ki} := c_{ki} \\ z_{ki} := c_{ki} \end{cases} \tag{4.22}$$

This means that the single matrix multiplication algorithm we are looking for will be presented with the use of variables corresponding to the AB product. In the essence, the choice of which of the AB, UV or XY products to choose to represent the multiplication algorithm, is arbitrary.

Using the assignment (4.22) we can rewrite the trilinear representation of the AB product (4.20) in the following form:

$$\sum_{i,j,k=0}^{n-1} a_{ij} b_{jk} c_{ki} = \sum_{i=0}^{n-1} a_{ii} b_{ii} c_{ii} + \sum_{(i,j,k)\in S} \left(a_{ij} b_{jk} c_{ki} + u_{jk} v_{ki} w_{ij} + x_{ki} y_{ij} z_{jk} \right) \tag{4.23}$$

Now the question is how to apply the $(2n) \times (2n)$ block scheme from Sect. 4.1 in the single disjoint case. We show this using the example of the c_{ki^+} block. Observe the construction, which consists of the sum of all products AB, UV and XY after assignment (4.22). Taking into account the trilinear equality (4.23), we obtain:

$$\sum_S \left(a_{i^+j}b_{jk} + a_{i^+j^+}b_{j^+k}\right)c_{ki^+} + \sum_S \left(a_{j^+k}b_{ki} + a_{j^+k^+}b_{k^+i}\right)c_{ij^+}$$

$$+ \sum_S \left(a_{k^+i}b_{ij} + a_{k^+i^+}b_{i^+j}\right)c_{jk^+} = \sum_S \left(a_{i^+j}b_{jk} + a_{j^+k}b_{ki} + a_{k^+i}b_{ij}\right)c_{ki^+}$$

$$+ \sum_S \left(a_{i^+j^+}b_{j^+k} + a_{j^+k^+}b_{k^+i} + a_{k^+i^+}b_{i^+j}\right)c_{ki^+}$$

$$= \left[\sum_{i,j,k=0}^{n-1} a_{i^+j}b_{jk} + \sum_{i,j,k=0}^{n-1} a_{i^+j^+}b_{j^+k}\right]c_{ki^+} - \left[\sum_{i=0}^{n-1} a_{i^+i}b_{ii} + \sum_{i=0}^{n-1} a_{i^+i^+}b_{i^+i}\right]c_{ii^+}$$

The equality obtained shows that the sum of the c_{ki^+}, w_{ij^+} and z_{jk^+} blocks of the triple disjoint algorithm, over all the indices (i, j, k) in the set S (4.21), after the variable rearrangement (4.22) produces the desired single matrix multiplication except for the terms of the form $a_{i^+i}b_{ii}$ and $a_{i^+i^+}b_{i^+i}$. These two missing series of products can simply be calculated directly, without the use of aggregates, or by uniting them with one of these aggregates. More on this later.

To adapt the respective sums from the previous sub-section for the case when the set over which we are summing is not a cube, we do not need to change the main aggregate P_{ijk} and the auxiliary products $P_{ij}^{(0)}$, $P_{jk}^{(1)}$ and $P_{ki}^{(2)}$. It is enough to call it with new indeterminates defined by assignment (4.22), but with the condition that the index triple (i, j, k) belongs to the set S (4.21). However, the 6 series of products $Q_{..}^{(\cdot)}$, formula (4.9), can be combined into 2 series. To do this, they must first be redefined as follows (we assume that a_{ij} stands for each respective element of the matrix A etc.):

$$\begin{cases} Q_{ij}^{(0')}(A,B) = a_{ij}\left[\sum\limits_{k\in K(i,j)} \left(b_{jk}+b_{ki}\right)\right] \\ Q_{jk}^{(0')}(A,B) = a_{jk}\left[\sum\limits_{i\in I(j,k)} \left(b_{ki}+b_{ij}\right)\right] \\ Q_{ki}^{(0')}(A,B) = a_{ki}\left[\sum\limits_{j\in J(i,k)} \left(b_{ij}+b_{jk}\right)\right] \end{cases}, \quad \begin{cases} Q_{ij}^{(1')}(A,B) = \left[\sum\limits_{k\in K(i,j)} \left(a_{jk}+a_{ki}\right)\right]b_{ij} \\ Q_{jk}^{(1')}(A,B) = \left[\sum\limits_{i\in I(j,k)} \left(a_{ki}+a_{ij}\right)\right]b_{jk} \\ Q_{ki}^{(1')}(A,B) = \left[\sum\limits_{j\in J(i,k)} \left(a_{ij}+a_{jk}\right)\right]b_{ki} \end{cases} \tag{4.24}$$

where $K(i, j) = \{k : (i, j, k) \in S\}$ etc. Let us denote the unwanted terms $\hat{f}$ as follows:

$$T_{ijk}^{(0)} = a_{ij}b_{ki}, \quad T_{jki}^{(1)} = a_{jk}b_{ij}, \quad T_{kij}^{(2)} = a_{ki}b_{jk}$$

Applying the trilinear representation of a single matrix multiplication (4.23) to the base trilinear equality (4.2), we can deduce, that we have to sum up the *respective* terms expressed by the equalities analogously to $\left(\widehat{AB}\right)_{ik}$, $\left(\widehat{UV}\right)_{ji}$ and $\left(\widehat{XY}\right)_{kj}$:

$$\begin{cases} (AB)_{ik} = \sum\limits_{j \in J(i,k)} \left[P_{ijk} - \left(P_{ij}^{(0)} + P_{jk}^{(1)} + P_{ki}^{(2)}\right) - T_{jki}^{(1)}\right] - \left(Q_{ki}^{(0')} + Q_{ki}^{(1')}\right) \\ (UV)_{ji} = \sum\limits_{k \in K(i,j)} \left[P_{ijk} - \left(P_{ij}^{(0)} + P_{jk}^{(1)} + P_{ki}^{(2)}\right) - T_{kij}^{(2)}\right] - \left(Q_{ij}^{(0')} + Q_{ij}^{(1')}\right) \\ (XY)_{kj} = \sum\limits_{i \in I(j,k)} \left[P_{ijk} - \left(P_{ij}^{(0)} + P_{jk}^{(1)} + P_{ki}^{(2)}\right) - T_{ijk}^{(0)}\right] - \left(Q_{jk}^{(0')} + Q_{jk}^{(1')}\right) \end{cases} \tag{4.25}$$

We can observe that among the sums that aggregate the respective terms (i.e. correspond to the same matrix entry) we have to sum up the following triples of aggregates:

$$Q_{ki}^{(0')},\ Q_{ij}^{(0')},\ Q_{jk}^{(0')}$$

and

$$Q_{ki}^{(1')},\ Q_{ij}^{(1')},\ Q_{jk}^{(1')}$$

It is possible to notice their analogous structure. In the triple of products $Q_{ki}^{(0')}, Q_{ij}^{(0')}, Q_{jk}^{(0')}$ we multiply a given, single element of the A matrix of the AB product by a sum of elements of a second matrix. Correspondingly in the triple $Q_{ki}^{(1')}, Q_{ij}^{(1')}, Q_{jk}^{(1')}$ the single element of the B matrix is on the right. Thus, for a given matrix element, the sum of those product triples we can unite into a single product and for the entire $(0, 1, ..., n-1)^2$ square we get n^2 of products for $Q_{ki}^{(0')}, Q_{ij}^{(0')}, Q_{jk}^{(0')}$ sum and another n^2 of products for $Q_{ki}^{(1')}, Q_{ij}^{(1')}, Q_{jk}^{(1')}$ sum. The unwanted terms $T_{ijk}^{(0)}, T_{jki}^{(1)}$ and $T_{kij}^{(2)}$ disappear after applying the 2×2 block scheme, according to Table 4.1.

Secondly, to the sum obtained we have to add the missing diagonal products of a general form $a_{i/i^+} b_{i/i^+}$, induced by the trilinear identity (4.23). The 2×2 block distribution of the terms to be added is shown in the diagram below:

$$\hat{R}_{2n}$$

$$= \left[\begin{array}{c|c} \ddots & \ddots \\ a_{ii}b_{ii} + a_{i(i+n)}b_{(i+n)i} & a_{ii}b_{i(i+n)} + a_{i(i+n)}b_{(i+n)(i+n)} \\ \ddots & \ddots \\ \hline \ddots & \ddots \\ a_{(i+n)i}b_{ii} + a_{(i+n)(i+n)}b_{(i+n)i} & a_{(i+n)i}b_{i(i+n)} + a_{(i+n)(i+n)}b_{(i+n)(i+n)} \\ \ddots & \ddots \end{array}\right] \tag{4.26}$$

Now let us count the bilinear products required by the presented single-matrix multiplication algorithm. To do this, let's first determine the cardinality of the set S (4.21). Considering that it induces two pairwise disjoint sets of corresponding cyclic shifts of the triples (i, j, k), covering a cube of n^3 elements except its main diagonal, the cardinality in question satisfies the equation $3|S| + n = n^3$. So $|S| = (n^3 - n)/3$ and we need $(n^3 - n)/3$ products P_{ijk}.

The three series of products $P^{(0)}_{ij}$, $P^{(1)}_{jk}$ and $P^{(2)}_{ki}$ are of the form: $a_{ij}b_{ij}$, $a_{jk}b_{jk}$ and $a_{ki}b_{ki}$, respectively. To prove that these three series of terms produce n^2 different products for $(i, j, k) \in S$ it is enough to analyse the structure of the first two. All of these three series of terms have a general structure $a_{i_1 i_2}b_{i_1 i_2}$. The S's subset $0 \leq j < k \leq n - 1$ applied to the second term $a_{jk}b_{jk}$ covers all cases where the first index is lower than the second. Another S's set subset $0 < j \leq i \leq n - 1$ applied to the first term $a_{ij}b_{ij}$ covers the cases where the first index is greater than or equal to the second one, with the exception that the second index is not equal to zero. The remaining case is when the second index is equal to zero. This last group of terms is generated by inequalities $0 \leq k < j \leq n - 1$ taken with $k = 0$ applied to the term $a_{jk}b_{jk}$ and $0 \leq i \leq j < n - 1$ taken with $i = j = 0$ applied to the term $a_{ij}b_{ij}$.

Considering that the sum of the respective products of the triple $Q^{(0')}_{ki}$, $Q^{(0')}_{ij}$, $Q^{(0')}_{jk}$ generates n^2 products and the same applies to the product triple $Q^{(1')}_{ki}$, $Q^{(1')}_{ij}$, $Q^{(1')}_{jk}$ the total number of P_{ijk}, $P^{(\cdot)}_{..}$ and $Q^{(\cdot')}_{..}$ products and $a_{(i+\cdot)(i+\cdot)}b_{(i+\cdot)(i+\cdot)}$ (reminder matrix (4.26)) products is:

$$8[\underbrace{(n^3 - n)/3}_{P_{ijk}} + \underbrace{n^2}_{P^{(\cdot)}_{..}} + \underbrace{2n^2}_{Q^{(\cdot')}_{..}} + \underbrace{n}_{a_{..}b_{..}}] = \frac{1}{3}(2n)^3 + 6(2n)^2 + \frac{8}{3}(2n)$$

It is also possible to get rid of the $8n$ terms $a_{i/i^+}b_{i/i^+}$. Namely, we can unite them with either the sum of $Q^{(0')}_{ki}$, $Q^{(0')}_{ij}$, $Q^{(0')}_{jk}$ or $Q^{(1')}_{ki}$, $Q^{(1')}_{ij}$, $Q^{(1')}_{jk}$ products triple (if necessary, see the example at the end of this section), or even obtain from the $P^{(\cdot)}_{..}$ series. This step does not affect their amount, n^2 each. In this way we obtain the following, final number of bilinear steps:

$$M_{\text{single}}(2n) = 8[\underbrace{(n^3-n)/3}_{P_{ijk}} + \underbrace{n^2}_{P_{..}^{(\cdot)}} + \underbrace{2n^2}_{Q_{..}^{(\cdot')}}] = \frac{1}{3}(2n)^3 + 6(2n)^2 - \frac{4}{3}(2n)$$

with the exponent equation:

$$\omega \leq \log_{2n}\left[M_{\text{single}}(2n)\right] = \log_{2n}\left[\frac{1}{3}(2n)^3 + 6(2n)^2 - \frac{4}{3}(2n)\right]$$

producing minimum $\omega \leq 2.79512$ for $2n = 70$. For the algorithm considered, the inequality:

$$\frac{1}{3}(2n)^3 + 6(2n)^2 - \frac{4}{3}(2n) < (2n)^3$$

is fulfilled from $2n = 10$. The algorithm from this section was first presented in the article Pan (1978). It was then repeated in the article Pan (1980), but in this article it was refined in a way that reduced the number of quadratic terms from 6 to 4.5. A further improvement is made in the article Pan (1981) which reduces the number of quadratic terms to 4. Finally, Pan (1982) reduced the number of quadratic terms to 3.75. Another article that uses both exact and approximate methods is Pan (1979). The article Kaporin (2004) showed that another fast disjoint matrix multiplication algorithm from the article Laderman et al. (1992) can be implemented as efficient computer code.

4.2.2 Example—Single Matrix Multiplication

The example we give for the algorithm with dimensionality 6×6, thus $n = 3$. Let us start with the form of the set S in this case, which contains the following $(n^3 - n)/3 = (3^3 - 3)/3 = 8$ triples (i, j, k):

$$S|_{n=3} = \{(0,0,1), (0,0,2), (0,1,2), (1,1,0), (1,1,2), (2,1,0), (2,2,0), (2,2,1)\}$$

Table 4.2 shows the distribution of triples (i, j, k) between their cyclic shifts. By convention, 1 stands for set S, 2 for $S^{(2)}$ and 3 for $S^{(3)}$. In each of the three diagrams there are horizontally $i = 0,\ 1,\ 2$ and $j = 0,\ 1,\ 2$ vertically. We can observe the empty main (in 3D) diagonal.

We show how the single matrix multiplication works for a generic $n \times n = 3 \times 3$ block, ignoring unwanted $T_{ijk}^{(0)}$, $T_{jki}^{(1)}$ and $T_{kij}^{(2)}$ terms. We calculate the $Q_{ki}^{(0')}$, $Q_{ij}^{(0')}$, $Q_{jk}^{(0')}$ terms from formula (4.24), taking into account the form of the S set (4.21). Notice how it works for $Q_{ij}^{(0')}$:

Table 4.2 Distribution of triples between their cyclic shifts

$k = 1$			$k = 2$			$k = 3$		
–	3	3	1	2	3	1	2	2
2	1	1	3	–	3	1	1	2
2	2	1	3	2	1	3	3	–

$$\left\{\begin{array}{l|l|l} Q_{00}^{(0')} = a_{0,0}(b_{0,1} + b_{0,2} + b_{1,0} + b_{2,0}) & Q_{01}^{(0')} = a_{0,1}(b_{1,2} + b_{2,0}) & Q_{02}^{(0')} = 0 \\ Q_{10}^{(0')} = 0 & Q_{11}^{(0')} = a_{1,1}(b_{0,1} + b_{1,0} + b_{1,2} + b_{2,1}) & Q_{12}^{(0')} = 0 \\ Q_{20}^{(0')} = 0 & Q_{21}^{(0')} = a_{2,1}(b_{0,2} + b_{1,0}) & Q_{22}^{(0')} = a_{2,2}(b_{0,2} + b_{1,2} + b_{2,0} + b_{2,1}) \end{array}\right.$$

For $Q_{jk}^{(0')}$:

$$\left\{\begin{array}{l|l|l} Q_{00}^{(0')} = 0 & Q_{01}^{(0')} = a_{0,1}(b_{0,0} + b_{1,0}) & Q_{02}^{(0')} = a_{0,2}(b_{0,0} + b_{2,0}) \\ Q_{10}^{(0')} = a_{1,0}(b_{0,1} + b_{0,2} + b_{1,1} + b_{2,1}) & Q_{11}^{(0')} = 0 & Q_{12}^{(0')} = a_{1,2}(b_{0,1} + b_{1,1} + b_{2,0} + b_{2,1}) \\ Q_{20}^{(0')} = a_{2,0}(b_{0,2} + b_{2,2}) & Q_{21}^{(0')} = a_{2,1}(b_{1,2} + b_{2,2}) & Q_{22}^{(0')} = 0 \end{array}\right.$$

And for $Q_{ki}^{(0')}$:

$$\left\{\begin{array}{l|l|l} Q_{00}^{(0')} = 0 & Q_{01}^{(0')} = a_{0,1}(b_{1,0} + b_{1,1}) & Q_{02}^{(0')} = a_{0,2}(b_{1,0} + b_{2,0} + b_{2,1} + b_{2,2}) \\ Q_{10}^{(0')} = a_{1,0}(b_{0,0} + b_{0,1}) & Q_{11}^{(0')} = 0 & Q_{12}^{(0')} = a_{1,2}(b_{2,1} + b_{2,2}) \\ Q_{20}^{(0')} = a_{2,0}(b_{0,0} + b_{0,1} + b_{0,2} + b_{1,2}) & Q_{21}^{(0')} = a_{2,1}(b_{1,1} + b_{1,2}) & Q_{22}^{(0')} = 0 \end{array}\right.$$

As a result of *uniting* the above three 3×3 groups of terms, we obtain the following matrix:

$$Q^{(0')} = \left[\begin{array}{l|l|l} a_{0,0}(b_{0,1} + b_{0,2} + b_{1,0} + b_{2,0}) & a_{0,1}(b_{0,0} + 2b_{1,0} + b_{1,1} + b_{1,2} + b_{2,0}) & a_{0,2}(b_{0,0} + b_{1,0} + 2b_{2,0} + b_{2,1} + b_{2,2}) \\ a_{1,0}(b_{0,0} + 2b_{0,1} + b_{0,2} + b_{1,1} + b_{2,1}) & a_{1,1}(b_{0,1} + b_{1,0} + b_{1,2} + b_{2,1}) & a_{1,2}(b_{0,1} + b_{1,1} + b_{2,0} + 2b_{2,1} + b_{2,2}) \\ a_{2,0}(b_{0,0} + b_{0,1} + 2b_{0,2} + b_{1,2} + b_{2,2}) & a_{2,1}(b_{0,2} + b_{1,0} + b_{1,1} + 2b_{1,2} + b_{2,2}) & a_{2,2}(b_{0,2} + b_{1,2} + b_{2,0} + b_{2,1}) \end{array}\right]$$

By analogy, we proceed with the product triple $Q_{ki}^{(1')}, Q_{ij}^{(1')}, Q_{jk}^{(1')}$ and after their *uniting* we obtain:

$$Q^{(1')} = \left[\begin{array}{l|l|l} (a_{0,1} + a_{0,2} + a_{1,0} + a_{2,0})b_{0,0} & (a_{0,0} + 2a_{1,0} + a_{1,1} + a_{1,2} + a_{2,0})b_{0,1} & (a_{0,0} + a_{1,0} + 2a_{2,0} + a_{2,1} + a_{2,2})b_{0,2} \\ (a_{0,0} + 2a_{0,1} + a_{0,2} + a_{1,1} + a_{2,1})b_{1,0} & (a_{0,1} + a_{1,0} + a_{1,2} + a_{2,1})b_{1,1} & (a_{0,1} + a_{1,1} + a_{2,0} + 2a_{2,1} + a_{2,2})b_{1,2} \\ (a_{0,0} + a_{0,1} + 2a_{0,2} + a_{1,2} + a_{2,2})b_{2,0} & (a_{0,2} + a_{1,0} + a_{1,1} + 2a_{1,2} + a_{2,2})b_{2,1} & (a_{0,2} + a_{1,2} + a_{2,0} + a_{2,1})b_{2,2} \end{array}\right]$$

Let us sum up the remaining terms to obtain the matrix product. To do this we use formula (4.25). These terms are distributed among the expressions $(AB)_{ik}$, $(UV)_{ji}$ and $(XY)_{kj}$ linear combinations. Let us show how it works using the example of the following subset of $(AB)_{ik}$:

$$
\left[\sum_{j \in J(i,k)} \left[P_{ijk} - \left(P^{(0)}_{ij} + P^{(1)}_{jk} + P^{(2)}_{ki} \right) \right] \right]_{ik;\ i=0,1;j=0,1}
$$

$$
= \begin{bmatrix}
0 & P_{001} - \left(P^{(0)}_{00} + P^{(1)}_{01} + P^{(2)}_{10} \right) & P_{002} + P_{012} - \left(P^{(0)}_{00} + P^{(0)}_{01} + P^{(1)}_{02} + P^{(1)}_{12} + 2P^{(2)}_{20} \right) \\
P_{110} - \left(P^{(0)}_{11} + P^{(1)}_{10} + P^{(2)}_{01} \right) & 0 & P_{112} - \left(P^{(0)}_{11} + P^{(1)}_{12} + P^{(2)}_{21} \right) \\
P_{210} + P_{220} - \left(P^{(0)}_{21} + P^{(0)}_{22} + P^{(1)}_{10} + P^{(1)}_{20} + 2P^{(2)}_{02} \right) & P_{221} - \left(P^{(0)}_{22} + P^{(1)}_{21} + P^{(2)}_{12} \right) & 0
\end{bmatrix}
$$

$$
= \begin{bmatrix}
0 & \begin{array}{c} (a_{0,0} + a_{0,1} + a_{1,0})(b_{0,0} + b_{0,1} + b_{1,0}) - a_{0,0}b_{0,0} - a_{0,1}b_{0,1} - a_{1,0}b_{1,0} \end{array} & \begin{array}{l} (a_{0,0} + a_{0,2} + a_{2,0})(b_{0,0} + b_{0,2} + b_{2,0}) + (a_{0,1} + a_{1,2} + a_{2,0})(b_{0,1} + b_{1,2} + b_{2,0}) \\ -a_{0,0}b_{0,0} - a_{0,1}b_{0,1} - a_{0,2}b_{0,2} - a_{1,2}b_{1,2} - 2a_{2,0}b_{2,0} \end{array} \\
(a_{0,1} + a_{1,0} + a_{1,1})(b_{0,1} + b_{1,0} + b_{1,1}) - a_{1,1}b_{1,1} - a_{1,0}b_{1,0} - a_{0,1}b_{0,1} & 0 & (a_{1,1} + a_{1,2} + a_{2,1})(b_{1,1} + b_{1,2} + b_{2,1}) - a_{1,1}b_{1,1} - a_{1,2}b_{1,2} - a_{2,1}b_{2,1} \\
\begin{array}{l} (a_{0,2} + a_{1,0} + a_{2,1})(b_{0,2} + b_{1,0} + b_{2,1}) + (a_{0,2} + a_{2,0} + a_{2,2})(b_{0,2} + b_{2,0} + b_{2,2}) \\ -a_{2,1}b_{2,1} - a_{2,2}b_{2,2} - a_{1,0}b_{1,0} - a_{2,0}b_{2,0} - 2a_{0,2}b_{0,2} \end{array} & (a_{1,2} + a_{2,1} + a_{2,2})(b_{1,2} + b_{2,1} + b_{2,2}) - a_{2,2}b_{2,2} - a_{2,1}b_{2,1} - a_{1,2}b_{1,2} & 0
\end{bmatrix}
$$

For the remaining two of the three subsets of $(UV)_{ji}$ and $(XY)_{kj}$ linear combinations we can obtain by analogy. Now we sum up three 3×3 matrices, corresponding to respective subsets of $(AB)_{ik}$, $(UV)_{ji}$ and $(XY)_{kj}$ linear combinations (4.25), with the $Q^{(0')}$ and $Q^{(1')}$ ones and we have almost obtained a general expression for a $n \times n$ block of the desired matrix product AB (without the undesired terms $T^{(\cdot)}_{\dots}$). All that remains is to add the reminder matrix (4.26) and apply the formula obtained 8 times, in accordance with the diagrams (4.3)–(4.5). In this way we get rid of the undesired terms and obtain the algorithm for $(2n, 2n, 2n)$ square matrix multiplication AB in $(2n)^3/3 + 6(2n)^2 + 8 \cdot (2n)/3$ bilinear products.

Nevertheless, we have predicted that it is possible to avoid separately computing of $8n$ terms $a_{i/i^+}b_{i/i^+}$ existing in the reminder matrix in question. To do this, we can add the diagonal matrix:

$$\hat{R}_3 = \begin{bmatrix} a_{0,0}b_{0,0} & & \\ & a_{1,1}b_{1,1} & \\ & & a_{2,2}b_{2,2} \end{bmatrix}$$

either to $Q^{(0')}$ or to $Q^{(1')}$ matrix. In the first case, we move the term $a_{i,i}$ before the common parenthesis, and in the second case, we move $b_{i,i}$ after the parenthesis, respectively. To make the example complete, let us show how it works using the example of the $Q^{(0')}$ matrix:

$$\begin{aligned} & Q^{(0')} + \hat{R}_3 \\ &= \left[\begin{array}{l|l|l} a_{0,0}(b_{0,1} + b_{0,2} + b_{1,0} + b_{2,0}) & a_{0,1}(b_{0,0} + 2b_{1,0} + b_{1,1} + b_{1,2} + b_{2,0}) & a_{0,2}(b_{0,0} + b_{1,0} + 2b_{2,0} + b_{2,1} + b_{2,2}) \\ a_{1,0}(b_{0,0} + 2b_{0,1} + b_{0,2} + b_{1,1} + b_{2,1}) & a_{1,1}(b_{0,1} + b_{1,0} + b_{1,2} + b_{2,1}) & a_{1,2}(b_{0,1} + b_{1,1} + b_{2,0} + 2b_{2,1} + b_{2,2}) \\ a_{2,0}(b_{0,0} + b_{0,1} + 2b_{0,2} + b_{1,2} + b_{2,2}) & a_{2,1}(b_{0,2} + b_{1,0} + b_{1,1} + 2b_{1,2} + b_{2,2}) & a_{2,2}(b_{0,2} + b_{1,2} + b_{2,0} + b_{2,1}) \end{array}\right] \\ &\quad + \begin{bmatrix} a_{0,0}(b_{0,0}) & & \\ & a_{1,1}(b_{1,1}) & \\ & & a_{2,2}(b_{2,2}) \end{bmatrix} \\ &= \left[\begin{array}{l|l|l} a_{0,0}\left[b_{0,1} + b_{0,2} + b_{1,0} + b_{2,0} + (b_{0,0})\right] & a_{0,1}(b_{0,0} + 2b_{1,0} + b_{1,1} + b_{1,2} + b_{2,0}) & a_{0,2}(b_{0,0} + b_{1,0} + 2b_{2,0} + b_{2,1} + b_{2,2}) \\ a_{1,0}(b_{0,0} + 2b_{0,1} + b_{0,2} + b_{1,1} + b_{2,1}) & a_{1,1}\left[b_{0,1} + b_{1,0} + b_{1,2} + b_{2,1} + (b_{1,1})\right] & a_{1,2}(b_{0,1} + b_{1,1} + b_{2,0} + 2b_{2,1} + b_{2,2}) \\ a_{2,0}(b_{0,0} + b_{0,1} + 2b_{0,2} + b_{1,2} + b_{2,2}) & a_{2,1}(b_{0,2} + b_{1,0} + b_{1,1} + 2b_{1,2} + b_{2,2}) & a_{2,2}\left[b_{0,2} + b_{1,2} + b_{2,0} + b_{2,1} + (b_{2,2})\right] \end{array}\right] \end{aligned}$$

In this way we reduce the number of products just to the announced $(2n)^3/3 + 6(2n)^2 - 4 \cdot (2n)/3$ ones.

4.3 Second Step to a Fast Triple Disjoint Matrix Multiplication Algorithm

In this section we show how to improve the triple disjoint matrix multiplication algorithm by reducing the number of quadratic terms in its complexity from 6 to 4 per a single matrix product of the three ones. The results presented in this section come from two articles, namely Laderman et al. (1992) and Respondek (2023). The notation and dimensions of the matrices used are the same as in Sect. 4.1. In particular, in all sums the variables range from 0 to $n - 1$ (which means that the set S (4.21) does not apply in this section).

As a preliminary, let us sum both sides of base equality (4.2) with respect to i, j and k. We obtain:

$$
\begin{aligned}
\text{Trace}(ABC + UVW + XYZ) &= \sum_{i,j,k} (a_{ij}b_{jk}c_{ki} + u_{jk}v_{ki}w_{ij} + x_{ki}y_{ij}z_{jk}) \\
&= \sum_{i,j,k} (a_{ij} + u_{jk} + x_{ki})(b_{jk} + v_{ki} + y_{ij})(c_{ki} + w_{ij} + z_{jk})
\end{aligned} \tag{4.27}
$$

$$
\left.\begin{aligned}
&-\sum_{i,j} a_{ij}y_{ij}\sum_{k}(c_{ki} + w_{ij} + z_{jk}) - \sum_{j,k} u_{jk}b_{jk}\sum_{i}(w_{ij} + z_{jk} + c_{ki}) \\
&-\sum_{k,i} x_{ki}v_{ki}\sum_{j}(z_{jk} + c_{ki} + w_{ij})
\end{aligned}\right\} \text{(A)}
$$

$$
\left.\begin{aligned}
&-\sum_{i,j} a_{ij}\left[\sum_{k}\left(b_{jk} + v_{ki}\right)\right]w_{ij} - \sum_{j,k} u_{jk}\left[\sum_{i}\left(v_{ki} + y_{ij}\right)\right]z_{jk} \\
&-\sum_{k,i} x_{ki}\left[\sum_{j}\left(y_{ij} + b_{jk}\right)\right]c_{ki} - \sum_{k,i}\left[\sum_{j}\left(a_{ij} + u_{jk}\right)\right]v_{ki}c_{ki} \\
&-\sum_{i,j}\left[\sum_{k}\left(u_{jk} + x_{ki}\right)\right]y_{ij}w_{ij} - \sum_{j,k}\left[\sum_{i}\left(x_{ki} + a_{ij}\right)\right]b_{jk}z_{jk}
\end{aligned}\right\} \text{(B)}
$$

$$
\left.-\sum_{i,j,k} a_{ij}v_{ki}z_{jk} - \sum_{i,j,k} u_{jk}y_{ij}c_{ki} - \sum_{i,j,k} x_{ki}b_{jk}w_{ij}\right\} \text{(C)}
$$

The article Laderman et al. (1992) on pages 565–566 of Sect. 3 therein proposes the following trilinear identity, with the q parameter equal to 1, representing three disjoint matrix products[2]:

$$
\begin{aligned}
&\text{Trace}(ABC + UVW + XYZ) \\
&= \sum_{i,j,k} (a_{ij}b_{jk}c_{ki} + u_{jk}v_{ki}w_{ij} + x_{ki}y_{ij}z_{jk}) \\
&= \sum_{i,j,k} \big[(a_{ij} + u_{jk} + x_{ki})(b_{jk} + v_{ki} + y_{ij})(c_{ki} + w_{ij} + z_{jk}) \\
&-T_0(i,j,k) - T_1(j,k,i) - T_2(k,i,j)\big] \\
&+ T_0' + T_1' + T_2' +
\end{aligned}
$$

[2] We changed the notation from $T_0(i)$, $T_1(j)$ and $T_2(k)$ to T_0', T_1' and T_2', because these expressions. do not depend on i, j and k, respectively.

$$\left.\begin{aligned}
&-\sum_{i,j}\left(a_{ij}+\frac{1}{g}\sum_k x_{ki}\right)\left(y_{ij}+\frac{1}{g}\sum_k v_{ki}\right)\left(gw_{ij}+\sum_k c_{ki}\right)\\
&-\sum_{i,j}\left(a_{ij}+\frac{1}{h}\sum_k u_{jk}\right)\left(y_{ij}+\frac{1}{h}\sum_k b_{jk}\right)\left(hw_{ij}+\sum_k z_{jk}\right)\\
&-\sum_{j,k}\left(u_{jk}+\frac{1}{g}\sum_i a_{ij}\right)\left(b_{jk}+\frac{1}{g}\sum_i y_{ij}\right)\left(gz_{jk}+\sum_i w_{ij}\right)\\
&-\sum_{j,k}\left(u_{jk}+\frac{1}{h}\sum_i x_{ki}\right)\left(b_{jk}+\frac{1}{h}\sum_i v_{ki}\right)\left(hz_{jk}+\sum_i c_{ki}\right)\\
&-\sum_{k,i}\left(x_{ki}+\frac{1}{g}\sum_j u_{jk}\right)\left(v_{ki}+\frac{1}{g}\sum_j b_{jk}\right)\left(gc_{ki}+\sum_j z_{jk}\right)\\
&-\sum_{k,i}\left(x_{ki}+\frac{1}{h}\sum_j a_{ij}\right)\left(v_{ki}+\frac{1}{h}\sum_j y_{ij}\right)\left(hc_{ki}+\sum_j w_{ij}\right)
\end{aligned}\right\}\quad (*)\qquad (4.28)$$

for $g + h = n$. The sub-expressions are the following:

$$T_0(i,j,k) = a_{ij}v_{ki}z_{jk}, \qquad T_1(j,k,i) = u_{jk}y_{ij}c_{ki}, \qquad T_2(k,i,j) = x_{ki}b_{jk}w_{ij}$$

$$\begin{aligned}
T_0' = \sum_i \Bigg[&\left(\frac{1}{h}\sum_j a_{ij}+\frac{1}{g}\sum_k x_{ki}\right)\left(\frac{1}{h}\sum_j y_{ij}+\frac{1}{g}\sum_k v_{ki}\right)\left(g\sum_j w_{ij}+h\sum_k c_{ki}\right)\\
&+\frac{q-g}{h^2}\sum_j a_{ij}\sum_j y_{ij}\sum_j w_{ij}+\frac{q-h}{g^2}\sum_k x_{ki}\sum_k v_{ki}\sum_k c_{ki}\Bigg]
\end{aligned}$$

$$\begin{aligned}
T_1' = \sum_j \Bigg[&\left(\frac{1}{g}\sum_i a_{ij}+\frac{1}{h}\sum_k u_{jk}\right)\left(\frac{1}{g}\sum_i y_{ij}+\frac{1}{h}\sum_k b_{jk}\right)\left(h\sum w_{ij}+g\sum_k z_{jk}\right)\\
&+\frac{q-h}{g^2}\sum_i a_{ij}\sum_i y_{ij}\sum_i w_{ij}+\frac{q-g}{h^2}\sum_k u_{jk}\sum_k b_{jk}\sum_k z_{jk}\Bigg]
\end{aligned}$$

$$\begin{aligned}
T_2' = \sum_k \Bigg[&\left(\frac{1}{h}\sum_i x_{ki}+\frac{1}{g}\sum_j u_{jk}\right)\left(\frac{1}{h}\sum_i v_{ki}+\frac{1}{g}\sum_j b_{jk}\right)\left(g\sum_i c_{ki}+h\sum_j z_{jk}\right)\\
&+\frac{q-g}{h^2}\sum_i x_{ki}\sum_i v_{ki}\sum_i c_{ki}+\frac{q-h}{g^2}\sum_j u_{jk}\sum_j b_{jk}\sum_j z_{jk}\Bigg]
\end{aligned}$$

To prove the above trilinear equality, we will show that the trilinear formula (4.28) to be proved reduces to the simpler identity (4.27). For this purpose we divide factors in (4.28) into separate, complementary groups.

- As the first group of factors let us have a look at a series of sums indicated by (*) of (4.28), the following three sums of triple products:

$$\begin{aligned}
&\sum_{i,j} a_{ij}y_{ij}\left(gw_{ij} + \sum_k c_{ki}\right) + \sum_{i,j} a_{ij}y_{ij}\left(hw_{ij} + \sum_k z_{jk}\right)\\
&\sum_{j,k} u_{jk}b_{jk}\left(gz_{jk} + \sum_i w_{ij}\right) + \sum_{j,k} u_{jk}b_{jk}\left(hz_{jk} + \sum_i c_{ki}\right)\\
&\sum_{k,i} x_{ki}v_{ki}\left(gc_{ki} + \sum_j z_{jk}\right) + \sum_{k,i} x_{ki}v_{ki}\left(hc_{ki} + \sum_j w_{ij}\right)
\end{aligned} \tag{4.29}$$

Considering $g + h = n$ we obtain, by the example of the first sum:

$$\begin{aligned}
&\sum_{i,j} a_{ij}y_{ij}\left(gw_{ij} + \sum_k c_{ki}\right) + \sum_{i,j} a_{ij}y_{ij}\left(hw_{ij} + \sum_k z_{jk}\right)\\
&\quad = \sum_{i,j} a_{ij}y_{ij}\left(gw_{ij} + hw_{ij} + \sum_k c_{ki} + \sum_k z_{jk}\right)\\
&\quad = \sum_{i,j} a_{ij}y_{ij}\left(nw_{ij} + \sum_k c_{ki} + \sum_k z_{jk}\right)
\end{aligned}$$

We can rewrite nw_{ij} in the form $\sum_k w_{ij}$. So:

$$\sum_{i,j} a_{ij}y_{ij}\left(nw_{ij} + \sum_k c_{ki} + \sum_k z_{jk}\right) = \sum_{i,j} a_{ij}y_{ij} \sum_k (c_{ki} + w_{ij} + z_{jk})$$

Thus this part of proven equality (4.28) is reduced to the first sum of part (A) in (4.27). The remaining two expressions in (4.29) analogically represent the second and third sum of part (A) in (4.27).

- For the second group of terms we again look at the series (*) in (4.28) six triple product sums of the following form:

$$\sum_{i,j}\left(\frac{1}{g}\sum_{k}x_{ki}\right)\left(\frac{1}{g}\sum_{k}v_{ki}\right)gw_{ij},\quad \sum_{i,j}\left(\frac{1}{h}\sum_{k}u_{jk}\right)\left(\frac{1}{h}\sum_{k}b_{jk}\right)hw_{ij}$$

$$\sum_{j,k}\left(\frac{1}{g}\sum_{i}a_{ij}\right)\left(\frac{1}{g}\sum_{i}y_{ij}\right)gz_{jk},\quad \sum_{j,k}\left(\frac{1}{h}\sum_{i}x_{ki}\right)\left(\frac{1}{h}\sum_{i}v_{ki}\right)hz_{jk} \tag{4.30}$$

$$\sum_{k,i}\left(\frac{1}{g}\sum_{j}u_{jk}\right)\left(\frac{1}{g}\sum_{j}b_{jk}\right)gc_{ki},\quad \sum_{k,i}\left(\frac{1}{h}\sum_{j}a_{ij}\right)\left(\frac{1}{h}\sum_{j}y_{ij}\right)hc_{ki}$$

To get to know which of the above terms (4.30) corresponds to the terms in (4.27) and/or is reduced by the correction terms T_0', T_1', T_2' and $T_0(i,j,k)$, $T_1(j,k,i)$, $T_2(k,i,j)$, we look at the example of a selected expression from (4.30):

$$\sum_{k,i}\left(\frac{1}{h}\sum_{j}a_{ij}\right)\left(\frac{1}{h}\sum_{j}y_{ij}\right)hc_{ki}=\sum_{i}\left(\frac{1}{h}\sum_{j}a_{ij}\right)\left(\frac{1}{h}\sum_{j}y_{ij}\right)h\sum_{k}c_{ki}$$

It is now clear that each of the expressions (4.30) is reduced by its matching part in the terms T_0', T_1' or T_2' just in the trilinear equality (4.28).

- Another group of terms includes expressions from the series (*) in (4.28) of the form:

$$\sum_{i,j}a_{ij}\left(\frac{1}{h}\sum_{k}b_{jk}\right)\left(hw_{ij}+\sum_{k}z_{jk}\right)+\sum_{j,k}\left(\frac{1}{g}\sum_{i}a_{ij}\right)b_{jk}\left(gz_{jk}+\sum_{i}w_{ij}\right)$$

$$\sum_{j,k}u_{jk}\left(\frac{1}{h}\sum_{i}v_{ki}\right)\left(hz_{jk}+\sum_{i}c_{ki}\right)+\sum_{k,i}\left(\frac{1}{g}\sum_{j}u_{jk}\right)v_{ki}\left(gc_{ki}+\sum_{j}z_{jk}\right)$$

$$\sum_{k,i}x_{ki}\left(\frac{1}{g}\sum_{j}b_{jk}\right)\left(gc_{ki}+\sum_{j}z_{jk}\right)+\sum_{j,k}\left(\frac{1}{h}\sum_{i}x_{ki}\right)b_{jk}\left(hz_{jk}+\sum_{i}c_{ki}\right)$$

$$\sum_{i,j}a_{ij}\left(\frac{1}{g}\sum_{k}v_{ki}\right)\left(gw_{ij}+\sum_{k}c_{ki}\right)+\sum_{k,i}\left(\frac{1}{h}\sum_{j}a_{ij}\right)v_{ki}\left(hc_{ki}+\sum_{j}w_{ij}\right)$$

$$\sum_{j,k}u_{jk}\left(\frac{1}{g}\sum_{i}y_{ij}\right)\left(gz_{jk}+\sum_{i}w_{ij}\right)+\sum_{i,j}\left(\frac{1}{h}\sum_{k}u_{jk}\right)y_{ij}\left(hw_{ij}+\sum_{k}z_{jk}\right)$$

$$\sum_{k,i}x_{ki}\left(\frac{1}{h}\sum_{j}y_{ij}\right)\left(hc_{ki}+\sum_{j}w_{ij}\right)+\sum_{i,j}\left(\frac{1}{g}\sum_{k}x_{ki}\right)y_{ij}\left(gw_{ij}+\sum_{k}c_{ki}\right) \tag{4.31}$$

By the example of the first expression in series (4.31) we give a general rule:

$$\sum_{i,j} a_{ij}\left(\frac{1}{h}\sum_{k} b_{jk}\right)\left(hw_{ij}+\sum_{k} z_{jk}\right)+\sum_{j,k}\left(\frac{1}{g}\sum_{i} a_{ij}\right)b_{jk}\left(gz_{jk}+\sum_{i} w_{ij}\right)$$
$$=\sum_{i,j,k}\left[a_{ij}b_{jk}\left(w_{ij}+\frac{1}{h}\sum_{k_1} z_{jk_1}\right)+a_{ij}b_{jk}\left(z_{jk}+\frac{1}{g}\sum_{i_1} w_{i_1j}\right)\right]$$
$$=\sum_{i,j,k}\left[\left(a_{ij}b_{jk}w_{ij}+a_{ij}b_{jk}z_{jk}\right)+a_{ij}b_{jk}\left(\frac{1}{h}\sum_{k_1} z_{jk_1}+\frac{1}{g}\sum_{i_1} w_{i_1j}\right)\right]$$

We can see that each of the six sums (4.31) produces two groups of expressions:

- Sum of the expressions of the form $a_{ij}b_{jk}w_{ij}+a_{ij}b_{jk}z_{jk}$. It is easy to see that they are desired terms, which summed over all six expressions (4.31) forms part (B) of equality (4.27).
- Sum of the expressions of the form $a_{ij}b_{jk}\left(1/h\sum_{k_1} z_{jk_1}+1/g\sum_{i_1} w_{i_1j}\right)$ which are undesired. They need to be reduced by correction terms of trilinear equality (4.28). Let us find a corresponding expression in (4.28), giving terms of a general form $\sum a_{ij}\sum b_{jk}\sum w_{ij}$ and $\sum a_{ij}\sum b_{jk}\sum z_{jk}$.

We can find a proper expression in the correction term T_1':

$$\sum_{j}\left(\frac{1}{g}\sum_{i} a_{ij}\right)\left(\frac{1}{h}\sum_{k} b_{jk}\right)\left(h\sum_{i} w_{ij}+g\sum_{k} z_{jk}\right)$$
$$=\sum_{j}\sum_{i}\sum_{k}\frac{1}{gh}a_{ij}b_{jk}\left(h\sum_{i_1} w_{i_1j}+g\sum_{k_1} z_{jk_1}\right)$$
$$=\sum_{i,j,k} a_{ij}b_{jk}\left(\frac{1}{h}\sum_{k_1} z_{jk_1}+\frac{1}{g}\sum_{i_1} w_{i_1j}\right)$$

To sum up, the group of six terms of the form (4.31) create group (B) of equality (4.27) and the undesired expressions they produce are corrected by the corresponding correction terms of T_0', T_1', T_2'.

It is also worth to notice that all the terms in (4.27) are covered by the so-far presented groups of expressions from trilinear identity (4.28). It means that the remaining terms in (4.28) reduce themselves within that identity.

- The last group of terms in the series (*) in (4.28) comprises six sums of triple products, of a common form:

$$\sum_{i,j}\left(\frac{1}{g}\sum_k x_{ki}\right)\left(\frac{1}{g}\sum_k v_{ki}\right)\left(\sum_k c_{ki}\right),\quad \sum_{i,j}\left(\frac{1}{h}\sum_k u_{jk}\right)\left(\frac{1}{h}\sum_k b_{jk}\right)\left(\sum_k z_{jk}\right)$$
$$\sum_{j,k}\left(\frac{1}{g}\sum_i a_{ij}\right)\left(\frac{1}{g}\sum_i y_{ij}\right)\left(\sum_i w_{ij}\right),\quad \sum_{j,k}\left(\frac{1}{h}\sum_i x_{ki}\right)\left(\frac{1}{h}\sum_i v_{ki}\right)\left(\sum_i c_{ki}\right)$$
$$\sum_{k,i}\left(\frac{1}{g}\sum_j u_{jk}\right)\left(\frac{1}{g}\sum_j b_{jk}\right)\left(\sum_j z_{jk}\right),\quad \sum_{k,i}\left(\frac{1}{h}\sum_j a_{ij}\right)\left(\frac{1}{h}\sum_j y_{ij}\right)\left(\sum_j w_{ij}\right) \tag{4.32}$$

Let us analyse the first trilinear product in (4.32):

$$\sum_{i,j}\left(\frac{1}{g}\sum_k x_{ki}\right)\left(\frac{1}{g}\sum_k v_{ki}\right)\left(\sum_k c_{ki}\right)$$

The inside, summed expression $(1/g\sum_k x_{ki})(1/g\sum_k v_{ki})(\sum_k c_{ki})$ does not depend on the j variable. So we can claim:

$$\sum_{i,j}\left(\frac{1}{g}\sum_k x_{ki}\right)\left(\frac{1}{g}\sum_k v_{ki}\right)\left(\sum_k c_{ki}\right) = n\cdot\sum_{i}\left(\frac{1}{g}\sum_k x_{ki}\right)\left(\frac{1}{g}\sum_k v_{ki}\right)\left(\sum_k c_{ki}\right) \tag{4.33}$$

Thus the corresponding term in T_0' reduces (4.33) iff $q = n$. For the remaining sums of triple products (4.32) there exist corresponding five reducing expressions in T_0', T_1' and T_2'.

- The last group of terms are in T_0', T_1' and T_2'. These terms balance themselves in the given term, in the outside sums. The terms in question are the following:

$$T_0' : \frac{g}{h^2}\sum_j a_{ij}\sum_j y_{ij}\sum_j w_{ij},\quad \frac{h}{g^2}\sum_k x_{ki}\sum_k v_{ki}\sum_k c_{ki}$$
$$T_1' : \frac{h}{g^2}\sum_i a_{ij}\sum_i y_{ij}\sum_i w_{ij},\quad \frac{g}{h^2}\sum_k u_{jk}\sum_k b_{jk}\sum_k z_{jk}$$
$$T_2' : \frac{g}{h^2}\sum_i x_{ki}\sum_i v_{ki}\sum_i c_{ki},\quad \frac{h}{g^2}\sum_j u_{jk}\sum_j b_{jk}\sum_j z_{jk}$$

To sum up, in the trilinear identity in Sect. 4.3 of the article Laderman et al. (1992), pages 565–566, its part included in the correction terms T_0', T_1' and T_2' of a general form:

$$\frac{1-g}{h^2}\sum f_{i_1 i_2}^{(1)}\sum f_{i_1 i_2}^{(2)}\sum f_{i_1 i_2}^{(3)}, \quad \text{and} \quad \frac{1-h}{g^2}\sum f_{i_1 i_2}^{(4)}\sum f_{i_1 i_2}^{(5)}\sum f_{i_1 i_2}^{(6)}$$

requires to be corrected to, respectively, with the q parameter equal to n:

$$\frac{n-g}{h^2}\sum f_{i_1 i_2}^{(1)}\sum f_{i_1 i_2}^{(2)}\sum f_{i_1 i_2}^{(3)}, \quad \text{and} \quad \frac{n-h}{g^2}\sum f_{i_1 i_2}^{(4)}\sum f_{i_1 i_2}^{(5)}\sum f_{i_1 i_2}^{(6)}$$

Now let us present the algorithm in the bilinear form. As usual it is obtained by equating the coefficients of c_{ki}, w_{ij} and z_{jk} on both sides. Now we decided to omit detailed calculations because they are similar and, also importantly, no more sophisticated than in previous sections. Like in Laderman et al. (1992) we choose $g = 1$ and $h = n - 1$ as well as we omit the $T_0(i, j, k)$, $T_1(j, k, i)$ and $T_2(k, i, j)$ terms.

At first let us define the non-scalar products. We have to perform n^3 of P_{ijk} products:

$$P_{ijk} = (a_{ij} + u_{jk} + x_{ki})(b_{jk} + v_{ki} + y_{ij}) \tag{4.34}$$

Next come six series with n^2 products in each of them, of the form:

$$\begin{cases} P_{ij}^{(0)} = \left(a_{ij} + \sum_k x_{ki}\right)\left(y_{ij} + \sum_k v_{ki}\right), P_{ij}^{(1)} = \left(a_{ij} + \frac{1}{n-1}\sum_k u_{jk}\right)\left(y_{ij} + \frac{1}{n-1}\sum_k b_{jk}\right) \\ P_{jk}^{(2)} = \left(u_{jk} + \sum_i a_{ij}\right)\left(b_{jk} + \sum_i y_{ij}\right), P_{jk}^{(3)} = \left(u_{jk} + \frac{1}{n-1}\sum_i x_{ki}\right)\left(b_{jk} + \frac{1}{n-1}\sum_i v_{ki}\right) \\ P_{ki}^{(4)} = \left(x_{ki} + \sum_j u_{jk}\right)\left(v_{ki} + \sum_j b_{jk}\right), P_{ki}^{(5)} = \left(x_{ki} + \frac{1}{n-1}\sum_j a_{ij}\right)\left(v_{ki} + \frac{1}{n-1}\sum_j y_{ij}\right) \end{cases} \tag{4.35}$$

Finally from T_0', T_1' and T_2' we have nine groups with n products in each, of the form:

$$\begin{cases} P_i^{(0)} = \left(\frac{1}{n-1}\sum_j a_{ij} + \sum_k x_{ki}\right)\left(\frac{1}{n-1}\sum_j y_{ij} + \sum_k v_{ki}\right) \\ P_i^{(1)} = \frac{1}{n-1}\sum_j a_{ij}\sum_j y_{ij}, \quad P_i^{(2)} = \sum_k x_{ki}\sum_k v_{ki} \end{cases}$$

$$\begin{cases} P_j^{(3)} = \left(\sum_i a_{ij} + \frac{1}{n-1}\sum_k u_{jk}\right)\left(\sum_i y_{ij} + \frac{1}{n-1}\sum_k b_{jk}\right) \\ P_j^{(4)} = \sum_i a_{ij}\sum_i y_{ij}, \quad P_j^{(5)} = \frac{1}{n-1}\sum_k u_{jk}\sum_k b_{jk} \end{cases} \tag{4.36}$$

$$\begin{cases} P_k^{(6)} = \left(\frac{1}{n-1}\sum_i x_{ki} + \sum_j u_{jk}\right)\left(\frac{1}{n-1}\sum_i v_{ki} + \sum b_{jk}\right) \\ P_k^{(7)} = \frac{1}{n-1}\sum_i x_{ki}\sum_i v_{ki}, \quad P_k^{(8)} = \sum_j u_{jk}\sum_j b_{jk} \end{cases}$$

The desired three matrix products are expressed explicitly by linear combinations. of the above products:

$$\begin{cases} (AB)_{ik} = \sum_j \left(P_{ijk} - P_{ij}^{(0)} - P_{jk}^{(3)}\right) - \left[P_{ki}^{(4)} + (n-1)P_{ki}^{(5)}\right] + (n-1)P_i^{(0)} \\ \qquad +P_i^{(2)} + P_k^{(6)} + P_k^{(7)} \\ (UV)_{ji} = \sum_k \left(P_{ijk} - P_{jk}^{(2)} - P_{ki}^{(5)}\right) - \left[P_{ij}^{(0)} + (n-1)P_{ij}^{(1)}\right] + P_i^{(0)} + P_i^{(1)} \\ \qquad +(n-1)P_j^{(3)} + P_j^{(4)} \\ (XY)_{kj} = \sum_i \left(P_{ijk} - P_{ij}^{(1)} - P_{ki}^{(4)}\right) - \left[P_{jk}^{(2)} + (n-1)P_{jk}^{(3)}\right] + P_j^{(3)} + P_j^{(5)} \\ \qquad +(n-1)P_k^{(6)} + P_k^{(8)} \end{cases} \tag{4.37}$$

Without the $T_0(i, j, k)$, $T_1(j, k, i)$ and $T_2(k, i, j)$ terms, the trilinear identity (4.28) contains $n^3 + 6n^2 + 9n$ products. Applying the 2×2 block scheme for $n \times n$ matrices from Sect. 4 of the article Laderman et al. (1992) we get rid of the $T_0(i, j, k)$, $T_1(j, k, i)$ and $T_2(k, i, j)$ terms and obtain:

$$M^{(2)}_{\text{disjoint}}(2n) = 8(n^3 + 6n^2 + 9n) = (2n)^3 + 12(2n)^2 + 36(2n)$$

products necessary to calculate. The obtained exponent expression is:

$$\omega \le \log_{2n}\left[M^{(2)}_{\text{disjoint}}(2n)/3\right] = \log_{2n}\left[\frac{(2n)^3}{3} + 4(2n)^2 + 9(2n)\right]$$

For $2n = 48$ we obtain an algorithm with an efficiency exponent $\omega < 2.77706$.

4.3.1 Example—Second Step to Triple Disjoint Matrix Multiplication

We show how the second algorithm works in detail using the example of $2n = 4$. The $n^3 = 2^3 = 8$ of P_{ijk} products is calculated directly from formula (4.34):

$$\begin{cases} P_{000} = (a_{0,0} + u_{0,0} + x_{0,0})(b_{0,0} + v_{0,0} + y_{0,0}) & P_{100} = (a_{1,0} + u_{0,0} + x_{0,1})(b_{0,0} + v_{0,1} + y_{1,0}) \\ P_{001} = (a_{0,0} + u_{0,1} + x_{1,0})(b_{0,1} + v_{1,0} + y_{0,0}) & P_{101} = (a_{1,0} + u_{0,1} + x_{1,1})(b_{0,1} + v_{1,1} + y_{1,0}) \\ P_{010} = (a_{0,1} + u_{1,0} + x_{0,0})(b_{1,0} + v_{0,0} + y_{0,1}) & P_{110} = (a_{1,1} + u_{1,0} + x_{0,1})(b_{1,0} + v_{0,1} + y_{1,1}) \\ P_{011} = (a_{0,1} + u_{1,1} + x_{1,0})(b_{1,1} + v_{1,0} + y_{0,1}) & P_{111} = (a_{1,1} + u_{1,1} + x_{1,1})(b_{1,1} + v_{1,1} + y_{1,1}) \end{cases}$$

How to obtain the six n^2 product series, we show using the example of $P^{(0)}_{ij}$ (formula (4.35)):

$$P^{(0)}_{ij} = \left(a_{ij} + \sum_{k=0}^{n-1=1} x_{ki}\right)\left(y_{ij} + \sum_{k=0}^{n-1=1} v_{ki}\right) = \left[a_{ij} + (x_{0i} + x_{1i})\right]\left[y_{ij} + (v_{0i} + v_{1i})\right]$$

Thus:

$$\begin{cases} P^{(0)}_{00} = \left[a_{0,0} + \left(x_{0,0} + x_{1,0}\right)\right]\left[y_{0,0} + \left(v_{0,0} + v_{1,0}\right)\right] \\ P^{(0)}_{01} = \left[a_{0,1} + \left(x_{0,0} + x_{1,0}\right)\right]\left[y_{0,1} + \left(v_{0,0} + v_{1,0}\right)\right] \\ P^{(0)}_{10} = \left[a_{1,0} + \left(x_{0,1} + x_{1,1}\right)\right]\left[y_{1,0} + \left(v_{0,1} + v_{1,1}\right)\right] \\ P^{(0)}_{11} = \left[a_{1,1} + \left(x_{0,1} + x_{1,1}\right)\right]\left[y_{1,1} + \left(v_{0,1} + v_{1,1}\right)\right] \end{cases}$$

and so on for the remaining five n^2 product series, which we calculate by analogy::

$$\begin{cases} P^{(1)}_{00} = \left[a_{0,0} + \frac{1}{2-1}\left(u_{0,0} + u_{0,1}\right)\right]\left[y_{0,0} + \frac{1}{2-1}\left(b_{0,0} + b_{0,1}\right)\right] \\ \quad = \left[a_{0,0} + \left(u_{0,0} + u_{0,1}\right)\right]\left[y_{0,0} + \left(b_{0,0} + b_{0,1}\right)\right] \\ P^{(1)}_{01} = \left[a_{0,1} + \left(u_{1,0} + u_{1,1}\right)\right]\left[y_{0,1} + \left(b_{1,0} + b_{1,1}\right)\right] \\ P^{(1)}_{10} = \left[a_{1,0} + \left(u_{0,0} + u_{0,1}\right)\right]\left[y_{1,0} + \left(b_{0,0} + b_{0,1}\right)\right] \\ P^{(1)}_{11} = \left[a_{1,1} + \left(u_{1,0} + u_{1,1}\right)\right]\left[y_{1,1} + \left(b_{1,0} + b_{1,1}\right)\right] \end{cases}$$

$$\begin{cases} P^{(2)}_{00} = \left[u_{0,0} + \left(a_{0,0} + a_{1,0}\right)\right]\left[b_{0,0} + \left(y_{0,0} + y_{1,0}\right)\right] & P^{(3)}_{00} = \left[u_{0,0} + \left(x_{0,0} + x_{0,1}\right)\right]\left[b_{0,0} + \left(v_{0,0} + v_{0,1}\right)\right] \\ P^{(2)}_{01} = \left[u_{0,1} + \left(a_{0,0} + a_{1,0}\right)\right]\left[b_{0,1} + \left(y_{0,0} + y_{1,0}\right)\right] & P^{(3)}_{01} = \left[u_{0,1} + \left(x_{1,0} + x_{1,1}\right)\right]\left[b_{0,1} + \left(v_{1,0} + v_{1,1}\right)\right] \\ P^{(2)}_{10} = \left[u_{1,0} + \left(a_{0,1} + a_{1,1}\right)\right]\left[b_{1,0} + \left(y_{0,1} + y_{1,1}\right)\right] & P^{(3)}_{10} = \left[u_{1,0} + \left(x_{0,0} + x_{0,1}\right)\right]\left[b_{1,0} + \left(v_{0,0} + v_{0,1}\right)\right] \\ P^{(2)}_{11} = \left[u_{1,1} + \left(a_{0,1} + a_{1,1}\right)\right]\left[b_{1,1} + \left(y_{0,1} + y_{1,1}\right)\right] & P^{(3)}_{11} = \left[u_{1,1} + \left(x_{1,0} + x_{1,1}\right)\right]\left[b_{1,1} + \left(v_{1,0} + v_{1,1}\right)\right] \end{cases}$$

$$\begin{cases} P_{00}^{(4)} = [x_{0,0} + (u_{0,0} + u_{1,0})][v_{0,0} + (b_{0,0} + b_{1,0})] & P_{00}^{(5)} = [x_{0,0} + (a_{0,0} + a_{0,1})][v_{0,0} + (y_{0,0} + y_{0,1})] \\ P_{01}^{(4)} = [x_{0,1} + (u_{0,0} + u_{1,0})][v_{0,1} + (b_{0,0} + b_{1,0})] & P_{01}^{(5)} = [x_{0,1} + (a_{1,0} + a_{1,1})][v_{0,1} + (y_{1,0} + y_{1,1})] \\ P_{10}^{(4)} = [x_{1,0} + (u_{0,1} + u_{1,1})][v_{1,0} + (b_{0,1} + b_{1,1})] & P_{10}^{(5)} = [x_{1,0} + (a_{0,0} + a_{0,1})][v_{1,0} + (y_{0,0} + y_{0,1})] \\ P_{11}^{(4)} = [x_{1,1} + (u_{0,1} + u_{1,1})][v_{1,1} + (b_{0,1} + b_{1,1})] & P_{11}^{(5)} = [x_{1,1} + (a_{1,0} + a_{1,1})][v_{1,1} + (y_{1,0} + y_{1,1})] \end{cases}$$

And the last nine series we calculate from formula (4.36). For $P_i^{(0)}$, $P_i^{(1)}$ and $P_i^{(2)}$, we use the following rule:

$$\begin{cases} P_i^{(0)} = \left(\frac{1}{2-1}\sum_{j=0}^{1} a_{ij} + \sum_{k=0}^{1} x_{ki}\right)\left(\frac{1}{2-1}\sum_{j=0}^{1} y_{ij} + \sum_{k=0}^{1} v_{ki}\right) \\ \quad = [(a_{i0} + a_{i1}) + (x_{0i} + x_{1i})][(y_{i0} + y_{i1}) + (v_{0i} + v_{1i})] \\ P_i^{(1)} = \frac{1}{2-1}\sum_{j=0}^{1} a_{ij} \sum_{j=0}^{1} y_{ij} = (a_{i0} + a_{i1})(y_{i0} + y_{i1}) \\ P_i^{(2)} = \sum_{k=0}^{1} x_{ki} \sum_{k=0}^{1} v_{ki} = (x_{0i} + x_{1i})(v_{0i} + v_{1i}) \end{cases}$$

Thus:

$$\begin{cases} P_0^{(0)} = [(a_{0,0} + a_{0,1}) + (x_{0,0} + x_{1,0})][(y_{0,0} + y_{0,1}) + (v_{0,0} + v_{1,0})] \\ P_1^{(0)} = [(a_{1,0} + a_{1,1}) + (x_{0,1} + x_{1,1})][(y_{1,0} + y_{1,1}) + (v_{0,1} + v_{1,1})] \\ P_0^{(1)} = (a_{0,0} + a_{0,1})(y_{0,0} + y_{0,1}), \quad P_0^{(2)} = (x_{0,0} + x_{1,0})(v_{0,0} + v_{1,0}) \\ P_1^{(1)} = (a_{1,0} + a_{1,1})(y_{1,0} + y_{1,1}), \quad P_1^{(2)} = (x_{0,1} + x_{1,1})(v_{0,1} + v_{1,1}) \end{cases}$$

$$\begin{cases} P_0^{(3)} = [(a_{0,0} + a_{1,0}) + (u_{0,0} + u_{0,1})][(y_{0,0} + y_{1,0}) + (b_{0,0} + b_{0,1})] \\ P_1^{(3)} = [(a_{0,1} + a_{1,1}) + (u_{1,0} + u_{1,1})][(y_{0,1} + y_{1,1}) + (b_{1,0} + b_{1,1})] \\ P_0^{(4)} = (a_{0,0} + a_{1,0})(y_{0,0} + y_{1,0}), \quad P_0^{(5)} = (u_{0,0} + u_{0,1})(b_{0,0} + b_{0,1}) \\ P_1^{(4)} = (a_{0,1} + a_{1,1})(y_{0,1} + y_{1,1}), \quad P_1^{(5)} = (u_{1,0} + u_{1,1})(b_{1,0} + b_{1,1}) \end{cases}$$

$$\begin{cases} P_0^{(6)} = [(x_{0,0} + x_{0,1}) + (u_{0,0} + u_{1,0})][(v_{0,0} + v_{0,1}) + (b_{0,0} + b_{1,0})] \\ P_1^{(6)} = [(x_{1,0} + x_{1,1}) + (u_{0,1} + u_{1,1})][(v_{1,0} + v_{1,1}) + (b_{0,1} + b_{1,1})] \\ P_0^{(7)} = (x_{0,0} + x_{0,1})(v_{0,0} + v_{0,1}), \quad P_0^{(8)} = (u_{0,0} + u_{1,0})(b_{0,0} + b_{1,0}) \\ P_1^{(7)} = (x_{1,0} + x_{1,1})(v_{1,0} + v_{1,1}), \quad P_1^{(8)} = (u_{0,1} + u_{1,1})(b_{0,1} + b_{1,1}) \end{cases}$$

Let us now construct the matrix products we are looking for. The construction is a linear combination (4.37). As the first product of three, i.e. AB we obtain:

$$(AB)_{ik} = \sum_{j=0}^{1} \left(P_{ijk} - P_{ij}^{(0)} - P_{jk}^{(3)}\right) - \left(P_{ki}^{(4)} + P_{ki}^{(5)}\right) + P_{i}^{(0)}$$
$$+ P_{i}^{(2)} + P_{k}^{(6)} + P_{k}^{(7)} = \left[\left(P_{i0k} - P_{i0}^{(0)} - P_{0k}^{(3)}\right) + \left(P_{i1k} - P_{i1}^{(0)} - P_{1k}^{(3)}\right)\right]$$
$$- \left(P_{ki}^{(4)} + P_{ki}^{(5)}\right) + P_{i}^{(0)} + P_{i}^{(2)} + P_{k}^{(6)} + P_{k}^{(7)}$$

Thus:

$$AB = \left[\begin{array}{c|c} \left(P_{000} - P_{00}^{(0)} - P_{00}^{(3)}\right) + \left(P_{010} - P_{01}^{(0)} - P_{10}^{(3)}\right) - \left(P_{00}^{(4)} + P_{00}^{(5)}\right) + P_{0}^{(0)} + P_{0}^{(2)} + P_{0}^{(6)} + P_{0}^{(7)} & \left(P_{001} - P_{00}^{(0)} - P_{01}^{(3)}\right) + \left(P_{011} - P_{01}^{(0)} - P_{11}^{(3)}\right) - \left(P_{10}^{(4)} + P_{10}^{(5)}\right) + P_{0}^{(0)} + P_{0}^{(2)} + P_{1}^{(6)} + P_{1}^{(7)} \\ \left(P_{100} - P_{10}^{(0)} - P_{00}^{(3)}\right) + \left(P_{110} - P_{11}^{(0)} - P_{10}^{(3)}\right) - \left(P_{01}^{(4)} + P_{01}^{(5)}\right) + P_{1}^{(0)} + P_{1}^{(2)} + P_{0}^{(6)} + P_{0}^{(7)} & \left(P_{101} - P_{10}^{(0)} - P_{01}^{(3)}\right) + \left(P_{111} - P_{11}^{(0)} - P_{11}^{(3)}\right) - \left(P_{11}^{(4)} + P_{11}^{(5)}\right) + P_{1}^{(0)} + P_{1}^{(2)} + P_{1}^{(6)} + P_{1}^{(7)} \end{array}\right]$$

The construction of the remaining products UV and XY is analogous and we will skip it. In the above formula we have neglected the unwanted terms $T_0(i, j, k)$, $T_1(j, k, i)$ and $T_2(k, i, j)$ terms, because we have finally applied the 2×2 block scheme for $n \times n$ matrices from Sect. 4 (see e.g. diagram (4.5); Table 4.1) to obtain the desired three $(2n) \times (2n)$ products.

In the next chapter we will move on to another class of fast matrix multiplication, with arbitrary precision, opposite to exact considered in this and the previous chapter.

References

Kaporin I (2004) The aggregation and cancellation techniques as a practical tool for faster matrix multiplication. Theoret Comput Sci 315(2):469–510

Laderman J, Pan V, Sha XH (1992) On practical algorithms for accelerated matrix multiplication. Linear Algebra Appl 162–164:557–588

Pan V (1980) New fast algorithms for matrix operations. SIAM J Comput 9(2):321–342

Pan V (1981) New combinations of methods for the acceleration of matrix multiplications. Comput Math Appl 7(1):73–125

Pan V (1982) Trilinear aggregating with implicit canceling for a new acceleration of matrix multiplication. Comput Math Appl 8(1):23–34

Pan V (1984) How can we speed up matrix multiplication? SIAM Rev 26(3):393–415

Pan V (1978) Strassen's algorithm is not optimal—trilinear technique of aggregating uniting and canceling for constructing fast algorithms for matrix operations. In: Proceedings of the 19th annual symposium on foundations of computer science, Ann Arbor, Michigan, pp 166–176

Pan V (1979) Field extension and trilinear aggregating, uniting and canceling for the acceleration of matrix multiplications. In: 20th annual symposium on foundations of computer science (SFCS 1979), San Juan, PR, USA, pp 28–38

Respondek JS (2023) Correction of 'J. Laderman, V. Pan, X. H. Sha, On practical Algorithms for Accelerated Matrix Multiplication, Linear Algebra and its Applications. Vol. 162–164 (1992) pp. 557–588' arXiv:2302.07626

Chapter 5
Arbitrary Precision Approximating Algorithms

The notion of the so-called arbitrary precision approximating algorithms for matrix multiplication (abbreviated as APA-algorithms) was introduced in the article Bini et al. (1979). The idea was to make the algorithm faster rather than calculate the exact result of the multiplication. However, Bini later showed how to transform a given approximate algorithm into its exact counterpart. Nonetheless, this comes at an additional cost of a logarithmic factor in complexity. Arbitrary precision algorithms are also available in versions for disjoint matrix multiplication.

Moreover, the methods for constructing a single matrix multiplication algorithm based on disjoint algorithms also apply to the arbitrary precision algorithms. We describe it later in Sect. 6.1 for the equidimensional disjoint matrices to be multiplied, and in Sect. 6.2 for different sizes of matrices to be multiplied at once.

The chapter is organised as follows:

- In Sect. 5.1 we introduce the reader to the idea of arbitrary precision approximating algorithms by means of an understandable example. The example we have chosen is historically later than the seminal article by Bini et al., but is much easier to analyse and still reveals the crucial features of approximate algorithms.
- In Sect. 5.2 we show how to construct a corresponding exact algorithm from an approximate one, in the general case. We have illustrated the general results using the example of the 3×3 algorithm from Sect. 5.1.
- In Sect. 5.3 we showed that in the time complexity of the exact algorithm constructed on the basis of the approximate algorithm induces additional logarithmic factor. In addition, we illustrated the results using the example from Sect. 5.1.
- In Sect. 5.4 we presented the famous result of Bini et al. from 1979, which combines partial matrix multiplication and approximation methods with the duality of matrix multiplication (Sect. 3.7), giving an exponent equal to 2.7799.
- In Sect. 5.5 we introduced the approximate double disjoint matrix multiplication algorithm, which was described in the corresponding exact version in Sect. 3.10.

J. S. Respondek, *Fast Matrix Multiplication with Applications*, Studies in Big Data 166,
https://doi.org/10.1007/978-3-031-76930-6_5

- In Sect. 5.6 we presented the double disjoint approximation algorithm introduced by Schonhage (1981), which has a complexity of $\omega \leq 2.5479$.
- In Sect. 5.7 we show a more sophisticated algorithm proposed by Schonhage (1981), in the triple disjoint version, which gives complexity $\omega \leq 2.521\ 8127$.

5.1 Approximate Multiplication of 3 x 3 Problem by 21 Products

Let us present the idea of the general class of APA algorithms using the example of—once again—3×3 matrices, discovered by Schonhage in his article from 1981. The algorithm is presented there on page 438. It is worth recalling that the fastest exact, non-commutative algorithm for the 3×3 case so far was given by Laderman (1976), presented in Sect. 3.5, using 23 multiplications, and has not been broken despite much effort, e.g. by the quite contemporary work of Heule, Kauers et al. (2021). It gives a worse efficiency than the Strassen's result for 2×2, because $\log_3 23 > \log_2 7$. However, the APA algorithm example below requires only 21 multiplications.

To compute the *approximate* matrix product $D \approx AB$, we first compute the following 21 aggregated products for $i, j \in \{1, 2, 3\}$:

$$\begin{cases} u_{i,i} = \left(a_{i,1} + \varepsilon^2 a_{i,2}\right)\left(\varepsilon^2 b_{1,i} + b_{2,i}\right) \\ v_{i,i} = \left(a_{i,1} + \varepsilon^2 a_{i,3}\right) b_{3,i} \\ w_i = a_{i,1}\left(b_{2,i} + b_{3,i}\right) \\ u_{i,j} = \left(a_{i,1} + \varepsilon^2 a_{j,2}\right)\left(b_{2,i} - \varepsilon b_{1,j}\right), \quad (i \neq j) \\ v_{i,j} = \left(a_{i,1} + \varepsilon^2 a_{j,3}\right)\left(b_{3,i} + \varepsilon b_{1,j}\right), \quad (i \neq j) \end{cases} \tag{5.1}$$

Using the product series (5.1), we can express the desired matrix product $D \approx AB$ as the following linear combinations:

$$\begin{cases} d'_{i,i} = \dfrac{1}{\varepsilon^2}\left(u_{i,i} + v_{i,i} - w_i\right) \\ d'_{j,i} = \dfrac{1}{\varepsilon^2}\left(u_{i,j} + v_{i,j} - w_i\right) + \dfrac{1}{\varepsilon}\left(v_{j,i} - v_{j,j}\right), \quad (i \neq j) \end{cases} \tag{5.2}$$

Inserting (5.1) into (5.2), gives:

$$D = AB + R_1 \varepsilon + R_2 \varepsilon^2 \tag{5.3}$$

where, let's call them the constant, residual matrices have the form:

$$R_1 = \begin{bmatrix} 0 & a_{12}b_{11} - a_{13}b_{11} + a_{13}b_{31} - a_{23}b_{31} & a_{12}b_{11} - a_{13}b_{11} + a_{13}b_{31} - a_{33}b_{31} \\ a_{22}b_{12} - a_{23}b_{12} - a_{13}b_{32} + a_{23}b_{32} & 0 & a_{22}b_{12} - a_{23}b_{12} + a_{23}b_{32} - a_{33}b_{32} \\ a_{32}b_{13} - a_{33}b_{13} - a_{13}b_{33} + a_{33}b_{33} & a_{32}b_{13} - a_{33}b_{13} - a_{23}b_{33} + a_{33}b_{33} & 0 \end{bmatrix}$$

$$R_2 = \begin{bmatrix} -a_{12}b_{11} & -a_{23}b_{12} & -a_{33}b_{13} \\ -a_{13}b_{11} & -a_{22}b_{12} & -a_{33}b_{13} \\ -a_{13}b_{11} & -a_{23}b_{12} & -a_{32}b_{13} \end{bmatrix} \tag{5.4}$$

The crucial point in the APA algorithm is to select small ε, but simultaneously ε must not be equal to zero. So if $\varepsilon \to 0$ and $\varepsilon \neq 0$, then D converges to the exact value AB, but is never exact. In Sect. 5.2 we show how to obtain an exact algorithm based on the APA algorithm. Another example of reducing the number of non-scalar multiplications is given in the article Bini et al. (1980). In this case it is presented by the example of a problem other than matrix multiplication. To sum up, since $\log_3 21 < \log_2 7$, we obtain a non-exact non-commutative algorithm for the case 3×3, but in return we gain better efficiency than the Strassen's algorithm.

5.2 Relation Between Arbitrary Precision Algorithms and Exact Algorithms

Interestingly, it is possible to obtain an exact fast matrix multiplication algorithm from the corresponding arbitrary precision approximating algorithm of the same dimensionality. The theoretical basis for this was given by Dario Bini in an article from 1980.

The matrix product $D \approx AB$ calculated by the APA algorithm, contains additional terms with the ε parameter, of maximum degree denoted by d. It can be represented as:

$$D = AB + \sum_{i=1}^{d} R_i \varepsilon^i \tag{5.5}$$

where AB is the unknown exact value of the product and R_i are constant, residual matrices of appropriate size. The idea proposed by Bini (1980) is first to compute the product (5.5) $(d + 1)$ times, for pairwise different ε-s, i.e. with $\varepsilon_i \neq \varepsilon_j$ for $i \neq j$. We obtain a series of $(d + 1)$ matrix equalities:

$$D_k = AB + \sum_{i=1}^{d} R_i \varepsilon_k^i, \quad k = 1, \ldots, d + 1$$

Next, we sum the above results in a linear combination as follows:

$$S = \sum_{k=1}^{d+1} \alpha_k D_k = \sum_{k=1}^{d+1} \alpha_k \left[AB + \sum_{i=1}^{d} R_i \varepsilon_k^i \right] \tag{5.6}$$

Let us rearrange the terms as follows:

$$S = \sum_{k=1}^{d+1} \alpha_k \left[AB + \sum_{i=1}^{d} R_i \varepsilon_k^i \right] = AB \sum_{k=1}^{d+1} \alpha_k + \sum_{i=1}^{d} R_i \sum_{k=1}^{d+1} \alpha_k \varepsilon_k^i$$

It is easy to see that in order to obtain the exact value of the product AB, it is sufficient to set the coefficient associated with AB equal to one:

$$\sum_{k=1}^{d+1} \alpha_k = 1 \tag{5.7}$$

and make the remaining coefficients associated with the residual matrices $R_1, \ldots, R_d$ disappear:

$$\sum_{k=1}^{d+1} \alpha_k \varepsilon_k^i = 0, \quad i = 1, \ldots, d \tag{5.8}$$

It is convenient to rewrite the series of Eqs. (5.7), (5.8) in a matrix form (Bini 1980, page 92):

$$\begin{bmatrix} 1 & 1 & \cdots & 1 \\ \varepsilon_1 & \varepsilon_2 & \cdots & \varepsilon_{d+1} \\ \varepsilon_1^2 & \varepsilon_2^2 & \cdots & \varepsilon_{d+1}^2 \\ \vdots & \vdots & \ddots & \vdots \\ \varepsilon_1^d & \varepsilon_2^d & \cdots & \varepsilon_{d+1}^d \end{bmatrix} \begin{bmatrix} \alpha_1 \\ \alpha_2 \\ \vdots \\ \alpha_{d+1} \end{bmatrix} = \begin{bmatrix} 1 \\ 0 \\ \vdots \\ 0 \end{bmatrix} \tag{5.9}$$

Thus we obtained a system with the well-known Vandermonde special type matrix. It is worth noting that it is strongly discouraged to solve Vandermonde systems using generic matrix methods, both for reasons of accuracy and time efficiency.

Using a generic method to solve the Vandermonde systems yields a cubic time algorithm, which would obviously waste the time efficiency gain from fast matrix multiplication. We have a specialised algorithm to invert the Vandermonde matrix in quadratic time. More about this in the Appendix on page 239.

In the next section we show the Bini's, Vandermonde-based method using the example of the 3×3 case approximating algorithm from Sect. 5.1.

5.2.1 Example—Obtaining the Exact 3 × 3 Algorithm from the APA Algorithm

In the example, we show how to transform the arbitrary precision approximating algorithm for multiplying 3×3 matrices in 21 non-scalar multiplications into the corresponding exact one. The desired product AB must be obtained from the Eq. (5.3).

We can see that the exact value of the product is burdened by the terms of the order $d = 2$. So we calculate the equality (5.3) $(d + 1) = 3$ times, for pairwise different $\varepsilon_1,\ \varepsilon_2,\ \varepsilon_3$:

$$\begin{cases} D_1 = AB + R_1\varepsilon_1 + R_2\varepsilon_1^2 \\ D_2 = AB + R_1\varepsilon_2 + R_2\varepsilon_2^2 \\ D_3 = AB + R_1\varepsilon_3 + R_2\varepsilon_3^2 \end{cases} \tag{5.10}$$

with unknown product AB and elements of matrices $D_1,\ D_2,\ D_3$ given as a linear combinations (5.2) of the aggregated products (5.1). Now we combine $D_1,\ D_2,\ D_3$ linearly, according to formula (5.6), with the coefficients $\alpha_1,\ \alpha_2,\ \alpha_3$:

$$\begin{aligned} S = \alpha_1 D_1 + \alpha_2 D_2 + \alpha_3 D_3 = \alpha_1\left(AB + R_1\varepsilon_1 + R_2\varepsilon_1^2\right) \\ + \alpha_2\left(AB + R_1\varepsilon_2 + R_2\varepsilon_2^2\right) + \alpha_3\left(AB + R_1\varepsilon_3 + R_2\varepsilon_3^2\right) \end{aligned} \tag{5.11}$$

Let us rearrange the above sum with respect to the successive powers of the ε parameters. Observe:

$$S = (\alpha_1 + \alpha_2 + \alpha_3)AB + \left(\alpha_1\varepsilon_1 + \alpha_2\varepsilon_2 + \alpha_3\varepsilon_3\right)R_1 + \left(\alpha_1\varepsilon_1^2 + \alpha_2\varepsilon_2^2 + \alpha_3\varepsilon_3^2\right)R_2$$

In order to make the argument fully clear, we should emphasise that we know the (exact!) values of the parameters $\varepsilon_1,\ \varepsilon_2,\ \varepsilon_3$ (we choose them arbitrarily; we just have to make sure that they are pairwise distinct), and we also know the matrices R_1 and R_2, since they consist of the given exact entries of the matrices $A,\ B$ to be multiplied, as in the formula (5.4).

The crucial step in the Bini construction is to choose the values of the parameters of the linear combination $\alpha_1,\ \alpha_2,\ \alpha_3$ in such a way that the residual matrices R_1 and R_2 vanish. Clearly, they must satisfy the following set of equations:

$$\begin{cases} \alpha_1 \quad + \alpha_2 \quad + \alpha_3 \quad = 1 \\ \alpha_1\varepsilon_1 + \alpha_2\varepsilon_2 + \alpha_3\varepsilon_3 = 0 \\ \alpha_1\varepsilon_1^2 + \alpha_2\varepsilon_2^2 + \alpha_3\varepsilon_3^2 = 0 \end{cases}$$

Finally, the set of equations in matrix form, using the Vandermonde matrix, is given by:

$$\begin{bmatrix} 1 & 1 & 1 \\ \varepsilon_1 & \varepsilon_2 & \varepsilon_3 \\ \varepsilon_1^2 & \varepsilon_2^2 & \varepsilon_3^2 \end{bmatrix} \begin{bmatrix} \alpha_1 \\ \alpha_2 \\ \alpha_3 \end{bmatrix} = \begin{bmatrix} 1 \\ 0 \\ 0 \end{bmatrix} \tag{5.12}$$

The values of the linear combination coefficients are as follows:

$$\begin{bmatrix} \alpha_1 \\ \alpha_2 \\ \alpha_3 \end{bmatrix} = \begin{bmatrix} \frac{\varepsilon_2\varepsilon_3}{(\varepsilon_1-\varepsilon_2)(\varepsilon_1-\varepsilon_3)} \\ -\frac{\varepsilon_1\varepsilon_3}{(\varepsilon_1-\varepsilon_2)(\varepsilon_2-\varepsilon_3)} \\ \frac{\varepsilon_1\varepsilon_2}{(\varepsilon_1-\varepsilon_3)(\varepsilon_2-\varepsilon_3)} \end{bmatrix} \tag{5.13}$$

In summary, to obtain an exact value of the matrix product AB from the arbitrary precision approximating algorithm, the following steps are required (using the 3×3 Schonhage (1981), page 438, algorithm as an example):

- Arbitrary choose distinct ε_1, ε_2, ε_3 parameters.
- Calculate the aggregated products $u_{11}, \ldots, u_{33}, v_{11}, \ldots, v_{33}, w_1, \ldots, w_3$ (5.1), separately for each ε_i, $i = 1, 2, 3$.
- Calculate the approximate values D_1, D_2, D_3 of the matrix product AB by linearly combining the aggregated products according to formula (5.2).
- Find the values of the coefficients in the linear combination (5.11), i.e. the values of $\alpha_1, \alpha_2, \alpha_3$. They are calculated from the linear system of equations with the Vandermonde system matrix. For the 3×3 dimensional example considered, the values are given explicitly by formula (5.13) (the order of this and the previous step does not matter).
- Finally, find the *exact* value of the matrix product AB, which we obtain directly from the linear combination:

$$AB = \alpha_1 D_1 + \alpha_2 D_2 + \alpha_3 D_3$$

5.3 Complexity of Exact Algorithms Based on APA Algorithms

The natural question is what time complexity we obtain when we construct an exact algorithm based on the approximate algorithm in the way presented in Sect. 5.2. An analysis of this is given in Romani (1980), Sect. 5, pages 83–84. Some basis for this is also provided by the work Bini (1980), already mentioned.

Suppose we have an approximate algorithm for fast matrix multiplication, with base dimension $n_0 \times n_0$ requiring r multiplications with correction entries of ε with highest degree d. Its complexity is $O(n^\omega)$ with $\omega = \log_{n_0} r$.

If, on this basis, we want to derive a—still approximate—algorithm for the dimension $n \times n$, we apply this algorithm recursively q times and obtain equality, with dimensions of the form:

$$n_0^q \times n_0^q = n \times n$$

So we have to apply a given algorithm $q = \log_{n_0} n$ times recursively. Clearly, the degree of the correction polynomial of the ε variable is now:

$$\deg\left[\left(\varepsilon^{d}\right)^{q}\right] = \deg\left[\varepsilon^{d\cdot q}\right] = d \cdot q = d \cdot \log_{n_0} n$$

As we described in Sect. 5.2, to obtain an exact result from the approximate matrix multiplication algorithm, we need to apply it $d + 1$ times. So the total number of multiplications is:

$$\begin{aligned} T(n) &= r^{q}(1 + d \cdot q) = r^{\log_{n_0} n}\left(1 + d \cdot \log_{n_0} n\right) \\ &= n^{\omega}\left(1 + d \cdot \log_{n_0} n\right) = O(\log n \cdot n^{\omega}) \end{aligned}$$

since $r^{\log_{n_0} n} = n^{\log_{n_0} r}$.

In summary, transforming the approximate fast matrix multiplication algorithm into the corresponding exact algorithm adds a logarithmic factor to its—polynomial—$O(n^{\omega})$ efficiency class.

For example, in the approximate algorithm presented in Sect. 5.1 we have the base dimension $n_0 = 3$, the number of non-scalar multiplications $r = 21$ and the corrections terms of the highest degree $d = 2$. In formula (5.3) we have residual matrices R_1, R_2 multiplied by the correction terms, of which R_2 is multiplied by the highest degree of those correction terms. At the second level of recursion, the entries of R_1, R_2 are matrices of 3×3 dimensions, which are the linear combinations of the products $a_{.} \cdot b_{.}$, which we obtain from the formula (5.3) used for scalars. So the entries of R_1, R_2 are burdened by both correction terms of 2nd degree and the degree of $R_2\varepsilon^2$ term, which together give $2 + 2 = 4 = 2d$. Applying the algorithm once again recursively, the entries of R_2 contain the 4th order correction terms, multiplied again by ε^2, so that the degree of $R_2\varepsilon^2$ expression is now $4 + 2 = 6 = 3d$. We have illustrated the process in Table 5.1.

It can be seen that the size of the matrix increases exponentially, while the degree of the correction term increases only linearly. This ensures that the final efficiency of the exact algorithm is only burdened by a logarithmic factor.

However, we have to wait a few recursion levels to make the method profitable. In the example in question at the *qth* recursion level we need to run the approximate algorithm $1 + 2q$ times with $r = 21$ multiplications for the base dimension, so $(1 + 2q) \cdot 21^q$ multiplications are needed, with comparison to 27^q obtained from

Table 5.1 Correction term degree with respect to the recursion level

Recursion level	Matrix dimension	Correction term maximum degree
1	$3^1 \times 3^1 = 3 \times 3$	$2 = d$
2	$3^2 \times 3^2 = 9 \times 9$	$2 + 2 = 2d$
3	$3^3 \times 3^3 = 27 \times 27$	$2d + 2 = 3d$
4	$3^4 \times 3^4 = 81 \times 81$	$3d + 2 = 4d$
$\vdots$	$\vdots$	$\vdots$
q	$n_0^q \times n_0^q$	$q \cdot d$

the definition-based algorithm and 23^q by the Laderman (1976) result. We have $(1 + 2q) \cdot 21^q < 27^q$ starting from $q = 14$, so the 3×3 approximating algorithm, which is used for the exact calculation, requires less multiplications starting from the dimension $n = 3^{14} = 4\,,782,969$ than the definition-based algorithm.

A general form of the inequality when the Bini's construction to find an exact algorithm based on the APA algorithm require less non-scalar multiplications than the definitional algorithm have the form:

$$(1 + qd) \cdot r^q < \left(n_0^q\right)^3$$

The above described Bini's construction bases on the two main steps:

- Apply the base APA algorithm with d degree correction terms recursively q times to obtain the APA algorithm with $q \cdot d$ correction terms,
- on this basis, construct an exact algorithm by running the approximate algorithm $d + 1$ times.

In other words, during recursion we are still working with an approximate version of the algorithms, and only in the final step do we transform it into an exact one.

One can see a different method here; at each level of recursion we use an exact version of proper submatrices. Let us analyse whether the efficiency of such an approach is competitive. For an approximate algorithm with r non-commutative steps on $(q+1)th$ recursion level we need to perform r non-scalar multiplications of exact $n_0^q \times n_0^q$-dimensional matrices. To make them exact, we need to do this $d + 1$ times, at each level of recursion. Thus, at the recursion level $(q+1)th$ the number of non-scalar multiplications increases $(d + 1)r$ times, resulting in the total number of multiplications $[(d + 1)r]^q = O(n^{\omega+k})$, where $k = \log_{n_0}(d + 1)$. In summary, the approach of making the matrices exact at each recursion level is slower.

In the next section we show another classical result that combines the methods of the APA algorithms, some of the decomposition methods together with a new notion of so-called partial matrix multiplication.

5.4 Partial Matrix Multiplication

In this section we present the results of the classic article Bini et al. (1979) which, by combining of a few techniques, obtained the exponent $\omega_{12} = 2.7799$ of the complexity $O(n^\omega)$ for the $12 \times 12 \times 12$ problem.

Bini proposed the so-called partial matrix multiplication, for matrices of the form:

$$Z = XY = \begin{bmatrix} x_{11} & x_{12} \\ x_{21} & 0 \end{bmatrix} \begin{bmatrix} y_{11} & y_{12} \\ y_{21} & y_{22} \end{bmatrix}$$

In general, by definition, it takes 6 multiplications to multiply. Moreover, the article Leeuwen and Boas (1978) proved in proposition 5.4, pages 75–76, that 6 multiplications for two 2×2 matrices, one of which is triangular (which is just our case of 2×2 partial matrix multiplication), also is the minimum number of multiplications required by any (exact) algorithm. However, Bini et al. (1979) proposed an arbitrary precision algorithm using 5 multiplications. The aggregated products proposed by Bini have the following form:

$$\begin{cases} m_1 = (x_{21} + \varepsilon x_{11})y_{11} \\ m_2 = x_{12}y_{12} \\ m_3 = x_{21}(y_{11} + y_{12} + \varepsilon y_{21}) \\ m_4 = (x_{12} + x_{21})(y_{12} + \varepsilon y_{21}) \\ m_5 = (x_{12} + \varepsilon x_{11})(y_{12} + \varepsilon y_{22}) \end{cases}$$

We combine them linearly as follows:

$$\begin{cases} z_{11} = (m_1 - m_2 - m_3 + m_4)/\varepsilon \\ z_{21} = m_1 \\ z_{12} = (m_5 - m_2)/\varepsilon \\ z_{22} = -m_2 + m_4 \end{cases}$$

As a result, we obtain the approximate formula with $d = 1$ degrees of correction terms:

$$Z \approx XY + \varepsilon \begin{bmatrix} 0 & x_{11}y_{22} \\ x_{11}y_{11} & x_{12}y_{21} + x_{21}y_{21} \end{bmatrix}$$

The symmetry allows us to obtain the corresponding formulae for the product of the form:

$$Z' = X'Y = \begin{bmatrix} 0 & x'_{12} \\ x'_{21} & x'_{22} \end{bmatrix} \begin{bmatrix} y_{11} & y_{12} \\ y_{21} & y_{22} \end{bmatrix}$$

Observe the products and their respective linear combinations:

$$\begin{cases} m'_1 = \left(x'_{12} + \varepsilon x'_{22}\right)y_{21} \\ m'_2 = x'_{21}y_{22} \\ m'_3 = x'_{12}(y_{21} + y_{22} + \varepsilon y_{11}) \\ m'_4 = \left(x'_{21} + x'_{12}\right)(y_{22} + \varepsilon y_{11}) \\ m'_5 = \left(x'_{21} + \varepsilon x'_{22}\right)(y_{22} + \varepsilon y_{12}) \end{cases}, \quad \begin{cases} z'_{21} = \left(m'_1 - m'_2 - m'_3 + m'_4\right)/\varepsilon \\ z'_{11} = m'_1 \\ z'_{22} = \left(m'_5 - m'_2\right)/\varepsilon \\ z'_{12} = -m'_2 + m'_4 \end{cases}$$

Now we can combine the two symmetrical versions according to the following diagram::

$$\begin{bmatrix} x_{11} & x_{12} \\ x_{21} & x'_{12} \\ x'_{21} & x'_{22} \end{bmatrix} \begin{bmatrix} y_{11} & y_{12} \\ y_{21} & y_{22} \end{bmatrix} = \begin{bmatrix} z_{11} & z_{12} \\ z_{21} + z'_{11} & z_{22} + z'_{12} \\ z'_{21} & z'_{22} \end{bmatrix}$$

Thus, by combining two symmetric versions of the same algorithm of partial matrix multiplication of the size $(2, 2, 2)$, with zero one entry, we obtained non-partial algorithm[1] for the $(3, 2, 2)$ problem by 10 multiplications, with gain equal to 2 multiplication over the definitional $3 \cdot 2 \cdot 2 = 12$ multiplications. Now we use the duality Theorem 3.7.1 twice to obtain the algorithms for $(2, 3, 2)$ and $(2, 2, 3)$ problems, both in 10 multiplications.

Finally, we use the multiplicative (recursion) rule described in Sect. 3.8 twice, in any order. For example, by combining the $(3, 2, 2)$ problem with $(2, 3, 2)$ we obtain an algorithm for the $(3 \cdot 2, 2 \cdot 3, 2 \cdot 2) = (6, 6, 4)$ size problem in $10 \cdot 10 = 100$ multiplications. Now we use the multiplicative rule again, now for the problems $(6, 6, 4)$ and $(2, 2, 3)$ obtain the algorithm for the $(6 \cdot 2, 6 \cdot 2, 4 \cdot 3) =(12, 12, 12)$ size problem realised by in $100 \cdot 10 = 1000$ multiplications. We obtain a complexity of:

$$O\left(n^{\log_{12} 1000}\right) \approx O\left(n^{2.7799}\right)$$

In the next few sections, we will look in more detail at other results and techniques for approximate fast matrix multiplication.

5.5 Arbitrary Precision Double Disjoint Matrix Multiplication

In this section we show how to apply the idea of arbitrary precision algorithms (APA), introduced in Sects. 5.1–5.4, to double disjoint matrix multiplication. As in Sect. 3.10, our goal is to compute two matrix products XY and UV at once, with less number of non-scalar multiplications than an exact algorithm used twice would require for this problem. The results presented in this section were proposed by article Pan (1984). When constructing an algorithm, we start from elementary equality:

$$\begin{aligned} \left(x_{ij} + u_{jk}\right)\left(y_{jk} + \varepsilon v_{ki}\right)\left(w_{ij} + \varepsilon z_{ki}\right) &= \varepsilon\left(x_{ij}y_{jk}z_{ki} + u_{jk}v_{ki}w_{ij}\right) \\ &+ x_{ij}\left(y_{jk} + \varepsilon v_{ki}\right)w_{ij} + u_{jk}y_{jk}\left(w_{ij} + \varepsilon z_{ki}\right) + \varepsilon^2\left(x_{ij} + u_{jk}\right)v_{ki}z_{ki} \end{aligned} \tag{5.14}$$

[1] In Sect. 7.2 we show how to construct a non-partial (there called as 'total') algorithm on the base of a partial matrix multiplication algorithm in a general case. Respective tools proposed Schonhage (1981).

where by x_{ij}, y_{jk}, u_{jk}, v_{ki} we mean the respective real entries of X, Y, U, V matrices with indices varying in the ranges given by expression (3.69) for certain real ε. Similarly to equality (3.68), identity (5.14) holds without using the commutativity of all variables except ε. In particular, equality (5.14) holds even if we drop all sub-indexes.

Let us now sum the identity (5.14) with respect to the three variables i, j, k. After optimising with respect to the number of non-scalar multiplications by an analogous method as we used in expression (3.71), devoted to the exact double disjoint matrix multiplication, we obtain the arbitrary precision double disjoint matrix multiplication algorithm we are looking for, in trilinear form (Pan 1984, pages 411–412):

$$\begin{aligned}\varepsilon\sum_{i,j,k}\left(x_{ij}y_{jk}z_{ki}+u_{jk}v_{ki}w_{ij}\right) &= \sum_{i,j,k}(x_{ij}+u_{jk})(y_{jk}+\varepsilon v_{ki})(w_{ij}+\varepsilon z_{ki})\\ &-\sum_{i,j}x_{ij}\left[\sum_{k}(y_{jk}+\varepsilon v_{ki})\right]w_{ij}-\sum_{j,k}u_{jk}y_{jk}\sum_{i}(w_{ij}+\varepsilon z_{ki})\\ &-\varepsilon^2\sum_{k,i}\left[\sum_{j}(x_{ij}+u_{jk})v_{ki}\right]z_{ki}\end{aligned} \tag{5.15}$$

To transform it into the bilinear form we use the analogous method as in Sect. 3.10, which allows us to omit detailed calculations without losing the clarity of the derivation. We obtain the following expressions for the entries of the matrix products (Pan 1984, page 399):

$$\begin{cases}\sum_{j}x_{ij}y_{jk}\triangleq(XY)_{ik}=\sum_{j}\left(Q^{(0)}_{ijk}-Q^{(2)}_{jk}\right)-\varepsilon Q^{(3)}_{ki}\\ \sum_{k}u_{jk}v_{ki}\triangleq(UV)_{ji}=\frac{1}{\varepsilon}\left\{\sum_{k}\left(Q^{(0)}_{ijk}-Q^{(2)}_{jk}\right)-Q^{(1)}_{ij}\right\}\end{cases} \tag{5.16}$$

where the products $Q^{(0)}_{\cdot} \div Q^{(3)}_{\cdot}$ are now the following four series:

$$\begin{cases}Q^{(0)}_{ijk}=\left(x_{ij}+u_{jk}\right)\left(y_{jk}+\varepsilon v_{ki}\right)\\ Q^{(1)}_{ij}=x_{ij}\sum_{k}\left(y_{jk}+\varepsilon v_{ki}\right)\\ Q^{(2)}_{jk}=u_{jk}y_{jk}\\ Q^{(3)}_{ki}=\left[\sum_{j}\left(x_{ij}+u_{jk}\right)\right]v_{ki}\end{cases} \tag{5.17}$$

for $i = 0, 1, \ldots, m-1$; $j = 0, 1, \ldots, n-1$; $k = 0, 1, \ldots, p-1$. We can see that:

$$\lim_{\varepsilon\to 0}\varepsilon Q^{(3)}_{ki}=0$$

thus, to obtain the approximate value of the matrix product XY, we can omit the $\varepsilon Q_{ki}^{(3)}$ term and obtain the arbitrary precision algorithm in:

$$M_{m,n,p,\varepsilon} = mnp + mn + np$$

non-scalar products. When $m = p = n$ we have to perform $n^3 + 2n^2$ non-scalar multiplications. The inequality $n^3 + 2n^2 < 2n^3$ is satisfied for $n \geq 3$. In this square case, the inequality of the efficiency exponent is $\omega \leq \log_n [M_{n,n,n,\varepsilon}/2] = \log_n [(n^3 + 2n^2)/2]$ with minimum for $n = 7$ equal to $\omega \leq 2.77294$.

In Chap. 6 we show that by choosing the dimension values $m = p = 7,\ n = 1$ (Pan 1984, page 400), we can obtain an algorithm with the following complexity exponent for single, particularly square, matrix multiplication:

$$\omega = 3 \log_{mpn} (M_{m,n,p,\varepsilon}/2) = 3 \log_{7\cdot 7\cdot 1}(63/2) = 2.65941 \tag{5.18}$$

5.5.1 Example

We show the example of the APA algorithm for double disjoint matrix multiplication in the square case for $n = 2$, so that all indices i, j, k vary in the same range from 0 to 1. The $Q_{ijk}^{(0)}$ product series consists of $2^3 = 8$ elements:

$$\begin{cases} P_{000}^{(0)} = (x_{0,0} + u_{0,0})(y_{0,0} + \varepsilon v_{0,0}), & P_{100}^{(0)} = (x_{1,0} + u_{0,0})(y_{0,0} + \varepsilon v_{0,1}) \\ P_{010}^{(0)} = (x_{0,1} + u_{1,0})(y_{1,0} + \varepsilon v_{0,0}), & P_{110}^{(0)} = (x_{1,1} + u_{1,0})(y_{1,0} + \varepsilon v_{0,1}) \\ P_{001}^{(0)} = (x_{0,0} + u_{0,1})(y_{0,1} + \varepsilon v_{1,0}), & P_{101}^{(0)} = (x_{1,0} + u_{0,1})(y_{0,1} + \varepsilon v_{1,1}) \\ P_{011}^{(0)} = (x_{0,1} + u_{1,1})(y_{1,1} + \varepsilon v_{1,0}), & P_{111}^{(0)} = (x_{1,1} + u_{1,1})(y_{1,1} + \varepsilon v_{1,1}) \end{cases}$$

For $Q_{jk}^{(2)}$ product series we have $2 \cdot 2 = 4$ entries:

$$\begin{cases} Q_{00}^{(2)} = u_{0,0}y_{0,0}, \quad Q_{10}^{(2)} = u_{1,0}y_{1,0}, \quad Q_{01}^{(2)} = u_{0,1}y_{0,1}, \quad Q_{11}^{(2)} = u_{1,1}y_{1,1} \end{cases}$$

Finally, we can calculate the series $Q_{ij}^{(1)}$ according to (5.17):

$$Q_{ij}^{(1)} = x_{ij} \sum_{k=0}^{1} (y_{jk} + \varepsilon v_{ki}) = x_{ij}[(y_{j0} + \varepsilon v_{0i}) + (y_{j1} + \varepsilon v_{1i})]$$

So the series $Q_{ij}^{(1)}$ consists of the following $2 \cdot 2 = 4$ entries:

$$\begin{cases} Q_{00}^{(1)} = x_{0,0}\left[(y_{0,0} + \varepsilon v_{0,0}) + (y_{0,1} + \varepsilon v_{1,0})\right], & Q_{10}^{(1)} = x_{1,0}\left[(y_{0,0} + \varepsilon v_{0,1}) + (y_{0,1} + \varepsilon v_{1,1})\right] \\ Q_{01}^{(1)} = x_{0,1}\left[(y_{1,0} + \varepsilon v_{0,0}) + (y_{1,1} + \varepsilon v_{1,0})\right], & Q_{11}^{(1)} = x_{1,1}\left[(y_{1,0} + \varepsilon v_{0,1}) + (y_{1,1} + \varepsilon v_{1,1})\right] \end{cases}$$

To obtain the two matrix products we are looking for, we linearly combine the products according to formula (5.16), without the $\varepsilon Q_{ki}^{(3)}$ term:

$$\begin{cases} (XY)_\varepsilon = \left[\begin{array}{c|c} (Q_{000}^{(0)} - Q_{00}^{(2)}) + (Q_{010}^{(0)} - Q_{10}^{(2)}) & (Q_{001}^{(0)} - Q_{01}^{(2)}) + (Q_{011}^{(0)} - Q_{11}^{(2)}) \\ \hline (Q_{100}^{(0)} - Q_{00}^{(2)}) + (Q_{110}^{(0)} - Q_{10}^{(2)}) & (Q_{101}^{(0)} - Q_{01}^{(2)}) + (Q_{111}^{(0)} - Q_{11}^{(2)}) \end{array}\right] \\ UV \quad = \frac{1}{\varepsilon}\left[\begin{array}{c|c} (Q_{000}^{(0)} - Q_{00}^{(2)}) + (Q_{001}^{(0)} - Q_{01}^{(2)}) - Q_{00}^{(1)} & (Q_{100}^{(0)} - Q_{00}^{(2)}) + (Q_{101}^{(0)} - Q_{01}^{(2)}) - Q_{10}^{(1)} \\ \hline (Q_{010}^{(0)} - Q_{10}^{(2)}) + (Q_{011}^{(0)} - Q_{11}^{(2)}) - Q_{01}^{(1)} & (Q_{110}^{(0)} - Q_{10}^{(2)}) + (Q_{111}^{(0)} - Q_{11}^{(2)}) - Q_{11}^{(1)} \end{array}\right] \end{cases}$$

From (5.16) we can see that for the product UV we obtained an exact value. Exact value of XY is:

$$XY = (XY)_\varepsilon - \varepsilon\left[\begin{array}{c|c} \left[(x_{0,0} + u_{0,0}) + (x_{0,1} + u_{1,0})\right]v_{0,0} & \left[(x_{0,0} + u_{0,1}) + (x_{0,1} + u_{1,1})\right]v_{1,0} \\ \hline \left[(x_{1,0} + u_{0,0}) + (x_{1,1} + u_{1,0})\right]v_{0,1} & \left[(x_{1,0} + u_{0,1}) + (x_{1,1} + u_{1,1})\right]v_{1,1} \end{array}\right]$$

In summary, the total number of products is now $M_{2,2,2,\varepsilon} = 8 + 4 + 4 = 16$.

5.6 Schonhage's Vector-by-Vector Approximate Matrix Multiplication Algorithm

The article Schonhage (1981) in Sect. 6 proposed a relatively simple algorithm for double disjoint matrix multiplication, using two possible orders of vectors, which leads to a quite favourable ω complexity exponent just by using of so-called tau-theorem (shown later in Chap. 6). In fact, it calculates at once $s = 2$ matrix products of dimensionality $(m, 1, n)$ and $(1, (m-1)(n-1), 1)$ by $r = mn + 1$ steps (non-commutative multiplications), for certain integers $m > 1$ and $n > 1$:

$$AB = \begin{bmatrix} a_1 \\ \vdots \\ a_i \\ \vdots \\ a_m \end{bmatrix} \begin{bmatrix} b_1 & \cdots & b_j & \cdots & b_n \end{bmatrix} = \begin{bmatrix} a_i b_j \end{bmatrix}_{\substack{i=1,\dots,m \\ j=1,\dots,n}} \tag{5.19}$$

and:

$$UV = \begin{bmatrix} u_1 & \cdots & u_{(m-1)(n-1)} \end{bmatrix} \begin{bmatrix} v_1 \\ \vdots \\ v_{(m-1)(n-1)} \end{bmatrix} = \sum_{k=1}^{(m-1)(n-1)} u_k v_k \tag{5.20}$$

Originally Schonhage (1981) gave the algorithm in a trilinear form; we present it in a bilinear form. Let us calculate the AB matrix product of the two above. To do this, we define the following two-dimensional series of products:

$$P_{ij} = \begin{cases} (a_i + \varepsilon u_{ij})(b_j + \varepsilon v_{ij}), & 1 \leq i < m,\ 1 \leq j < n \\ \left(a_m - \varepsilon \sum\limits_{i'=1}^{m-1} u_{i'j}\right) b_j = P_{mj}, & 1 \leq j < n,\ i = m \\ a_i \left(b_n - \varepsilon \sum\limits_{j'=1}^{n-1} v_{ij'}\right) = P_{in}, & 1 \leq i < m,\ j = n \\ a_m b_n, & i = m,\ j = n \end{cases} \tag{5.21}$$

We can immediately observe the following obvious series of limits:

$$\lim_{\varepsilon \to 0} P_{ij} = a_i b_j \tag{5.22}$$

for $i = 1, \ldots, m$ and $j = 1, \ldots, n$. This means that the P_{ij} product series directly approximates all elements of the first matrix product AB [formula (5.19)]. To find the second matrix product UV, we sum up all these products. More precisely, we want to find the sum $\sum_{i,j=1}^{m,n} P_{ij}$. In finding it will help the following diagram:

$$\left[\begin{array}{ccc|c} P_{11} & \cdots & P_{1(n-1)} & P_{1n} \\ \vdots & \ddots & \vdots & \vdots \\ P_{(m-1)1} & \cdots & P_{(m-1)(n-1)} & P_{(m-1)n} \\ \hline P_{m1} & \cdots & P_{m(n-1)} & P_{mn} \end{array}\right] \tag{5.23}$$

Thus, the desired sum consists of four groups and we can sum it up as follows:

$$\begin{aligned} \sum_{i,j=1}^{m,n} P_{ij} &= \sum_{i,j=1}^{m-1,n-1} P_{ij} + \sum_{j=1}^{n-1} P_{mj} + \sum_{i=1}^{m-1} P_{in} + P_{mn} \\ &= \sum_{i,j=1}^{m-1,n-1} (a_i + \varepsilon u_{ij})(b_j + \varepsilon v_{ij}) + \sum_{j=1}^{n-1} \left(a_m - \varepsilon \sum_{i=1}^{m-1} u_{ij}\right) b_j \\ &\quad + \sum_{i=1}^{m-1} a_i \left(b_n - \varepsilon \sum_{j=1}^{n-1} v_{ij}\right) + a_m b_n \end{aligned}$$

We can see that the pairs of the terms of the form $\sum_{i,j=1}^{m-1,n-1} a_i(\varepsilon v_{ij})$ and $\sum_{i,j=1}^{m-1,n-1} (\varepsilon u_{ij}) b_j$ cancel each other. Using the diagram (5.23) in the opposite direction, adapted to $a_i b_j$ terms, we obtain:

$$\sum_{i,j=1}^{m,n} P_{ij} = \left[\sum_{i,j=1}^{m-1,n-1} a_i b_j + a_m\left(\sum_{j=1}^{n-1} b_j\right) + \left(\sum_{i=1}^{m-1} a_i\right) b_n + a_m b_n\right] + \varepsilon^2 \sum_{i,j=1}^{m-1,n-1} u_{ij} v_{ij}$$
$$= \sum_{i,j=1}^{m,n} a_i b_j + \varepsilon^2 \sum_{i,j=1}^{m-1,n-1} u_{ij} v_{ij}$$

Using—rather not so simple—identity $\sum_{i,j=1}^{m,n} a_i b_j = \left(\sum_{i=1}^{m} a_i\right) \cdot \left(\sum_{j=1}^{n} b_j\right)$, it is possible to get the expression for the scalar product $\sum_{i,j=1}^{m-1,n-1} u_{ij} v_{ij}$ of the $\sum_{i,j=1}^{m,n} P_{ij}$ sum:

$$\sum_{i,j=1}^{m-1,n-1} u_{ij} v_{ij} = \frac{1}{\varepsilon^2}\left[\sum_{i,j=1}^{m,n} P_{ij} - \left(\sum_{i=1}^{m} a_i\right)\left(\sum_{j=1}^{n} b_j\right)\right] \tag{5.24}$$

To see that the sum $\sum_{i,j=1}^{m-1,n-1} u_{ij} v_{ij}$ expresses the UV matrix product, it is sufficient to arrange the u_{ij} variable series as follows:

$$U = \left[\, u_{11} \cdots u_{1(n-1)} \,\middle|\, \cdots \,\middle|\, u_{i1} \cdots u_{i(n-1)} \,\middle|\, \cdots \,\middle|\, u_{(m-1)1} \;\; u_{(m-1)(n-1)} \,\right]$$

and the series v_{ij} can be represented analogously.

It is worth noting that to calculate the second matrix product in the pair AB, UV, it is sufficient to perform only one (!) additional multiplication:

$$(a_1 + a_2 + \cdots + a_m)(b_1 + b_2 + \cdots + b_n)$$

rather than the definitional $(m-1)(n-1)$ multiplications. In Chap. 6 we show how this algorithm can be converted into a single matrix multiplication algorithm of complexity $\omega \leq 2.5479$.

5.6.1 Example—Schonhage Double Disjoint (4,1,4), (1,9,1) Problem

To be precise, our aim is to compute two matrix products simultaneously. The first of the two AB has dimensionality $(4, 1, 4)$, and the second, denoted UV, is of dimensionality $(1, (4-1)\cdot(4-1), 1) = (1, 9, 1)$. The definitional approach here leads to $4\cdot1\cdot4 + 1\cdot9\cdot1 = 25$ products to perform, whereas the Schonhage double disjoint algorithm requires $r = mn + 1 = 4\cdot4 + 1 = 17$ products. The entries of AB products are directly approximated by the P_{ij} products (5.21). For $1 \leq i < 4$, $1 \leq j < 4$ we have:

$$P_{11} = (a_1 + \varepsilon u_{11})(b_1 + \varepsilon v_{11}),\ P_{12} = (a_1 + \varepsilon u_{12})(b_2 + \varepsilon v_{12}),\ P_{13} = (a_1 + \varepsilon u_{13})(b_3 + \varepsilon v_{13})$$
$$P_{21} = (a_2 + \varepsilon u_{21})(b_1 + \varepsilon v_{21}),\ P_{22} = (a_2 + \varepsilon u_{22})(b_2 + \varepsilon v_{22}),\ P_{23} = (a_2 + \varepsilon u_{23})(b_3 + \varepsilon v_{23})$$
$$P_{31} = (a_3 + \varepsilon u_{31})(b_1 + \varepsilon v_{31}),\ P_{32} = (a_3 + \varepsilon u_{32})(b_2 + \varepsilon v_{32}),\ P_{33} = (a_3 + \varepsilon u_{33})(b_3 + \varepsilon v_{33})$$

$$\text{For } 1 \le j < 4,\ i = 4: \quad \begin{aligned} P_{41} &= (a_4 - \varepsilon u_{11} - \varepsilon u_{21} - \varepsilon u_{31})b_1 \\ P_{42} &= (a_4 - \varepsilon u_{12} - \varepsilon u_{22} - \varepsilon u_{32})b_2 \\ P_{43} &= (a_4 - \varepsilon u_{13} - \varepsilon u_{23} - \varepsilon u_{33})b_3 \end{aligned}$$

$$\text{For } 1 \le i < 4, j = 4: \quad \begin{aligned} P_{14} &= a_1(b_4 - \varepsilon v_{11} - \varepsilon v_{12} - \varepsilon v_{13}) \\ P_{24} &= a_2(b_4 - \varepsilon v_{21} - \varepsilon v_{22} - \varepsilon v_{23}) \\ P_{34} &= a_3(b_4 - \varepsilon v_{31} - \varepsilon v_{32} - \varepsilon v_{33}) \end{aligned}$$

$$\text{For } i = 4,\ j = 4: \quad P_{44} = a_4 b_4$$

The matrix product AB, according to formula (5.19) in combination with the limit series (5.22), can approximated as:

$$(AB)_{ij} \approx P_{ij}, \quad i = 1, \ldots, m, \quad j = 1, \ldots, n$$

As we have seen, to find the second matrix product UV we only need to do one more multiplication. From formula (5.24) we obtain:

$$\sum_{i,j=1}^{3,3} u_{ij} v_{ij} = \frac{1}{\varepsilon^2} \left[\left(\sum_{i,j=1}^{4,4} P_{ij} \right) - (a_1 + a_2 + a_3 + a_4)(b_1 + b_2 + b_3 + b_4) \right]$$

and the multiplication in question is $(a_1 + a_2 + a_3 + a_4)(b_1 + b_2 + b_3 + b_4)$. On the other hand, the definition-based formula (5.20) results in a single element matrix 1×1:

$$UV = \left[\sum_{k=1}^{9} u_k v_k \right]$$

The desired scalar product is obtained by assignment:

$$\sum_{k=1}^{9} u_k v_k := \sum_{i,j=1}^{3,3} u_{ij} v_{ij}$$

In the next section we show another, more sophisticated disjoint algorithm proposed by Schonhage (1981), which works on three matrix products simultaneously and gives a slightly better efficiency exponent.

5.7 Triple Disjoint Matrix Multiplication Algorithm by Schonhage

In this section we present the triple disjoint matrix multiplication algorithm proposed in Section 8 of the article Schonhage (1981), pages 453–454. The algorithm in question calculates three matrix products at once, where in each of the products at least one of the multiplication factors is a horizontal or vertical vector.

Let us consider $1 \times k$ matrices $A,\ B,\ C$, $k \times n$ matrices X, Y, Z and $n \times 1$ matrices U, V, W. We want to obtain the following three matrix products at once:

$$\begin{cases} AY \\ UB \\ XV \end{cases} \tag{5.25}$$

They can be represented in a trilinear form:

$$F = Trace(AYW + UBZ + XVC) = \sum_{i,j} \big(a_i y_{ij} w_j + u_j b_i z_{ij} + x_{ij} v_j c_i\big) \tag{5.26}$$

where the summation variables vary in the ranges $1 \le i \le k$ and $1 \le j \le n$. It represents three disjoint matrix products of dimensionality $(1, k, n)$, $(n, 1, k)$ and $(k, n, 1)$. Schonhage (1981) introduces six auxiliary variables for a more concise form of the algorithm for the vectors:

$$\begin{aligned} a &= \sum_i a_i, \quad b = \sum_i b_i, \quad c = \sum_i c_i \\ u &= \sum_j u_j, \quad v = \sum_j v_j, \quad w = \sum_j w_j \end{aligned} \tag{5.27}$$

And for the matrices, we each denote three auxiliary series of variables:

$$x_i = \sum_j x_{ij}, \quad y_i = \sum_j y_{ij}, \quad z_i = \sum_j z_{ij} \tag{5.28}$$

The algorithm proposed by Schonhage is based on the following trilinear identity:

$$\begin{aligned} &\varepsilon^{12} \sum_{i,j} \big(a_i + \varepsilon^4 u_j + \varepsilon^6 x_{ij}\big)\big(b_i + \varepsilon^2 v_j + \varepsilon^5 y_{ij}\big)\big(c_i + \varepsilon^3 w_j + \varepsilon^4 z_{ij}\big) \\ &- \varepsilon^8 \sum_i \big(a_i + \varepsilon^8 u + \varepsilon^{10} x_i\big)\big(b_i + \varepsilon^6 v + \varepsilon^9 y_i\big)\big(c_i + \varepsilon^7 w + \varepsilon^8 z_i\big) \\ &- \varepsilon^4 \sum_j \big(a + \varepsilon^8 u_j\big)\big(b + \varepsilon^6 v_j\big)\big(c + \varepsilon^7 w_j\big) \end{aligned}$$

$$
\begin{aligned}
&+ \left(a+\varepsilon^{12}u\right)\left(b+\varepsilon^{10}v\right)\left(c+\varepsilon^{11}w\right) \\
&+ p\varepsilon^{8}\sum_{i} a_i b_i c_i \\
&- p\cdot abc \\
&= \varepsilon^{20}F + \varepsilon^{18}\sum_{i,j} a_i v_j z_{ij} + O^{+}\left(\varepsilon^{21}\right)
\end{aligned}
\tag{5.29}
$$

where:

$$p = 1 - n\varepsilon^{4}$$

and $O^{+}(\varepsilon^{q})$ denotes a polynomial in the variable ε of degree equal to or greater than q. In the identity (5.29) the term $O^{+}\left(\varepsilon^{21}\right)$ contains terms of degree up to 35.

Let us introduce the following product series, which are convenient to represent the algorithm in bilinear form:

$$
\begin{cases}
P_{ij}\left(a_i, u_j, x_{ij}, b_i, v_j, y_{ij}\right) = \left(a_i + \varepsilon^{4}u_j + \varepsilon^{6}x_{ij}\right)\left(b_i + \varepsilon^{2}v_j + \varepsilon^{5}y_{ij}\right) \\
Q_i^{(1)}(a_i, u, x_i, b_i, v, y_i) = \left(a_i + \varepsilon^{8}u + \varepsilon^{10}x_i\right)\left(b_i + \varepsilon^{6}v + \varepsilon^{9}y_i\right) \\
Q_j^{(2)}\left(a, u_j, b, v_j\right) = \left(a + \varepsilon^{8}u_j\right)\left(b + \varepsilon^{6}v_j\right) \\
R^{(1)}(a, u, b, v) = \left(a + \varepsilon^{12}u\right)\left(b + \varepsilon^{10}v\right) \\
Q_i^{(3)}(a_i, b_i) = a_i b_i \\
R^{(2)}(a, b) = ab
\end{cases}
\tag{5.30}
$$

Using the product series (5.30), the trilinear identity (5.29) can be written down as follows (for brevity we have denoted $P_{ij}\left(a_i, u_j, x_{ij}, b_i, v_j, y_{ij}\right)$ as P_{ij}, $Q_i^{(1)}(a_i, u, x_i, b_i, v, y_i)$ as $Q_i^{(1)}$ etc.):

$$
\begin{aligned}
&\varepsilon^{12}\sum_{i,j} P_{ij}\cdot\left(c_i + \varepsilon^{3}w_j + \varepsilon^{4}z_{ij}\right) - \varepsilon^{8}\sum_{i} Q_i^{(1)}\cdot\left(c_i + \varepsilon^{7}w + \varepsilon^{8}z_i\right) \\
&- \varepsilon^{4}\sum_{j} Q_j^{(2)}\cdot\left(c + \varepsilon^{7}w_j\right) + R^{(1)}\cdot\left(c + \varepsilon^{11}w\right) + p\varepsilon^{8}\sum_{i} Q_i^{(3)}\cdot c_i \\
&- p\cdot R^{(2)}\cdot c = \varepsilon^{20}F + \varepsilon^{18}\sum_{i,j} a_i v_j z_{ij} + O^{+}\left(\varepsilon^{21}\right)
\end{aligned}
\tag{5.31}
$$

In the trilinear identity in question, all the terms in the representation of the matrix products (5.25) that we want to calculate are desired, except the terms $\sum_{i,j} a_i v_j z_{ij}$ and $O^{+}\left(\varepsilon^{21}\right)$. The last one leads to the approximate matrix multiplication algorithm. To get rid of the undesired term $\sum_{i,j} a_i v_j z_{ij}$ Schonhage (1981), page 454, suggests to set up a second copy of the trilinear form (5.31), with the same matrices A, V, Z but with another matrices $\tilde{B}$, $\tilde{C}$, $\tilde{U}$, $\tilde{W}$, $\tilde{X}$, $\tilde{Y}$ and with the Z matrix taken with the negative sign. This notation applies respectively to the auxiliary variables (5.27) and (5.28). To ensure that changing the sign of the Z matrix does not affect the trilinear representation of (another) triple disjoint matrix multiplication, we also change the sign of the $\tilde{U}$ matrix. The corresponding trilinear representation of the second matrix multiplication has the following form:

$$\begin{aligned}\tilde{F} &= Trace\left[A\tilde{Y}\tilde{W} + (-\tilde{U})\tilde{B}(-Z) + \tilde{X}V\tilde{C}\right] \\ &= Trace\left[A\tilde{Y}\tilde{W} + \tilde{U}\tilde{B}Z + \tilde{X}V\tilde{C}\right] \\ &= \sum_{i,j}\left[a_i\tilde{y}_{ij}\tilde{w}_j + \tilde{u}_j\tilde{b}_i z_{ij} + \tilde{x}_{ij}v_j\tilde{c}_i\right]\end{aligned} \tag{5.32}$$

The second copy of the trilinear form (5.31) has the form:

$$\begin{aligned}&\varepsilon^{12}\sum_{i,j}\tilde{P}_{ij}\cdot\left(\tilde{c}_i + \varepsilon^3\tilde{w}_j - \varepsilon^4 z_{ij}\right) - \varepsilon^8\sum_i \tilde{Q}_i^{(1)}\cdot\left(\tilde{c}_i + \varepsilon^7\tilde{w} - \varepsilon^8 z_i\right) \\ &- \varepsilon^4\sum_j \tilde{Q}_j^{(2)}\cdot\left(\tilde{c} + \varepsilon^7\tilde{w}_j\right) + \tilde{R}^{(1)}\cdot\left(\tilde{c} + \varepsilon^{11}\tilde{w}\right) + p\varepsilon^8\sum_i \tilde{Q}_i^{(3)}\cdot\tilde{c}_i \\ &- p\cdot\tilde{R}^{(2)}\cdot\tilde{c} = \varepsilon^{20}\tilde{F} - \varepsilon^{18}\sum_{i,j} a_i v_j z_{ij} + O^+\left(\varepsilon^{21}\right)\end{aligned} \tag{5.33}$$

where the new product series is defined using series (5.30):

$$\begin{cases}\tilde{P}_{ij} = P_{ij}\left(a_i, -\tilde{u}_j, \tilde{x}_{ij}, \tilde{b}_i, v_j, \tilde{y}_{ij}\right) \\ \tilde{Q}_i^{(1)} = Q_i^{(1)}\left(a_i, -\tilde{u}, \tilde{x}_i, \tilde{b}_i, v, \tilde{y}_i\right) \\ \tilde{Q}_j^{(2)} = Q_j^{(2)}\left(a, -\tilde{u}_j, \tilde{b}, v_j\right) \\ \tilde{R}^{(1)} = R^{(1)}\left(a, -\tilde{u}, \tilde{b}, v\right) \\ \tilde{Q}_i^{(3)} = Q_i^{(3)}\left(a_i, \tilde{b}_i\right) \\ \tilde{R}^{(2)} = R^{(2)}\left(a, \tilde{b}\right)\end{cases} \tag{5.34}$$

Now we add by sides two trilinear forms (5.31) and (5.33):

$$\begin{aligned}&\varepsilon^{12}\sum_{i,j}P_{ij}\cdot\left(c_i + \varepsilon^3 w_j\right) + \varepsilon^{12}\sum_{i,j}\tilde{P}_{ij}\cdot\left(\tilde{c}_i + \varepsilon^3\tilde{w}_j\right) \\ &+ \varepsilon^{16}\sum_{i,j}\left(P_{ij} - \tilde{P}_{ij}\right)z_{ij} - \varepsilon^8\sum_i Q_i^{(1)}\cdot\left(c_i + \varepsilon^7 w_i\right) \\ &- \varepsilon^8\sum_i \tilde{Q}_i^{(1)}\cdot\left(\tilde{c}_i + \varepsilon^7\tilde{w}\right) - \varepsilon^{16}\sum_i\left(Q_i^{(1)} - \tilde{Q}_i^{(1)}\right)z_i \\ &- \varepsilon^4\sum_j Q_j^{(2)}\cdot\left(c + \varepsilon^7 w_j\right) - \varepsilon^4\sum_j \tilde{Q}_j^{(2)}\cdot\left(\tilde{c} + \varepsilon^7\tilde{w}_j\right)\end{aligned}$$

$$+ R^{(1)} \cdot \left(c + \varepsilon^{11} w\right) + \tilde{R}^{(1)} \cdot \left(\tilde{c} + \varepsilon^{11} \tilde{w}\right)$$
$$+ p\varepsilon^8 \sum_i Q_i^{(3)} \cdot c_i + p\varepsilon^8 \sum_i \tilde{Q}_i^{(3)} \cdot \tilde{c}_i$$
$$- p \cdot R^{(2)} \cdot c - p \cdot \tilde{R}^{(2)} \cdot \tilde{c} = \varepsilon^{20}\left(F + \tilde{F}\right) + O^+\left(\varepsilon^{21}\right) \tag{5.35}$$

We can see that the term $\sum_{i,j} a_i v_j z_{ij}$ no longer exists in the obtained identity (5.35). Let us transform the trilinear identity (5.35) into a bilinear form. To do this, we will start the work by comparing the expressions corresponding to the c_i terms. Considering the trilinear form F [formula (5.26)], they represent the values of the matrix product XV. We insert here the auxiliary variable c as $\sum_i c_i$ (5.27):

$$\varepsilon^{20} \sum_i \left[\sum_j x_{ij} v_j\right] c_i = \varepsilon^{12} \sum_{i,j} P_{ij} c_i - \varepsilon^8 \sum_i Q_i^{(1)} c_i - \varepsilon^4 \sum_j Q_j^{(2)} \sum_i c_i$$
$$+ R^{(1)} \sum_i c_i + p\varepsilon^8 \sum_i Q_i^{(3)} c_i - pR^{(2)} \sum_i c_i + O^+\left(\varepsilon^{21}\right)$$

One of the above terms requires the summation order to be changed:

$$\varepsilon^4 \sum_j Q_j^{(2)} \sum_i c_i = \varepsilon^4 \sum_{i,j} Q_j^{(2)} c_i = \varepsilon^4 \sum_i \left[\sum_j Q_j^{(2)}\right] c_i$$

Thus:

$$\varepsilon^{20} \sum_i \left[\sum_j x_{ij} v_j\right] c_i = \varepsilon^{12} \sum_i \left[\sum_j P_{ij}\right] c_i - \varepsilon^8 \sum_i Q_i^{(1)} c_i$$
$$- \varepsilon^4 \sum_i \left[\sum_j Q_j^{(2)}\right] c_i + R^{(1)} \sum_i c_i + p\varepsilon^8 \sum_i Q_i^{(3)} c_i - pR^{(2)} \sum_i c_i$$
$$+ O^+\left(\varepsilon^{21}\right) = \sum_i \left[\varepsilon^{12} \sum_j P_{ij} - \varepsilon^8 Q_i^{(1)} - \varepsilon^4 \sum_j Q_j^{(2)} + R^{(1)} + p\varepsilon^8 Q_i^{(3)} - pR^{(2)}\right] c_i$$
$$+ O^+\left(\varepsilon^{21}\right)$$

The values of the entries of the matrix XV in the bilinear form are:

$$(XV)_i = \sum_j x_{ij} v_j = \sum_j \left(\frac{1}{\varepsilon^8} P_{ij} - \frac{1}{\varepsilon^{16}} Q_j^{(2)}\right) - \frac{1}{\varepsilon^{12}} Q_i^{(1)}$$
$$+ \frac{p}{\varepsilon^{12}} Q_i^{(3)} + \frac{1}{\varepsilon^{20}} R^{(1)} - \frac{p}{\varepsilon^{20}} R^{(2)} + O^+(\varepsilon) \tag{5.36}$$

The structure of the trilinear identity (5.35) is analogous for the $\tilde{c}_i$ variables. Thus, the formula (5.36) also expresses the respective entries of the matrix product $\tilde{X}V$ after replacing the $P_{..}$, $Q_{.}$, $R_{.}$ products by the respective $\tilde{P}_{..}$, $\tilde{Q}_{.}$, $\tilde{R}_{.}$ products defined by formula (5.35).

By analogy with the matrix product AY, we obtain:

$$(AY)_j = \sum_i a_i y_{ij} = \frac{1}{\varepsilon^5} \sum_i \left(P_{ij} - Q_i^{(1)} \right) - \frac{1}{\varepsilon^9} Q_j^{(2)} + \frac{1}{\varepsilon^9} R^{(1)} + O^+(\varepsilon) \tag{5.37}$$

After replacing the products $P_{..}$, $Q_{.}$, $R_{.}$ by $\tilde{P}_{..}$, $\tilde{Q}_{.}$, $\tilde{R}_{.}$, formula (5.37) also expresses entries of the matrix product $A\tilde{Y}$. The procedure is slightly different for the matrix product corresponding to the z_{ij} variable. Let us collect from the trilinear identity (5.35) those terms, which contain z_{ij} entries:

$$\begin{aligned} &\varepsilon^{16} \sum_{i,j} (P_{ij} - \tilde{P}_{ij}) z_{ij} - \varepsilon^{16} \sum_i \left(Q_i^{(1)} - \tilde{Q}_i^{(1)} \right) z_i \\ &= \varepsilon^{20} \left(\sum_{i,j} u_j b_i z_{ij} + \sum_{i,j} \tilde{u}_j \tilde{b}_i z_{ij} \right) + O^+(\varepsilon^{2}) \end{aligned} \tag{5.38}$$

Let us replace z_i by $\sum_j z_{ij}$. Note the left part of the inequality (5.38):

$$\begin{aligned} &\varepsilon^{16} \sum_{i,j} (P_{ij} - \tilde{P}_{ij}) z_{ij} - \varepsilon^{16} \sum_i \left(Q_i^{(1)} - \tilde{Q}_i^{(1)} \right) \sum_j z_{ij} \\ &= \varepsilon^{16} \sum_{i,j} \left[\left(P_{ij} - Q_i^{(1)} \right) - \left(\tilde{P}_{ij} - \tilde{Q}_i^{(1)} \right) \right] z_{ij} \end{aligned} \tag{5.39}$$

And the right side of the equality (5.38) have the form:

$$\varepsilon^{20} \left(\sum_{i,j} u_j b_i z_{ij} + \sum_{i,j} \tilde{u}_j \tilde{b}_i z_{ij} \right) + O^+(\varepsilon^{21}) = \varepsilon^{20} \sum_{i,j} (u_j b_i + \tilde{u}_j \tilde{b}_i) z_{ij} + O^+(\varepsilon^{21}) \tag{5.40}$$

Equating (5.39) with (5.40) we obtain:

$$\varepsilon^4 \sum_{i,j} (u_j b_i + \tilde{u}_j \tilde{b}_i) z_{ij} = \sum_{i,j} \left[\left(P_{ij} - Q_i^{(1)} \right) - \left(\tilde{P}_{ij} - \tilde{Q}_i^{(1)} \right) \right] z_{i,} + O^+(\varepsilon^5) \tag{5.41}$$

Arriving at the bilinear form, we obtain the sum of two products, for i, j in their respective ranges:

$$u_j b_i + \tilde{u}_j \tilde{b}_i = \frac{1}{\varepsilon^4}\Big[\Big(P_{ij} - Q_i^{(1)}\Big) - \Big(\tilde{P}_{ij} - \tilde{Q}_i^{(1)}\Big)\Big] + O^+(\varepsilon) \tag{5.42}$$

To cancel the unwanted term $\sum_{i,j} a_i v_j z_{ij}$ in the trilinear form (5.29), we had to set up another version of that trilinear form [equality (5.33)], but we can also neatly use it to speed up the fast matrix multiplication algorithm. Let us start with the product XV and its second copy $\tilde{X}V$. We can arrange them into a single matrix multiplication using the following diagram:

$$\left[\frac{X}{\tilde{X}}\right] V \tag{5.43}$$

Single product XV has dimensionality $(k, n, 1)$, so it follows from diagram (5.43) that the dimensionality of coupled XV and $\tilde{X}V$ products is $(2k, n, 1)$. By analogy, we can smartly connect the pair of products AY and $A\tilde{Y}$:

$$A\big[\,Y \,\big|\, \tilde{Y}\,\big] \tag{5.44}$$

Now the dimensionality of a "source" single product AY is $(1, k, n)$, so for the above construction (5.44) we obtain the $(1, k, 2n)$ dimensionality.

The situation is slightly different for the pair of products corresponding to the z_{ij} variable. The equality (5.42), which expresses the value of the $u_j b_i + \tilde{u}_j \tilde{b}_i$ sum, can be arranged in a single block matrix multiplication by the following construction:

$$\big[\,U \,\big|\, \tilde{U}\,\big]\left[\frac{B}{\tilde{B}}\right] = \big(u_j b_i + \tilde{u}_j \tilde{b}_i\big)_{\substack{j=1,\ldots,n \\ i=1,\ldots,k}} \tag{5.45}$$

with the dimensionality of the product $(n, 2, k)$.

In summary, the pair of trilinear identities (5.29) and (5.33) represents a triple disjoint matrix multiplication algorithm with dimensionalities $(2k, n, 1)$, $(1, k, 2n)$ and $(n, 2, k)$. The amount of non-linear products [equalities (5.30) and (5.34)] is equal to $2(kn + k + n + 1 + k + 1) = 2(k + 1)(n + 2)$. In Chap. 6 we show how to use this, triple disjoint algorithm, to construct an algorithm for single matrix multiplication with the efficiency exponent $\omega \leq 2.5218127$.

5.7.1 Example—Triple Disjoint Matrix Multiplication Algorithm by Schonhage

In this example we show how to linearly combine 30 products, defined by the series (5.30) and (5.34), to obtain triple disjoint matrix multiplication algorithm for problems of dimensionality $(4, 3, 1)$, $(1, 2, 6)$ and $(3, 2, 2)$. To do this, we use

Schonhage's algorithm with dimension-related parameters $n = 3,\ k = 2$, in particular formulae (5.36), (5.37) and (5.42) expressing the 'individual' matrix products, together with diagrams (5.43), (5.44) and (5.45), showing how to combine them in pairs.

We calculate matrix products $A\left[\,Y \,\middle|\, \tilde{Y}\,\right]$, $\left[\,X^T \,\middle|\, \tilde{X}^T\,\right]^T V$ at once, and $\left[\,U \,\middle|\, \tilde{U}\,\right]\left[\,B^T \,\middle|\, \tilde{B}^T\,\right]^T$ with 1×2 matrices A, B, $\tilde{B}$, 2×3 matrices $X, \tilde{X}, Y, \tilde{Y}$ and 3×1 matrices $U, \tilde{U}, V$. The indices i, j are in the following sets:

$$i = 1, 2; \quad j = 1, 2, 3$$

Let us start with the product series P_{ij} and $\tilde{P}_{ij}$, which depend on two parameters. For P_{ij} we have:

$$\begin{cases}
P_{11} = \left(a_1 + \varepsilon^4 u_1 + \varepsilon^6 x_{11}\right)\left(b_1 + \varepsilon^2 v_1 + \varepsilon^5 y_{11}\right), & P_{12} = \left(a_1 + \varepsilon^4 u_2 + \varepsilon^6 x_{12}\right)\left(b_1 + \varepsilon^2 v_2 + \varepsilon^5 y_{12}\right) \\
P_{13} = \left(a_1 + \varepsilon^4 u_3 + \varepsilon^6 x_{13}\right)\left(b_1 + \varepsilon^2 v_3 + \varepsilon^5 y_{13}\right), & P_{21} = \left(a_2 + \varepsilon^4 u_1 + \varepsilon^6 x_{21}\right)\left(b_2 + \varepsilon^2 v_1 + \varepsilon^5 y_{21}\right) \\
P_{22} = \left(a_2 + \varepsilon^4 u_2 + \varepsilon^6 x_{22}\right)\left(b_2 + \varepsilon^2 v_2 + \varepsilon^5 y_{22}\right), & P_{23} = \left(a_2 + \varepsilon^4 u_3 + \varepsilon^6 x_{23}\right)\left(b_2 + \varepsilon^2 v_3 + \varepsilon^5 y_{23}\right)
\end{cases}$$

and for the second copy of the trilinear identity:

$$\begin{cases}
\tilde{P}_{11} = \left(a_1 - \varepsilon^4 \tilde{u}_1 + \varepsilon^6 \tilde{x}_{11}\right)\left(\tilde{b}_1 + \varepsilon^2 v_1 + \varepsilon^5 \tilde{y}_{11}\right), & \tilde{P}_{12} = \left(a_1 - \varepsilon^4 \tilde{u}_2 + \varepsilon^6 \tilde{x}_{12}\right)\left(\tilde{b}_1 + \varepsilon^2 v_2 + \varepsilon^5 \tilde{y}_{12}\right) \\
\tilde{P}_{13} = \left(a_1 - \varepsilon^4 \tilde{u}_3 + \varepsilon^6 \tilde{x}_{13}\right)\left(\tilde{b}_1 + \varepsilon^2 v_3 + \varepsilon^5 \tilde{y}_{13}\right), & \tilde{P}_{21} = \left(a_2 - \varepsilon^4 \tilde{u}_1 + \varepsilon^6 \tilde{x}_{21}\right)\left(\tilde{b}_2 + \varepsilon^2 v_1 + \varepsilon^5 \tilde{y}_{21}\right) \\
\tilde{P}_{22} = \left(a_2 - \varepsilon^4 \tilde{u}_2 + \varepsilon^6 \tilde{x}_{22}\right)\left(\tilde{b}_2 + \varepsilon^2 v_2 + \varepsilon^5 \tilde{y}_{22}\right), & \tilde{P}_{23} = \left(a_2 - \varepsilon^4 \tilde{u}_3 + \varepsilon^6 \tilde{x}_{23}\right)\left(\tilde{b}_2 + \varepsilon^2 v_3 + \varepsilon^5 \tilde{y}_{23}\right)
\end{cases}$$

For the products dependent on one parameter, i.e. $Q_i^{(1)}$, $Q_j^{(2)}$ and $Q_i^{(3)}$, we obtain:

$$\begin{cases}
Q_1^{(1)} = \left(a_1 + \varepsilon^8 (u_1 + u_2 + u_3) + \varepsilon^{10}\left(x_{1,1} + x_{1,2} + x_{1,3}\right)\right)\left(b_1 + \varepsilon^6 (v_1 + v_2 + v_3) + \varepsilon^9\left(y_{1,1} + y_{1,2} + y_{1,3}\right)\right) \\
Q_2^{(1)} = \left(a_2 + \varepsilon^8 (u_1 + u_2 + u_3) + \varepsilon^{10}\left(x_{2,1} + x_{2,2} + x_{2,3}\right)\right)\left(b_2 + \varepsilon^6 (v_1 + v_2 + v_3) + \varepsilon^9\left(y_{2,1} + y_{2,2} + y_{2,3}\right)\right) \\
\hline
\tilde{Q}_1^{(1)} = \left(a_1 - \varepsilon^8 (\tilde{u}_1 + \tilde{u}_2 + \tilde{u}_3) + \varepsilon^{10}\left(\tilde{x}_{1,1} + \tilde{x}_{1,2} + \tilde{x}_{1,3}\right)\right)\left(\tilde{b}_1 + \varepsilon^6 (v_1 + v_2 + v_3) + \varepsilon^9\left(\tilde{y}_{1,1} + \tilde{y}_{1,2} + \tilde{y}_{1,3}\right)\right) \\
\tilde{Q}_2^{(1)} = \left(a_2 - \varepsilon^8 (\tilde{u}_1 + \tilde{u}_2 + \tilde{u}_3) + \varepsilon^{10}\left(\tilde{x}_{2,1} + \tilde{x}_{2,2} + \tilde{x}_{2,3}\right)\right)\left(\tilde{b}_2 + \varepsilon^6 (v_1 + v_2 + v_3) + \varepsilon^9\left(\tilde{y}_{2,1} + \tilde{y}_{2,2} + \tilde{y}_{2,3}\right)\right)
\end{cases}$$

$$\begin{cases}
Q_1^{(2)} = \left(a_1 + a_2 + \varepsilon^8 u_1\right)\left(b_1 + b_2 + \varepsilon^6 v_1\right) \\
Q_2^{(2)} = \left(a_1 + a_2 + \varepsilon^8 u_2\right)\left(b_1 + b_2 + \varepsilon^6 v_2\right) \\
Q_3^{(2)} = \left(a_1 + a_2 + \varepsilon^8 u_3\right)\left(b_1 + b_2 + \varepsilon^6 v_3\right) \\
\hline
\tilde{Q}_1^{(2)} = \left(a_1 + a_2 - \varepsilon^8 \tilde{u}_1\right)\left(\tilde{b}_1 + \tilde{b}_2 + \varepsilon^6 v_1\right) \\
\tilde{Q}_2^{(2)} = \left(a_1 + a_2 - \varepsilon^8 \tilde{u}_2\right)\left(\tilde{b}_1 + \tilde{b}_2 + \varepsilon^6 v_2\right) \\
\tilde{Q}_3^{(2)} = \left(a_1 + a_2 - \varepsilon^8 \tilde{u}_3\right)\left(\tilde{b}_1 + \tilde{b}_2 + \varepsilon^6 v_3\right)
\end{cases}
\quad
\begin{cases}
Q_1^{(3)} = a_1 b_1 \\
Q_2^{(3)} = a_2 b_2 \\
\hline
\tilde{Q}_1^{(3)} = a_1 \tilde{b}_1 \\
\tilde{Q}_2^{(3)} = a_2 \tilde{b}_2
\end{cases}$$

And finally, we receive the $R^{(*)}$ products:

$$\begin{aligned}
R^{(1)} &= \left(a_1 + a_2 + \varepsilon^{12}(u_1 + u_2 + u_3)\right)\left(b_1 + b_2 + \varepsilon^{10}(v_1 + v_2 + v_3)\right) \\
R^{(2)} &= (a_1 + a_2)(b_1 + b_2)
\end{aligned}$$

$$\begin{aligned}
\tilde{R}^{(1)} &= \left(a_1 + a_2 - \varepsilon^{12}(\tilde{u}_1 + \tilde{u}_2 + \tilde{u}_3)\right)\left(\tilde{b}_1 + \tilde{b}_2 + \varepsilon^{10}(v_1 + v_2 + v_3)\right) \\
\tilde{R}^{(2)} &= (a_1 + a_2)\left(\tilde{b}_1 + \tilde{b}_2\right)
\end{aligned}$$

To obtain the matrix products we are looking for, all we have to do is linearly combine the 30 aggregated products. Let us start with the product XV, to be constructed by the linear combination (5.36):

$$\begin{cases} (XV)_1 \approx \left(\frac{1}{\varepsilon^8}P_{11} - \frac{1}{\varepsilon^{16}}Q_1^{(2)}\right) + \left(\frac{1}{\varepsilon^8}P_{12} - \frac{1}{\varepsilon^{16}}Q_2^{(2)}\right) + \left(\frac{1}{\varepsilon^8}P_{13} - \frac{1}{\varepsilon^{16}}Q_3^{(2)}\right) \\ \qquad - \frac{1}{\varepsilon^{12}}Q_1^{(1)} + \frac{p}{\varepsilon^{12}}Q_1^{(3)} + \frac{1}{\varepsilon^{20}}R^{(1)} - \frac{p}{\varepsilon^{20}}R^{(2)} \\ (XV)_2 \approx \left(\frac{1}{\varepsilon^8}P_{21} - \frac{1}{\varepsilon^{16}}Q_1^{(2)}\right) + \left(\frac{1}{\varepsilon^8}P_{22} - \frac{1}{\varepsilon^{16}}Q_2^{(2)}\right) + \left(\frac{1}{\varepsilon^8}P_{23} - \frac{1}{\varepsilon^{16}}Q_3^{(2)}\right) \\ \qquad - \frac{1}{\varepsilon^{12}}Q_2^{(1)} + \frac{p}{\varepsilon^{12}}Q_2^{(3)} + \frac{1}{\varepsilon^{20}}R^{(1)} - \frac{p}{\varepsilon^{20}}R^{(2)} \end{cases}$$

Analogously, we construct its second copy $\tilde{X}V$. We just have to remember that in the second series of products (5.34) we use the second copy, marked by a tilde, of all the matrices except A, V and the $\tilde{U}$ matrix with a negative sign. The formulas are analogous, but for didactic purposes we have decided to present them:

$$\begin{cases} \left(\tilde{X}V\right)_1 \approx \left(\frac{1}{\varepsilon^8}\tilde{P}_{11} - \frac{1}{\varepsilon^{16}}\tilde{Q}_1^{(2)}\right) + \left(\frac{1}{\varepsilon^8}\tilde{P}_{12} - \frac{1}{\varepsilon^{16}}\tilde{Q}_2^{(2)}\right) + \left(\frac{1}{\varepsilon^8}\tilde{P}_{13} - \frac{1}{\varepsilon^{16}}\tilde{Q}_3^{(2)}\right) \\ \qquad - \frac{1}{\varepsilon^{12}}\tilde{Q}_1^{(1)} + \frac{p}{\varepsilon^{12}}\tilde{Q}_1^{(3)} + \frac{1}{\varepsilon^{20}}\tilde{R}^{(1)} - \frac{p}{\varepsilon^{20}}\tilde{R}^{(2)} \\ \left(\tilde{X}V\right)_2 \approx \left(\frac{1}{\varepsilon^8}\tilde{P}_{21} - \frac{1}{\varepsilon^{16}}\tilde{Q}_1^{(2)}\right) + \left(\frac{1}{\varepsilon^8}\tilde{P}_{22} - \frac{1}{\varepsilon^{16}}\tilde{Q}_2^{(2)}\right) + \left(\frac{1}{\varepsilon^8}\tilde{P}_{23} - \frac{1}{\varepsilon^{16}}\tilde{Q}_3^{(2)}\right) \\ \qquad - \frac{1}{\varepsilon^{12}}\tilde{Q}_2^{(1)} + \frac{p}{\varepsilon^{12}}\tilde{Q}_2^{(3)} + \frac{1}{\varepsilon^{20}}\tilde{R}^{(1)} - \frac{p}{\varepsilon^{20}}\tilde{R}^{(2)} \end{cases}$$

The $\left[X^T \,\middle|\, \tilde{X}^T\right]^T V$ matrix product is constructed as follows:

$$\left[\begin{array}{c} X \\ \hline \tilde{X} \end{array}\right] V \approx \left[\begin{array}{c} (XV)_1 \\ (XV)_2 \\ \hline \left(\tilde{X}V\right)_1 \\ \left(\tilde{X}V\right)_2 \end{array}\right]$$

with dimensionality $(2k, n, 1) = (4, 3, 1)$. The pair of products AY and $A\tilde{Y}$ we build in accordance with the linear combination (5.37). And so, for the product AY, we have (explicit formulae for $A\tilde{Y}$ will be omitted at this point):

$$\begin{cases} (AY)_1 \approx \frac{1}{\varepsilon^5}\left[\left(P_{11}-Q_1^{(1)}\right)+\left(P_{21}-Q_2^{(1)}\right)\right]-\frac{1}{\varepsilon^9}Q_1^{(2)}+\frac{1}{\varepsilon^9}R^{(1)} \\ (AY)_2 \approx \frac{1}{\varepsilon^5}\left[\left(P_{12}-Q_1^{(1)}\right)+\left(P_{22}-Q_2^{(1)}\right)\right]-\frac{1}{\varepsilon^9}Q_2^{(2)}+\frac{1}{\varepsilon^9}R^{(1)} \\ (AY)_3 \approx \frac{1}{\varepsilon^5}\left[\left(P_{13}-Q_1^{(1)}\right)+\left(P_{23}-Q_2^{(1)}\right)\right]-\frac{1}{\varepsilon^9}Q_3^{(2)}+\frac{1}{\varepsilon^9}R^{(1)} \end{cases}$$

So the $A\left[\,Y \mid \tilde{Y}\,\right]$ product is built as follows:

$$A\left[\,Y \mid \tilde{Y}\,\right] \approx \left[\,(AY)_1\ (AY)_2\ (AY)_3 \mid \left(A\tilde{Y}\right)_1\ \left(A\tilde{Y}\right)_2\ \left(A\tilde{Y}\right)_3\,\right]$$

with dimensionality $(1, k, 2n) = (1, 2, 6)$. The structure of the last, third product $\left[\,U \mid \tilde{U}\,\right]\left[\,B^T \mid \tilde{B}^T\,\right]^T$, expressed by the linear combination (5.42), is different from the previous two, so we have decided to present it in full detail. The below diagram illustrates how we construct it:

$$\left[\,U \mid \tilde{U}\,\right]\begin{bmatrix} B \\ \hline \tilde{B} \end{bmatrix} = \left[\begin{array}{c|c} u_1 & \tilde{u}_1 \\ u_2 & \tilde{u}_2 \\ u_3 & \tilde{u}_3 \end{array}\right]\left[\begin{array}{cc} b_1 & b_2 \\ \hline \tilde{b}_1 & \tilde{b}_2 \end{array}\right]$$

$$= \frac{1}{\varepsilon^4}\left[\begin{array}{c|c} \left(P_{11}-Q_1^{(1)}\right)-\left(\tilde{P}_{11}-\tilde{Q}_1^{(1)}\right) & \left(P_{21}-Q_2^{(1)}\right)-\left(\tilde{P}_{21}-\tilde{Q}_2^{(1)}\right) \\ \left(P_{12}-Q_1^{(1)}\right)-\left(\tilde{P}_{12}-\tilde{Q}_1^{(1)}\right) & \left(P_{22}-Q_2^{(1)}\right)-\left(\tilde{P}_{22}-\tilde{Q}_2^{(1)}\right) \\ \left(P_{13}-Q_1^{(1)}\right)-\left(\tilde{P}_{13}-\tilde{Q}_1^{(1)}\right) & \left(P_{23}-Q_2^{(1)}\right)-\left(\tilde{P}_{23}-\tilde{Q}_2^{(1)}\right) \end{array}\right]$$

with dimensionality $(n, 2, k) = (3, 2, 2)$.

References

Bini D (1980) Relations between exact and approximate bilinear algorithms. Appl Calcolo 17:87–97
Bini D, Lotti G (1980) Stability of fast algorithms for matrix multiplication. Numer Math 36:63–72
Bini D, Lotti G, Romani F (1980) Approximate solutions for the bilinear form computational problem. SIAM J Comput 9(4):692–697
Bini D, Capovani M, Lotti G, Romani F (1979) O($n^{2.7799}$) Complexity for approximate matrix multiplication. Inform Process Lett 8(5):234–235
Heule MJH, Kauers M, Seidl M (2021) New ways to multiply 3x3-matrices J Symb Comput 104:899–916
Laderman JD (1976) A noncommutative algorithm for multiplying 3x3 matrices using 23 multiplications. Bull Am Math Soc 82(1):126–128
Leeuven J, Boas P-E (1978) Some elementary proofs of lower bounds in complexity theory. Linear Algebra Appl 19:63–80
Pan V (1984) How can we speed up matrix multiplication? SIAM Rev 26(3):393–415
Romani F (1980) Complexity measures for matrix multiplication algorithms. Calcolo 17:77–86
Schonhage A (1981) Partial and total matrix multiplication. SIAM J Comput 10(3):434–455

Chapter 6
From Disjoint to Single Matrix Multiplication

One can ask a question, whether it is possible to use the disjoint matrix multiplication algorithms for efficient single matrix multiplication. One of the possible solutions was proposed by Pan by the method of dispatching the entries of a single matrix product into the entries of separate, disjoint products. This method is described in detail in Chap. 5. But there are also general methods developed that allow any s-disjoint matrix multiplication algorithm to be combined into a single multiplication, for any s. Methods are described in Pan (1984) and Schonhage (1981). In general, two cases can be distinguished:

- The s-disjoint algorithm on the input accepts s different matrix products, but each one with the same dimensionality, in particular rectangular. We presented three methods for this case in Sect. 6.1.
- The s-disjoint algorithm works on s matrix products, where the dimensionality of each product is arbitrary. To do this, we use the famous tau-theorem formulated by Schonhage (1981), which we describe in Sect. 6.2.

At the end of the chapter we return to the disjoint algorithms of Sects. 5.5–5.7. For each of the algorithms presented there, we determine the efficiency using the tau-theorem.

6.1 From Disjoint to Single Matrix Multiplication—Equidimensional Case

In general, this section shows how to apply disjoint matrix multiplication algorithms to a single matrix multiplication.

More specifically, this section has shown that general methods can be used in conjunction with any disjoint matrix multiplication algorithm, designed to compute an arbitrary number of square or rectangular matrix products of equal dimensionality

J. S. Respondek, *Fast Matrix Multiplication with Applications*, Studies in Big Data 166,
https://doi.org/10.1007/978-3-031-76930-6_6

at once, to obtain an efficient algorithm for a single matrix multiplication problem. The results of this section apply in particular to those of Sects. 4.1, 4.3 and Sects. 3.10, 5.5 in the square case. We propose three different methods; two already known in the literature and we proposed a new one.

6.1.1 Block Method

The method presented in this section is the simplest, and was mentioned by Schonhage (1981, Section 7.2, page 451). Suppose we have to multiply two square matrices of $ns \times ns$ dimensions, where s is the number of disjoint matrix multiplications computed at once by the selected disjoint-type algorithm. To explain the idea, we start from the simple case with $s = 2$, where we multiply two square matrices of dimension $2n \times 2n$:

$$\left[\begin{array}{c|c} A_{11} & A_{12} \\ \hline A_{21} & A_{22} \end{array}\right] \cdot \left[\begin{array}{c|c} B_{11} & B_{12} \\ \hline B_{21} & B_{22} \end{array}\right] = \left[\begin{array}{c|c} A_{11}B_{11} + A_{12}B_{21} & A_{11}B_{12} + A_{12}B_{22} \\ \hline A_{21}B_{11} + A_{22}B_{21} & A_{21}B_{12} + A_{22}B_{22} \end{array}\right] \tag{6.1}$$

The method is as follows: in total we have to perform $s^3 = 8$ matrix products, each of dimension $n \times n$. We group all these 8 sub-products $A_{..} \cdot B_{..}$ into distinct pairs, and next to each pair we apply the double disjoint matrix multiplication algorithm, e.g. the one presented in Sect. 3.10. The number of possible such a partitions is $(s^3 - 1)!! = 7!! = 105$ (Chen and Koh 1992, Example 1.4.4, pages 21–22). However, each of them provides an equally efficient algorithm, so it does not matter which partition we choose. For example we can perform double disjoint matrix multiplications on the following 4 two-elementary subsets of matrix products, each in $n^3 + 3n^2$ multiplications:

$$\{A_{11}B_{11},\ A_{12}B_{21}\},\ \{A_{11}B_{12},\ A_{12}B_{22}\},\ \{A_{21}B_{11},\ A_{22}B_{21}\},\ \{A_{21}B_{12},\ A_{22}B_{22}\}$$

This way of pairing of the set of 8 matrix products seems to be the most natural, since each pair corresponds to a single entry of the desired $2n \times 2n$ block matrix product.

In a general case, the idea is illustrated below by the block matrix Eq. (6.2):

$$\begin{bmatrix} A_{11} & \cdots & A_{1s} \\ & \ddots & \\ \vdots & A_{ik} & \vdots \\ & \ddots & \\ A_{s1} & \cdots & A_{ss} \end{bmatrix} \begin{bmatrix} B_{11} & \cdots & B_{1s} \\ & \ddots & \\ \vdots & B_{kj} & \vdots \\ & \ddots & \\ B_{s1} & \cdots & B_{ss} \end{bmatrix} = \begin{bmatrix} \sum_{k=1}^{s} A_{1k}B_{k1} & \cdots & \sum_{k=1}^{s} A_{1k}B_{ks} \\ & \ddots & \\ \vdots & \sum_{k=1}^{s} A_{ik}B_{kj} & \vdots \\ & \ddots & \\ \sum_{k=1}^{s} A_{sk}B_{k1} & \cdots & \sum_{k=1}^{s} A_{sk}B_{ks} \end{bmatrix} \tag{6.2}$$

where $\dim A_{ik} = \dim B_{kj} = n \times n$. Now the set of s^3 matrix products can be divided into s^2 separate subsets of s elements, whose union is the set of all these products. We can do this in $(s^3)!/\left[(s^2)! \cdot (s!)^{s^2}\right]$ ways (Chen and Koh 1992, problem 1.43, page 22). In particular, for $s = 3$ we have possible $(3^3)!/((3^2)! \cdot (3!)^{3^2})$ =2, 977, 546, 171, 600, 000 groups of s^3 elements. From this variety, we choose such a group in which each s-element subset is in each of s^2 block entries. Next to each of these subsets, we apply a given s-disjoint matrix multiplication algorithm, i.e. to compute the $(ij)th$ block matrix entry we apply the s-disjoint algorithm to the matrix products subset $\{A_{i1}B_{2j}, A_{i2}B_{2j}, \ldots, A_{is}B_{s_j}\}$. The cost of this block method is $s^2 \cdot r(n, s)$, where $r(n, s)$ is the number of non-scalar multiplications required by a used disjoint matrix multiplication algorithm for s products of $n \times n$ matrices. For an efficient disjoint matrix multiplication algorithm, we can find such n_0, that $s^2 r(n, s) < (sn)^3$ for every $n > n_0$, thus obtaining a lower number of multiplications than in the direct method. In particular, for the double disjoint algorithm from Sect. 3.10 we have $2^2(n^3 + 3n^2) < (2n)^3$ for $n \geq 4$, and for the triple disjoint algorithm from Sect. 4.3 the inequality $3^2((2n)^3 + 12(2n)^2 + 36(2n)) < (3 \cdot (2n))^3$ is fulfilled for $2n \geq 10$.

If $r(n, s) = O(n^3)$ the limit for the complexity exponent of the block method in question is the following (we model $O(n^3)$ as Cn^3):

$$
\begin{aligned}
\omega &= \lim_{n\to\infty} \log_{sn}\left[s^2 r(s, n)\right] = \lim_{n\to\infty} \log_{sn}\left[s^2 O(n^3)\right] \\
&= \lim_{n\to\infty} \left[\log_{sn}(s^2) + \log_{sn} O(n^3)\right] \\
&= \lim_{n\to\infty} \left[\frac{\log(s^2)}{\log(sn)}\right] + \lim_{n\to\infty} \left[\frac{\log(Cn^3)}{\log(sn)}\right] = 0 + \lim_{n\to\infty} \left[\frac{\log C}{\log(sn)} + 3\frac{\log n}{\log s + \log n}\right] \\
&= [0 + 3 \cdot 1] = 3
\end{aligned}
$$

In Table 6.1 we have compared the block method with the definitional matrix multiplication (in n^3 steps) in the case of $s = 2$ (algorithm in Sect. 3.10). We give the complexity exponent obtained as well as the ratio of the number of multiplications in the block method with respect to the trivial algorithm for increasing matrix dimensions. We have separated the threshold dimension, which is better than the trivial algorithm, by the bolded entries.

In Table 6.2 we give an analogous overview for $s = 3$ (see Sect. 4.3).

6.1.2 Recursive Method

The method presented in this section was proposed in article Pan (1984), Section 5, pages 400–402. Its general idea is to recursively apply a given disjoint matrix multiplication algorithm, but the problem here is that these algorithms, by their very nature, multiply more than one pair of matrices at a time. So, given that our goal is to perform

Table 6.1 The block method—multiplication number survey (double disjoint algorithm)

	Matrix dimension	Number of *(def)	Number of * (block method)	Exponent ω	Ratio (disjoint/def)
n	$2n$	$(2n)^3$	$2^2(n^3+3n^2)$	$\log_{2n} 2^2(n^3+3n^2)$	$\frac{2^2(n^3+3n^2)}{(2n)^3}$
1	2	8	16	4.00000	200.00%
2	4	64	80	3.16096	125.00%
3	6	216	216	3.00000	100.00%
4	**8**	**512**	**448**	**2.93578**	**87.50%**
5	10	1000	800	2.90309	80.00%
6	12	1728	1296	2.88423	75.00%
7	14	2744	1960	2.87250	71.43%
8	16	4096	2816	2.86486	68.75%
$\vdots$	$\vdots$	$\vdots$	$\vdots$	$\vdots$	$\vdots$
∞	∞	∞	∞	3.00000	50.00%

a single matrix product, the question is how to provide more than one matrix product at the first recursion level?

The answer to this is an ingenious idea of Pan, described in the last paragraph of page 400 of his article (1984). He suggested dividing the multiplication into two phases. First, we multiply s matrix products at once, where one is the desired (single) product we want to obtain, and the remaining $s-1$ are "dummy" products. The values of the entries of the matrices in these "dummy matrix products" do not matter; in particular we can use $s-1$ copies of the input matrices (in which case we calculate the same desired product s times), zero matrices or even matrices with random entries. This initial phase requires more steps (non-scalar multiplications) than the definitional n^3 (e.g. with $s=2$ for algorithm from Sect. 3.10 the additional cost is $3n^2$), but in return we obtain a series of matrices with lower dimension. This series can be partitioned into a series of subsets of s elements,[1] for which we can apply the same disjoint matrix multiplication algorithm.

After the disjoint multiplication, we simply ignore all the $s-1$ matrix products except the desired one. The initial phase of the method in question is illustrated on Fig. 6.1.

Before showing how we proceed in the second phase in a general case, we give some concrete examples using the double disjoint algorithm (Sect. 3.10). In the square case, this algorithm requires $r(n, s=2) = n^3 + 3n^2$ steps. Importantly, $n^3 + 3n^2$ is even for any natural n. We assume that we are multiplying a pair of matrices of $n^q \times n^q$ dimension each.

[1] With the assumption, that the applied disjoint matrix multiplication algorithm works in number of steps $r(n, s)$ which is divisible by s.

Table 6.2 The block method—multiplication number survey (triple disjoint algorithm)

	Matrix dimension	Number of * (def)	Number of * (block method)	Exponent ω	Ratio (disjoint/def)
n	$3 \cdot (2n)$	$(6n)^3$	$3^2\left[(2n)^3 + 12(2n)^2 + 36(2n)\right]$	$\log_{6n} 3^2\left[(2n)^3 + 12(2n)^2 + 36(2n)\right]$	$\frac{3^2\left[(2n)^3+12(2n)^2+36(2n)\right]}{(6n)^3}$
1	6	216	1152	3.00000	533.33%
2	12	1728	3600	3.29537	208.33%
3	18	5832	7776	3.09953	133.33%
4	24	13,824	14,112	3.00649	102.08%
5	**30**	**27,000**	**23,040**	**2.95337**	**85.33%**
6	36	46,656	34,992	2.91972	75.00%
7	42	74,088	50,400	2.89692	68.03%
8	48	110,592	69,696	2.88073	63.02%
⋮	⋮	⋮	⋮	⋮	⋮
∞	∞	∞	∞	3.00000	33.33%

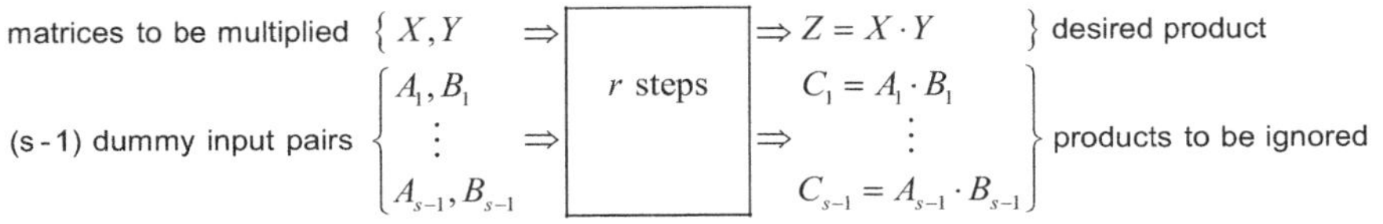

Fig. 6.1 First phase of the recursive method

- Let us take $n = 7$ and $q = 2$. In the first phase we have a product of $7^2 \times 7^2$ dimensional matrices. Applying the scheme from Fig. 6.1, we arrive at $7^3 + 3 \cdot 7^2 = 490$ products of 7×7 matrices. We arbitrarily group them into $490/2 = 245$ pairs of products. For each of these pairs we apply again the same double disjoint algorithm, again working in 490 steps. So in total we need $490 \cdot (490/2) = 120,050$ steps against $\left(7^2\right)^3 = 117,649$, as required by the definition of matrix multiplication. So for $q = 2$ this approach is inefficient in terms of the number of non-scalar multiplications required.

But suppose we want to obtain a single product of $7^3 \times 7^3$ dimensional matrices. In the first phase of the algorithm, we have 490 products of $7^2 \times 7^2$ dimensional matrices, i.e. $490/2 = 245$ pairs. We can multiply any pair of $7^2 \times 7^2$ matrices by 490 products of 7×7 matrices. Again, these 490 products of 7×7 matrices can be grouped into 245 pairs, each requiring 490 multiplications. We are on the last level of recursion, so these 490 multiplications are on scalars. In total we need $(490/2) \cdot (490/2) \cdot 490 = (490/2)^{2(=q-1)} \cdot 490 = 29,412,250$ steps against $\left(7^3\right)^3 = 40,353,607$, so for $q = 3$ the recursive algorithm needs fewer steps than the definition.

- Let us repeat the analysis for $n = 8$ and $q = 2$. We have $r(n, s) = 8^3 + 3 \cdot 8^2 = 704$. For an $8^2 \times 8^2$ dimensional single matrix product (inevitably, paired with the additional "dummy" matrix product) we need $704/2$ pairs of 8×8 matrix products, each requiring 704 multiplications. The total number of steps required is $(704/2)^1 \cdot 704 = 247,808 < \left(8^2\right)^3 = 262,144$.

The above examples show that the method leads to a lower number of required steps than the definition, despite the additional effort in the first phase induced by the requirement to calculate another $s - 1$ redundant matrix products. In the rest of this section, we obtain the necessary and sufficient conditions for the method to be efficient, and we determine its efficiency exponent.

In the general case, we have a disjoint matrix multiplication algorithm which computes s products at once in r steps (we assume that s divides r) and a pair of $n^q \times n^q$ dimensional matrices whose product is to be found.

We proceed as follows: in the initial phase we supplement a given pair of matrices to by multiplied by $s-1$ dummy $n^q \times n^q$ matrices and apply the disjoint multiplication algorithm, which requires to perform r products of $n^{q-1} \times n^{q-1}$ matrices. Now we move on to the second phase. To calculate these r products, we first divide them into r/s sets with s different products in each set. For each of the r/s sets we apply the same algorithm, obtaining $(r/s)r$ products of $n^{q-2} \times n^{q-2}$ matrices that can

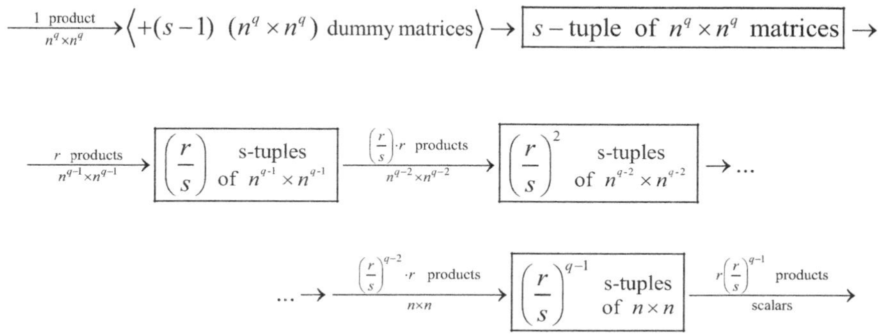

Fig. 6.2 Flowchart of the recursive method

be further divided into $((r/s)r))/s = (r/s)^2$ equinumerous sets. In the last step of the second phase we have s elementary sets of $n \times n$ matrices, each one to be computed in r multiplications on scalars. To summarise, the total number of non-scalar multiplications in the recursive method is

$$\underbrace{\left(\frac{r}{s}\right) \cdot \dots \cdot \left(\frac{r}{s}\right)}_{q-1} \cdot r = r\left(\frac{r}{s}\right)^{q-1} \tag{6.3}$$

Figure 6.2 is a flowchart showing how the recursive method works.[2]

The efficiency condition for the recursive method has the following form[3]:

$$r\left(\frac{r}{s}\right)^{q-1} < \left(n^q\right)^3 \Leftrightarrow q \geq q_{\min} = \left\lceil \frac{\log_n s}{3 - \log_n(r/s)} \right\rceil = \left\lceil \frac{\log_n s}{3 - \omega} \right\rceil \tag{6.4}$$

where $q_{\min}$ is a threshold for the minimum number of recursion levels at which the recursive method becomes profitable with respect to the matrix multiplication definition. From inequality (6.4) it follows that the necessary condition for the recursive method to be better than the definition-based matrix multiplication for a given base matrix dimension $n \times n$ is:

$$\frac{r}{s} < n^3$$

Furthermore, the lower the ratio $(r/s)/n^3$ of the applied disjoint matrix multiplication algorithm is, the lower the recursion level threshold $q_{\min}$. Another question is what is the computational complexity exponent for the recursive method. Observe:

[2] The number of possible partitionings of a given number of products into s-tuples is solved by problem 1.43 Chen and Koh (1992). However, which one we choose does not affect the complexity.

[3] The equality $\omega = \log_n(r/s)$ will be proved later in this section.

$$\omega = \lim_{q\to\infty} \log_{n^q} r\left(\frac{r}{s}\right)^{q-1} = \lim_{q\to\infty} \frac{\log r\left(\frac{r}{s}\right)^{q-1}}{q \log n}$$
$$= \lim_{q\to\infty} \frac{1}{q} \log_n r\left(\frac{r}{s}\right)^{q-1} = \lim_{q\to\infty}\left[\frac{1}{q}\log_n r + \frac{1}{q}(q-1)\log_n\left(\frac{r}{s}\right)\right]$$
$$= 0 + 1 \cdot \log_n\left(\frac{r}{s}\right) = \log_n\left(\frac{r}{s}\right) \tag{6.5}$$

where n^q is the number of rows and columns of the pair of matrices to be multiplied, and the number of necessary non-scalar multiplications is calculated according to Eq. (6.3).

We can see that as the number of recursion levels applied increases, the additional cost of $s - 1$ redundant products introduced in the initial phase (Fig. 6.1) becomes negligible. In Table 6.3 we have made a survey of a threshold $q_{\min}$ and a complexity exponent ω with respect to different base dimensions n (with $q \to \infty$) of the matrices we multiply, using the double disjoint matrix multiplication algorithm presented in Sect. 3.10. In particular, it is compatible with the results we obtained in the examples presented earlier in this section, for $n = 7,\ q = 2,\ 3$ and for $n = 8,\ q = 2$. We have also tabulated the ratios r/s and $(r/s)/n^3$.

We can see that the second ratio, i.e. $r/s < n^3$ becomes less than 1 from the base dimension $n = 4$. The complexity exponent is minimum for $n = 13$ and is equal to $\omega = \log_{13}(2704/2) = 2.810714$.

Table 6.3 The recursive method—efficiency exponent and $q_{\min}$ survey for $s = 2$

Base matrix dimension	Number of * (def)	Ratio	Ratio	Recursion level threshold	Limiting efficiency exponent
n	n^3	$\frac{r}{s}$	$\frac{r/s}{n^3}$	$q_{\min}$	$\omega = \log_n\left(\frac{r}{s}\right)$
2	8	10	125.00%	–	3.321928
3	27	27	100.00%		3.000000
4	64	56	87.50%	6	2.903677
5	125	100	80.00%	4	2.861353
6	216	162	75.00%	3	2.839442
7	343	245	71.43%		2.827087
8	512	352	68.75%	2	2.819811
⋮	⋮	⋮	⋮		⋮
12	1728	1080	62.50%		2.810857
13	**2197**	**1352**	**61.54%**		**2.810714**
14	2744	1666	60.71%		2.810921
⋮	⋮	⋮	⋮		⋮
∞	∞	∞	50.00%		3.000000

A similar analysis can be made with the recursive method using the triple disjoint matrix multiplication algorithm, which we describe in more detail in Sect. 4.3. However, we should note here that unlike in the double disjoint algorithm, where $r = n^3 + 3n^2$ is divisible by 2 for any natural n, the expression $r = (2n)^3 + 12(2n)^2 + 36(2n)$ is divisible by $s = 3$ iff $2n$ is divisible by 3. So the threshold for the recursion level here is $q_{\min} = 4$ for $2n = 12$, $q_{\min} = 3$ for $2n = 18$ and $q_{\min} = 2$ for $2n = 24$, respectively. The ratio inequality $r/s < (2n)^3$ is satisfied for $2n \geq 12$. We can see that, compared to the double disjoint multiplication algorithm, for $s = 3$ the threshold $q_{\min}$ receives corresponding values for greater dimensions. But in return we get a lower minimum of the complexity exponent equal to $\omega = \log_{2n}[r(2n)/3] = \log_{2n}(139, 968/3) = 2.777060$ for $2n = 48$.

The design idea of a recursive method for s-disjoint matrix multiplication in the rectangular case is analogous. Pan (1984, page 406) proved in proposition 7.1, that if we have an algorithm that computes s disjoint matrix multiplications of (m, n, p) dimensionality at once, working in r non-commutative steps, then by recursion we obtain an algorithm for a single (m, n, p) matrix multiplication problem with the following efficiency exponent:

$$\omega = 3 \log_{mnp}\left(\frac{r}{s}\right) \tag{6.6}$$

with an assumption that s divides r. In the same article from 1984 Pan proved in proposition 7.2 (page 407) that the complexity bound (6.6) is also valid if s does not divide r. We can see that if $m = n = p$ the efficiency limit for the square case (6.5) is a special case of a limit for the rectangular case (6.6). Indeed:

$$\omega = 3 \log_{mnp}\left(\frac{r}{s}\right)\Big|_{\substack{m=n \\ p=n}} = 3\frac{\log \frac{r}{s}}{\log n^3} = \frac{\log \frac{r}{s}}{\log n} = \log_n\left(\frac{r}{s}\right)$$

The result (6.6) will be used later in Sect. 6.2 to prove a special case of the famous Schonhage tau-theorem.

6.1.3 Hybrid Method

The main drawback of the recursive method, presented in the previous section, is that in the first recursion step we are forced to perform more products than just the one product we want to obtain (Fig. 6.1) of the same dimensionality. On the other hand, the block method described in Sect. 6.1.1 is free from this drawback, but does not yield from recursion. In this section we have proposed a new method that combines the advantages of both block and recursive methods. Namely, the calculation of a single square matrix product divided into 2 stages is proposed—we can call it a process:

- At first stage, we dispatch the $n^q \times n^q$ dimensional matrices into n^2 number of $n^{q-1} \times n^{q-1}$ dimensional blocks, as in the block method from Sect. 6.1.1. Assuming that s divides n, in each of n^2 blocks we have n/s s-tuples of matrix products. This stage is inspired by the block method by Schonhage (1981, Section 7.2, page 451).
- In the second stage we use recursion, as in the recursion method described in Sect. 6.1.2.

So after the first stage we get:

$$n^2 \cdot \frac{n}{s} = \frac{n^3}{s}$$

s-tuples of $\left(n^{q-1}, n^{q-1}, n^{q-1}\right)$ square matrix products, for $q > 1$. Then, in the second stage, we apply the s-disjoint matrix multiplication algorithm to each of these products, working in r steps and repeating this process recursively. We assume that s divides r. Observe (MM is an abbreviation for Matrix Multiplication):

$$\begin{aligned}
\frac{n^3}{s}\left\langle s - \text{tuples } \left(n^{q-1}, n^{q-1}, n^{q-1}\right)\right\rangle &= \frac{n^3}{s} \cdot r\left\langle MM \text{ of } \left(n^{q-2}, n^{q-2}, n^{q-2}\right)\right\rangle \\
&= \frac{n^3}{s} \cdot \frac{r}{s}\langle s - \text{tuples } (n^{q-2}, n^{q-2}, n^{q-2})\rangle \\
&= n^3\left(\frac{r}{s}\right)^2 \left\langle MM \text{ of } \left(n^{q-3}, n^{q-3}, n^{q-3}\right)\right\rangle \\
&\ \ \vdots \\
&= n^3\left(\frac{r}{s}\right)^{k-1} \left\langle MM \text{ of } \left(n^{q-k}, n^{q-k}, n^{q-k}\right)\right\rangle \\
&= n^3\left(\frac{r}{s}\right)^{k-1} \cdot \frac{1}{s}\left\langle s - \text{tuples } \left(n^{q-k}, n^{q-k}, n^{q-k}\right)\right\rangle \\
&\ \ \vdots \\
&= |k := q-1| = n^3\left(\frac{r}{s}\right)^{q-2} \langle MM \text{ of } (n, n, n)\rangle \\
&= n^3\left(\frac{r}{s}\right)^{q-2} \cdot \frac{1}{s}\langle s - \text{tuples } (n, n, n)\rangle \\
&= n^3\left(\frac{r}{s}\right)^{q-2} \cdot \frac{r}{s} = n^3\left(\frac{r}{s}\right)^{q-1}, \quad \text{for } q > 1
\end{aligned}$$

Here it is worth comparing the number of non-scalar multiplications obtained with that required by the recursion method, given by formula (6.3). The number of steps obtained in the proposed hybrid method is lower. In fact, we have an obvious inequality:

$$n^3\left(\frac{r}{s}\right)^{q-1} < r\left(\frac{r}{s}\right)^{q-1}$$

for $r > n^3$. Thus, in the case of the hybrid method, the minimum number of recursion levels $q_{\min}$ is constant and equal to 1. The question now is whether the computational complexity exponent obtained is also better than in the recursive method. The answer is negative. Observe:

$$
\begin{aligned}
\omega &= \lim_{q \to \infty} \log_{n^q} n^3 \left(\frac{r}{s}\right)^{q-1} = \lim_{q \to \infty} \frac{\log n^3 \left(\frac{r}{s}\right)^{q-1}}{q \log n} \\
&= \lim_{q \to \infty} \frac{3 \log n + (q-1) \log\left(\frac{r}{s}\right)}{q \log n} \\
&= \lim_{q \to \infty} \left[\frac{3}{q} + \frac{q-1}{q} \log_n \left(\frac{r}{s}\right)\right] = 0 + 1 \cdot \log_n \left(\frac{r}{s}\right) = \log_n \left(\frac{r}{s}\right)
\end{aligned}
$$

Thus, with the hybrid method we have reduced the number of necessary multiplications, but the complexity exponent, with the number of recursion levels going to infinity, remains unchanged. In Table 6.4 we have shown how the complexity exponent changes with increasing number of recursion levels as well as the ratio of the number of multiplications required by a given method with respect to the definition, for recursive and hybrid methods, for $s = 2$ using the disjoint algorithm from Sect. 3.10. For the base dimension we choose $n = 4$.

We can see that for a given recursion level the hybrid method gives a better exponent than the classical recursive method.

6.2 From Disjoint to Single Matrix Multiplication—General Case

In this section we show another way of using the disjoint matrix multiplication algorithm to obtain an efficient matrix multiplication algorithm.

However, what distinguishes the methods presented in this section from those presented in the previous Sect. 6.1, is that here we get rid of the significantly limiting assumption of equal dimensionality of a matrix product computed by a given disjoint matrix multiplication algorithm. In return, we can obtain a lower efficiency exponent.

For example, the approximate double disjoint algorithm presented in Sect. 5.5, used in combination with the recursive method which we described in Sect. 6.1, achieves its minimum for $m = p = n = 7$ equal to $\omega \leq \log_7 \lceil (7^3 + 2 \cdot 7^2)/2 \rceil = 2.772943$. On the contrary, if we choose $m = p = 7,\ n = 1$, we have double disjoint matrix multiplication algorithm for a pair of products of dimensions $(m, n, p) = (7, 1, 7)$ and $(n, p, m) = (1, 7, 7)$. After applying the method presented later in this section, we obtain the complexity $\omega \leq 2.65941$ (Eq. 5.18).

In general, the core of the method is to construct a smart recursive scheme tailored to the disjoint matrix multiplication algorithm used, allowing arbitrary dimensions of each matrix product that the disjoint algorithm computes at once, and also arbitrary

Table 6.4 Efficiency survey of recursive and hybrid method

Recursion level	Matrix dimension	ω (recursion method)	Ratio	ω (hybrid method)	Ratio (hybrid method)
q	$n^q \times n^q$	$\log_{n^q}\left[r\left(\frac{r}{s}\right)^{q-1}\right]$	$\frac{r\left(\frac{r}{s}\right)^{q-1}}{(n^q)^3}$	$\log_{n^q}\left[n^3\left(\frac{r}{s}\right)^{q-1}\right]$	$\frac{n^3\left(\frac{r}{s}\right)^{q-1}}{(n^q)^3}$
2	16	3.153677	153.13%	2.951838	87.50%
3	64	3.070344	133.98%	2.935784	76.56%
4	256	3.028677	117.24%	2.927758	66.99%
5	1024	3.003677	102.58%	2.922941	58.62%
6	**4096**	**2.987010**	**89.76%**	**2.919731**	**51.29%**
7	16,384	2.975106	78.54%	2.917437	44.88%
8	65,536	2.966177	68.72%	2.915717	39.27%
9	262,144	2.959233	60.13%	2.914379	34.36%
10	1,048,576	2.953677	52.62%	2.913309	30.07%
11	4,194,304	2.949132	46.04%	2.912434	26.31%
12	16,777,216	2.945344	40.28%	2.911704	23.02%
13	67,108,864	2.942138	35.25%	2.911086	20.14%
14	268,435,456	2.939391	30.84%	2.910557	17.62%
15	1,073,741,824	2.937010	26.99%	2.910098	15.42%
16	4,294,967,296	2.934927	23.61%	2.909697	13.49%
$\vdots$	$\vdots$	$\vdots$	$\vdots$	$\vdots$	$\vdots$
∞	∞	2.903677	0.00%	2.903677	0.00%

numbers of these products. Before we get to the general case, let us show how it works in the—not so much—special case.

6.2.1 *Particular Case—Triple Disjoint Algorithm with Cyclically Shifted Dimensions*

We will demonstrate the idea of the method by the often encountered in practice case of a triple disjoint matrix multiplication algorithm with cyclically shifted dimensions (e.g. Laderman et al. 1992, Section 5). Such an algorithm calculates in r steps at once three matrix products with the dimensions (m, n, p), (n, p, m) and (p, m, n). A single recursion phase, with triple recursive embedding, is shown in Fig. 6.3 On this diagram, blocks stand for a triple disjoint algorithm with given dimensions that works inside the blocks, and arrows stand for the matrix product dimensionality that we obtain in a particular recursion branch (in this example, we have as many as $3^3 = 27$

possible branches). The first block on the left stand for matrices with scalar elements i.e. simply for a scalar non-commutative multiplication. Observe the diagram:

Let us start analysing the recursion diagram from the second column of blocks. We use there three times the triple disjoint algorithm, but each time as the matrix elements we poke another rectangular matrices, but with scalar elements. The matrices we poke into the second column of triple disjoint algorithms form there three series of matrix subproducts, with (m, n, p), (n, p, m) and (p, m, n) dimensionality, respectively. Each of the triple disjoint products in the second column of the blocks requires r steps to be computed on these subproducts. If we use the definition-based matrix multiplication algorithm for them, we need to perform $r \cdot mnp + r \cdot npm + r \cdot pmn = 3mnp \cdot r$ scalar products.

But it is better to use the same triple disjoint matrix multiplication algorithm again for the submatrices, calculating at once three (m, n, p), (n, p, m) and (p, m, n) matrix

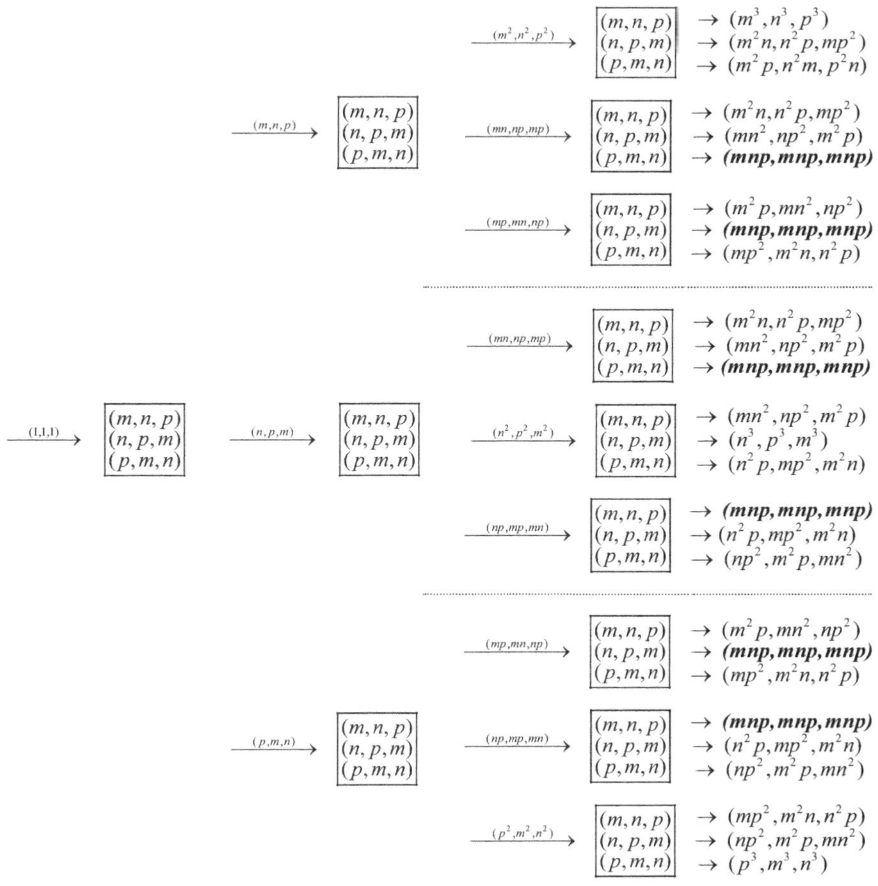

Fig. 6.3 Single phase of a recursion for a triple disjoint matrix multiplication algorithm

products (with scalar elements) in r steps. In this way, we reduce the number of steps required from $3mnp \cdot r$ to r^2. After running the recursion three times, which represents the whole diagram, we obtain 6 different products which we have bolded, but importantly of equal dimensionality (mnp, mnp, mnp) in r^3 steps.

In other words, after deleting[4] all the unwanted non-square products, the triple recursion scheme shown in Fig. 6.3 is nothing more than a 6-disjoint matrix multiplication algorithm for square matrices of equal dimension $mnp \times mnp$, working in r^3 steps. At this moment it is already possible to use it by the recursion method described in Sect. 6.1, and we have an algorithm for square matrices with the following efficiency (formula (6.6) or even formula (6.5) with $n := mnp$):

$$\begin{aligned}\omega_1 &= 3\log_{[(mnp)\cdot(mnp)\cdot(mnp)]}\left(\frac{r^3}{6}\right) = 3\log_{(mnp)^3}\left(\frac{r^3}{6}\right)\\ &= \frac{3\log r - \log 6}{\log(mnp)} \approx \frac{3\log r - 0.77815}{\log(mnp)}\end{aligned} \tag{6.7}$$

where ω_1 indicates that we have applied a single phase of the recursive scheme.

But it is still possible to improve this level of efficiency (6.7). In fact, let us repeat the recursive process shown in Fig. 6.3 To count how many square equidimensional problems we get by such a repeated recursion scheme, we note that each of the m, n, p dimensions must occur exactly q times on a particular recursion branch of the q-times composed diagram. For the first of the dimensions, say m, we have $\binom{3q}{q}$ possible distributions in the columns of the blocks in the q-times repeated diagram. For the second one dimension, say n, there remain $\binom{2q}{q}$ possible distributions and the location of the last dimension, here p, is determined by the first two. In summary, after repeating the diagram in Fig. 6.3 q times, we obtain algorithm for

$$s = \binom{3q}{q} \cdot \binom{2q}{q} = \frac{(3q)!}{(q!)^3} \tag{6.8}$$

disjoint problems of square and equal dimensionality $\left[(mnp)^q, (mnp)^q, (mnp)^q\right]$ computed in $\left(r^3\right)^q$ steps. Its efficiency exponent is obtained from (6.6) (compare with (6.7) and notice that $(3 \cdot 1)!/(1!)^3 = 6$):

$$\omega_q = 3\log_{[(mnp)^q\cdot(mnp)^q\cdot(mnp)^q]}\left(\frac{r^{3q}}{s}\right) = 3\log_{(mnp)^{3q}}\left(\frac{r^{3q}}{s}\right) \tag{6.9}$$

Let us now analyse what happens if $q \to \infty$. To do this, we transform the efficiency exponent formula (6.9) as follows:

[4] Some authors call this operation as 'pruning'.

$$\omega = \lim_{q\to\infty} \omega_q = \lim_{q\to\infty} 3\log_{(mnp)^{3q}}\left(\frac{r^{3q}}{s}\right) = 3\lim_{q\to\infty}\frac{1}{3q}\cdot\frac{\log\left(\frac{r^{3q}}{s}\right)}{\log(mnp)}$$
$$= 3\lim_{q\to\infty}\frac{\log\left(\frac{r^{3q}}{s}\right)^{\frac{1}{3q}}}{\log(mnp)} = 3\lim_{q\to\infty}\frac{\log\left(\frac{r}{s^{\frac{1}{3q}}}\right)}{\log(mnp)} \tag{6.10}$$

To find the limit for $s^{1/3q}$ with $q \to \infty$ we must return to the formula (6.8), which expresses the number s of disjoint problems we obtain after repeating the main recursion diagram q times. Observe:

$$s^{\frac{1}{3q}} = \left[\frac{(3q)!}{(q!)^3}\right]^{\frac{1}{3q}} = \frac{\left[(3q)!\right]^{\frac{1}{3q}}}{(q!)^{\frac{1}{q}}} \tag{6.11}$$

To find the value of ω, it is sufficient to use the following lemma:

Lemma

For any positive natural k the following limit is true:

$$\lim_{q\to\infty}\frac{\left[(kq)!\right]^{\frac{1}{kq}}}{(q!)^{\frac{1}{q}}} = k \tag{6.12}$$

Let us start the proof from the following, classical limit (e.g. Radulescu et al. 2009, problem 1.2.2 page 8):

$$\lim_{q\to\infty}\frac{\sqrt[q]{q!}}{q} = \frac{1}{e}$$

Let us define series $a_q = \sqrt[q]{q!}/q$ and an auxiliary series $b_q = a_{kq}$. Look at the construction:

$$\lim_{q\to\infty}\left(k\frac{b_q}{a_q}\right) = \lim_{q\to\infty}\left(k\frac{\frac{\sqrt[kq]{(kq)!}}{kq}}{\frac{\sqrt[q]{q!}}{q}}\right) = \lim_{q\to\infty}\frac{\left[(kq)!\right]^{\frac{1}{kq}}}{(q!)^{\frac{1}{q}}}$$

which is just the left side of the equality thesis of lemma (6.12). On the other hand, we have:

$$\lim_{q\to\infty}\left(k\frac{b_q}{a_q}\right) = k\cdot\frac{\lim_{q\to\infty} b_q}{\lim_{q\to\infty} a_q}$$

The last limit is clear if we notice that the series b_q, which is a subseries of a_q series, also converges to $1/e$. Thus:

$$\lim_{q\to\infty}\left(k\frac{b_q}{a_q}\right) = k\frac{\frac{1}{e}}{\frac{1}{e}} = k \quad Q.E.D$$

Let us return to the efficiency exponent. Applying the limit given by lemma (6.12), with $k = 3$, to the formula expressing the value of ω to be found (6.10) we finally arrive at the following value of the efficiency exponent:

$$\omega = 3\frac{\log\frac{r}{3}}{\log(mnp)} = 3\log_{mnp}\left(\frac{r}{3}\right) \tag{6.13}$$

It is worth noting that the efficiency exponent level which we obtained is the same as (6.6), which is valid for the case of s-disjoint matrix multiplication algorithms, but with equal dimensions of all the s sub-products that a given disjoint algorithm works on.

It is also worth comparing the value of $\omega = \omega_q$ with $q \to \infty$ (6.13) with the corresponding value if $q = 1$. Let us estimate (6.13) as:

$$\omega = 3\log_{mnp}\left(\frac{r}{3}\right) \approx \frac{3\log r - 1.43136}{\log(mnp)}$$

for any logarithm base. Comparing the last expression with the exponent (6.7), which is valid in the case of a single use of the triple recursion scheme shown in Fig. 6.3, it is clear that by using the recursion scheme several times we can obtain an algorithm with even better overall efficiency.

However, the above result is only valid for the case where the $s = 3$ disjoint algorithm with cyclically shifted dimensions is applied to the matrix multiplication problem it works on.

The above example is characterised by the fact that it leads directly to square matrices in both single and multiple repetitions of the recursion scheme from Fig. 6.3 This is not the case with the double disjoint algorithms in the exact and approximate versions presented in Sects. 3.10 and 5.5, which work on two matrix multiplication problems with cyclically shifted dimensions (m, n, p) and (n, p, m). Article Pan (1984) on page 407, just after proposition 7.2, shows how to deal with such case to obtain an algorithm for s-disjoint multiplication algorithm for rectangular matrix problem with $\left[(mn)^q, (np)^q, (pm)^q\right]$ dimensionality with efficiency $\omega = 3\log_{mnp}(r/2)$ and $s = (2q)!/(q!)^2$ [compare with formulae (6.8) and (6.13)].

However, from this result we can also derive the corresponding algorithm for square matrices, with the same efficiency exponent $\omega = 3\log_{mnp}(r/2)$. To achieve this, the first additional task[5] we need to do is to apply twice the duality theorem of Hopcroft and Musinski (1973), described in Sect. 3.7, to the $\left[(mn)^q, (np)^q, (pm)^q\right]$ problem. Next, we use the multiplicative (recursion) rule from Sect. 3.8 twice.

In particular, the above method explains how to achieve the complexity exponent $\omega \leq 2.659\,41$, announced in Sect. 5.5 also for square matrices by using the double disjoint approximate matrix multiplication algorithm.

6.2.2 General Case—Schonhage's Tau-Theorem

(Schonhage 1981, Section 7 page 449) *Suppose we have an exact or approximate algorithm for s-disjoint matrix multiplication of rank or border rank r, respectively, each of arbitrary size (k_i, m_i, n_i) for $i = 1, \ldots, s$. Then the time efficiency exponent is $\omega \leq 3\tau$, where τ satisfies the equation:*

$$\sum_{i=1}^{s} (k_i \cdot m_i \cdot n_i)^{\tau} = r \tag{6.14}$$

In the case of the problem we described in Sect. 6.2.1, we have $(k_1 = m,\ m_1 = n,\ n_1 = p)$, $(k_2 = n,\ m_2 = p,\ n_2 = m)$ and $(k_3 = p,\ m_3 = m,\ n_3 = n)$ thus $k_1 m_1 n_1 = k_2 m_2 n_2 = k_3 m_3 n_3 = mnp$ and the corresponding equation for the tau-theorem is of the form:

$$3(mnp)^{\tau} = r$$

with simple solution (6.13). In the general case, however, the Eq. (6.14) is a transcendental equation, so its solution cannot be expressed in an explicit form by a finite combination of elementary functions.

Nevertheless, Schonhage, in his seminal article from 1981, addressed the problem comprehensively and proposed an iterative numerical algorithm to solve Eq. (6.14) in a general form. Namely, Schonhage (1981) in Section 7.1, pages 449–450, proved that:

- Equation (6.14) has a unique solution $\tau \in [0, 1)$.
- The sequence:

$$\tau_0 = 1, \qquad \tau_{j+1} = \tau_j \cdot \frac{\ln r}{\ln\left[\sum_{i=1}^{s} (k_i \cdot m_i \cdot n_i)^{\tau_j}\right]}$$

[5] We also need to use the recursion method from Sect. 6.1 to shift from the s-disjoint algorithm for the rectangular matrix multiplication problem of $\left[(mn)^q, (np)^q, (pm)^q\right]$ dimensionality to a single matrix multiplication algorithm of the same dimensionality and efficiency.

is strictly decreasing and converges linearly to the solution of Eq. (6.14).

We can observe the following properties of the left and right parts of the tau-theorem Eq. (6.14):

- The inequality $2 < \omega \leq 3\tau$ implies that $\tau > 2/3$.
- For any non-trivial s-disjoint matrix multiplication algorithm $k_i m_i n_i \geq 1$ for $i = 1, ..., s$ but not all these terms are equal to one, so the function $\sum_{i=1}^{s} (k_i m_i n_i)^\tau$ is increasing with respect to the (real) τ parameter.
- For $\tau = 1$, the sum $\sum_{i=1}^{s} (k_i m_i n_i)^\tau = k_1 m_1 n_1 + \cdots + k_s m_s n_s$ expresses the number of multiplications required to calculate each of the s matrix products independently by the definition-based algorithm (Definition 1.1). Thus, any *fast* disjoint matrix product algorithm requires fewer multiplications, i.e. $r < \sum_{i=1}^{s} k_i m_i n_i$.

Now let us return to the time complexity analysis of disjoint matrix multiplication algorithms from the previous sections.

- For the arbitrary precision double disjoint matrix multiplication from Sect. 5.5, to obtain the complexity announced there, it is sufficient to apply equality (6.13) with $s = 2$.
- To determine the time complexity of Schonhage's vector-by-vector based matrix multiplication algorithm presented in Sect. 5.6 we apply the tau-theorem. The sum (6.14) leads to the transcendental equation:

$$(mn)^\tau + [(m-1)(n-1)]^\tau = mn + 1$$

Schonhage (1981) in Table 6.1, page 449, has tabulated the obtained τ-s for different m and n parameters. The best efficiency is obtained for $m = n = 4$. In this case, the tau-theorem equation takes the form:

$$(4 \cdot 4)^\tau + [(4-1)(4-1)]^\tau = 4 \cdot 4 + 1 \quad \Leftrightarrow \quad 16^\tau + 9^\tau = 17$$

whose (numerical!) solution is $\omega \leq 3\tau = 2.5479$.

- For the second, triple disjoint Schonhage's algorithm, which we showed in Sect. 5.7, we obtain from the tau-theorem [Sect. 6.2.2 Eq. (6.14)] a matrix multiplication algorithm of complexity:

$$\begin{aligned}&(2k \cdot n \cdot 1)^\tau + (1 \cdot k \cdot 2n)^\tau + (n \cdot 2 \cdot k)^\tau \\ &= 3(2kn)^\tau = 2(k+1)(n+2)\end{aligned}$$

So, in this case the τ-theorem equation does not have a transcendental form with an analytic solution:

$$\tau = \frac{\ln \frac{2}{3}(k+1)(n+2)}{\ln(2kn)}$$

Schonhage (1981), page 454, found that τ gets its minimum value for $k = 5,\ n = 11$ which leads to the following, final complexity:

$$\omega \leq 3\tau = 3\frac{\ln 52}{\ln 110} = 2.5218127...$$

References

Chen CC, Koh KM (1992) Principles and techniques in combinatorics. World Scientific Publishing, Singapore

Hopcroft J, Musinski J (1973) Duality applied to the complexity of matrix multiplication and other bi-linear forms. SIAM J Comput 2:159–173

Laderman J, Pan V, Sha XH (1992) On practical algorithms for accelerated matrix multiplication. Linear Algeb Appl 162–164:557–588

Pan V (1984) How can we speed up matrix multiplication? SIAM Rev 26(3):393–415

Radulescu TL, Radulescu VD, Andreescu T (2009) Problems in real analysis. Advanced calculus on the real axis. Springer, Dordrecht Heidelberg London New York

Schonhage A (1981) Partial and total matrix multiplication. SIAM J Comput 10(3):434–455

Chapter 7
From Partial to Total Matrix Multiplication

The notion of partial matrix multiplication was introduced by Bini et al. (1979), which we described in Sect. 5.4. This means that we are multiplying matrices where not all entries are filled; some of the entries in a given partial matrix multiplication algorithm are permanently set to zero. In this chapter we deal with the problem in a general form. Basically we have two problems:

- How to efficiently perform the partial matrix multiplication—example of an algorithm we presented in this chapter in Sect. 7.1.
- How to use any partial matrix multiplication to get a fast total matrix multiplication one. The word 'total' here means, that we consider 'normal' matrix multiplication, with allowed non-zero all the multiplied matrices entries. We presented a general method in Sect. 7.2.

7.1 Partial Matrix Multiplication

The concept of partial matrix multiplication was introduced by Bini et al. in 1979, resulting in an algorithm for total matrix multiplication with exponent 2.7799. We have described this result in Sect. 5.4. The technique of partial matrix multiplication and the construction of a total matrix multiplication algorithm based on it was further developed by Schonhage (1981). He developed both the general method and, in Sect. 5 of his article, gave some concrete examples of partial matrix multiplication algorithms. In this section we describe one of these examples in detail, and in the following Sect. 7.2, we discussed this general method and illustrated it with two examples.

Let us take a look at the algorithm for partial matrix multiplication presented by Schonhage on pages 445–446. We have a problem of dimension (k, m, n) and, what is specific to this algorithm of partial multiplication, we have an additional parameter q which is greater than or equal to 2. The dimensions of the problem are determined

J. S. Respondek, *Fast Matrix Multiplication with Applications*, Studies in Big Data 166,
https://doi.org/10.1007/978-3-031-76930-6_7

by this parameter as follows:

$$k = n = q + 1, \quad m = 2q$$

The algorithm can be illustrated in the following diagram, using the case $q = 2$:

$$A \cdot B = \begin{bmatrix} 0\,0 * * \\ 0\,0 * * \\ * * * * \end{bmatrix} \cdot \begin{bmatrix} * * * \\ * * * \\ 0\,0 * \\ 0\,0 * \end{bmatrix} \tag{7.1}$$

where the asterisks stand for non-zero entries. The problem has the following trilinear form:

$$F = \sum_{i=1}^{q} \sum_{j=1}^{q+1} a_{k,i} b_{i,j} c_{j,k} + \sum_{i=1}^{q+1} \sum_{j=1}^{q} a_{i,j+q} b_{j+q,k} c_{k,i}$$

The approximate algorithm proposed by Schonhage (1981) is:

$$\begin{aligned} &\sum_{i=1}^{q} \sum_{j=1}^{q} \left(a_{k,i} + \varepsilon^2 a_{i,j+q}\right)\left(b_{j+q,k} + \varepsilon^2 b_{i,j}\right)\left(\varepsilon c_{j,k} + \varepsilon c_{k,i} - c_{k,k}\right) \\ &\quad + \sum_{i=1}^{q} \left[a_{k,i} \left(\varepsilon^3 b_{i,k} + \sum_{j=1}^{q} \left(b_{j+q,k} + \varepsilon^2 b_{i,j}\right) \right) \right] \left(c_{k,k} - \varepsilon c_{k,i}\right) \\ &\quad + \sum_{j=1}^{q} \left[\left(\varepsilon^3 a_{k,j+q} + \sum_{i=1}^{q} \left(a_{k,i} + \varepsilon^2 a_{i,j+q}\right) \right) b_{j+q,k} \right] \left(c_{k,k} - \varepsilon c_{j,k}\right) \\ &\quad - \left(\sum_{i=1}^{q} a_{k,i} \right) \left(\sum_{j=1}^{q} b_{j+q,k} \right) c_{k,k} = \varepsilon^3 F + O^+\left(\varepsilon^4\right) \end{aligned}$$

We can distinguish between the following products in order to present them in the bilinear form we are introducing:

$$\begin{cases} P_{ij} = \left(a_{k,i} + \varepsilon^2 a_{i,j+q}\right)\left(b_{j+q,k} + \varepsilon^2 b_{i,j}\right) \\ Q_i^{(1)} = a_{k,i}\left[\varepsilon^3 b_{i,k} + \sum_{j=1}^{q}\left(b_{j+q,k} + \varepsilon^2 b_{i,j}\right)\right] \\ Q_j^{(2)} = \left[\varepsilon^3 a_{k,j+q} + \sum_{i=1}^{q}\left(a_{k,i} + \varepsilon^2 a_{i,j+q}\right)\right] b_{j+q,k} \\ R = \left(\sum_{i=1}^{q} a_{k,i}\right)\left(\sum_{j=1}^{q} b_{j+q,k}\right) \end{cases}$$

with i and j parameters ranging from 1 to q. Thus we have one series depending on two parameters, two depending on one parameter and the last, a single product. In total we have an algorithm that requires

$$l = q^2 + q + q + 1 = (q+1)^2$$

products (for partial matrix multiplication, Schonhage (1981) uses the variable l here to denote the number of products). From diagram (7.1) the form of the matrix product is given:

$$AB = \left[\begin{array}{cc|c} 0 & 0 & * \\ 0 & 0 & * \\ \hline * & * & * \end{array}\right] \tag{7.2}$$

The corresponding linear combination form can be found by analogy with the previous sections and we go straight to the result. The $c_{j,k}$ variable represents the last row of the product, except for the $c_{k,k}$ element:

$$(AB)_{k,j} = \frac{1}{\varepsilon^2}\left[\sum_{i=1}^{q} P_{ij} - Q_j^{(2)}\right] + O^+(\varepsilon), \quad j = 1, \ldots, q \tag{7.3}$$

The elements of the last column (also with the exception of its last element) of the desired partial matrix product are the following:

$$(AB)_{i,k} = \frac{1}{\varepsilon^2}\left[\sum_{j=1}^{q} P_{ij} - Q_i^{(1)}\right] + O^+(\varepsilon), \quad i = 1, \ldots, q \tag{7.4}$$

The lowest right matrix element is obtained from the following linear combination:

$$(AB)_{k,k} = \frac{1}{\varepsilon^3}\left[-\sum_{i,j=1}^{q} P_{ij} + \sum_{i=1}^{q} Q_i^{(1)} + \sum_{j=1}^{q} Q_j^{(2)} - R\right] + O^+(\varepsilon) \tag{7.5}$$

7.1.1 Example—Partial Matrix Multiplication Algorithm

In this section we show how the partial matrix multiplication algorithm works for $q = 2$. In the (k, m, n)-dimensional problem we have $k = n = q + 1 = 3$ and $m = 2q = 4$. The matrices to be multiplied are:

$$A \cdot B = \left[\begin{array}{cc|cc} 0 & 0 & a_{1,3} & a_{1,4} \\ 0 & 0 & a_{2,3} & a_{2,4} \\ \hline a_{3,1} & a_{3,2} & a_{3,3} & a_{3,4} \end{array}\right] \cdot \left[\begin{array}{cc|c} b_{1,1} & b_{1,2} & b_{1,3} \\ b_{2,1} & b_{2,2} & b_{2,3} \\ \hline 0 & 0 & b_{3,3} \\ 0 & 0 & b_{4,3} \end{array}\right] \tag{7.6}$$

We have $q^2 = 4$ of P_{ij} products:

$$\begin{cases} P_{11} = (a_{3,1} + \varepsilon^2 a_{1,3})(b_{3,3} + \varepsilon^2 b_{1,1}) & P_{12} = (a_{3,1} + \varepsilon^2 a_{1,4})(b_{4,3} + \varepsilon^2 b_{1,2}) \\ P_{21} = (a_{3,2} + \varepsilon^2 a_{2,3})(b_{3,3} + \varepsilon^2 b_{2,1}), & P_{22} = (a_{3,2} + \varepsilon^2 a_{2,4})(b_{4,3} + \varepsilon^2 b_{2,2}) \end{cases}$$

Next come two series, each with $q = 2$ products:

$$\begin{aligned} Q_i^{(1)} &= a_{3,i}\left[\varepsilon^3 b_{i,3} + \sum_{j=1}^{2}(b_{j+2,3} + \varepsilon^2 b_{i,j})\right] \\ &= a_{3,i}\left[\varepsilon^3 b_{i,3} + (b_{3,3} + \varepsilon^2 b_{i,1}) + (b_{4,3} + \varepsilon^2 b_{i,2})\right] \\ Q_j^{(2)} &= \left[\varepsilon^3 a_{3,j+2} + \sum_{i=1}^{2}(a_{3,i} + \varepsilon^2 a_{i,j+2})\right] b_{j+2,3} \\ &= \left[\varepsilon^3 a_{3,j+2} + (a_{3,1} + \varepsilon^2 a_{1,j+2}) + (a_{3,2} + \varepsilon^2 a_{2,j+2})\right] b_{j+2,3} \end{aligned}$$

So:

$$\begin{cases} Q_1^{(1)} = a_{3,1}\left[\varepsilon^3 b_{1,3} + (b_{3,3} + \varepsilon^2 b_{1,1}) + (b_{4,3} + \varepsilon^2 b_{1,2})\right] \\ Q_2^{(1)} = a_{3,2}\left[\varepsilon^3 b_{2,3} + (b_{3,3} + \varepsilon^2 b_{2,1}) + (b_{4,3} + \varepsilon^2 b_{2,2})\right] \end{cases}$$

$$\begin{cases} Q_1^{(2)} = \left[\varepsilon^3 a_{3,3} + (a_{3,1} + \varepsilon^2 a_{1,3}) + (a_{3,2} + \varepsilon^2 a_{2,3})\right] b_{3,3} \\ Q_2^{(2)} = \left[\varepsilon^3 a_{3,4} + (a_{3,1} + \varepsilon^2 a_{1,4}) + (a_{3,2} + \varepsilon^2 a_{2,4})\right] b_{4,3} \end{cases}$$

And the last, single product R has the following form:

$$R = \left(\sum_{i=1}^{2} a_{3,i}\right)\left(\sum_{j=1}^{2} b_{j+2,3}\right) = (a_{3,1} + a_{3,2})(b_{3,3} + b_{4,3})$$

Using the above P_{ij}, $Q_i^{(1)}$, $Q_j^{(2)}$ and R products, in the amount of $(q+1)^2 = (2+1)^2 = 9$, we construct the desired partial matrix product on the basis of linear combinations (7.3)–(7.5). Note:

$$AB = \left[\begin{array}{cc|c} 0 & 0 & \frac{1}{\varepsilon^2}\left(P_{11}+P_{12}-Q_1^{(1)}\right) \\ 0 & 0 & \frac{1}{\varepsilon^2}\left(P_{21}+P_{22}-Q_2^{(1)}\right) \\ \hline \frac{1}{\varepsilon^2}\left(P_{11}+P_{21}-Q_1^{(2)}\right) & \frac{1}{\varepsilon^2}\left(P_{12}+P_{22}-Q_2^{(2)}\right) & \begin{array}{l}\frac{1}{\varepsilon^3}(-P_{11}-P_{12}-P_{21}-P_{22}+ \\ +Q_1^{(1)}+Q_2^{(1)}+Q_1^{(2)}+Q_2^{(2)} \\ -R)\end{array} \end{array}\right] + \left(O^+(\varepsilon)\right)$$

where $\left(O^+(\varepsilon)\right)$ denotes a 3×3 matrix with a shape of the form (7.2).

7.2 From Partial to Total Matrix Multiplication

The article Schonhage (1981) in Sects. 3–4, pages 441–445, proposes a general method how to use any partial matrix multiplication algorithm to construct a total one. As Schonhage explains on page 441, a more precise description of partial matrix multiplication is the multiplication of matrices that are partially filled. We still have a problem of dimensionality (k, m, n) for the matrix product $A^{(1)} \cdot B^{(1)}$. The (1) indicates the first level of recursion, i.e. no recursion is used. Schonhage (1981) presented this method so precisely that in this lecture it is inevitable that the general formulas will be similar to those in Schonhage's work.

We assume that matrix $A^{(1)}$ has k_μ non-zero entries in its μ th column and matrix $B^{(1)}$ has n_μ non-empty entries in its μ th row, for $\mu = 1, 2, \ldots, m$ (k stands for the row number of the 'whole' A matrix, while k_μ denotes the number of non-zero rows (here scalars) in a selected μth column; analogous heuristics are between n and n_μ). We can represent them in block matrix form as follows (blocks of vectors would be a more appropriate term here):

$$A^{(1)} = \left[a_1^{(1)} \middle| \cdots \middle| a_m^{(1)}\right], \quad B^{(1)} = \left[\begin{array}{c} b_1^{(1)} \\ \hline \vdots \\ \hline b_m^{(1)} \end{array}\right]$$

Using the block matrix multiplication formula (3.2), we obtain their product:

$$A^{(1)} \cdot B^{(1)} = a_1^{(1)} b_1^{(1)} + a_2^{(1)} b_2^{(1)} + \cdots + a_m^{(1)} b_m^{(1)} \tag{7.7}$$

The number of multiplications to be performed using the definition base algorithm (1.1) is therefore:

$$f = k_1 n_1 + k_2 n_2 + \cdots + k_m n_m \tag{7.8}$$

because each of the sub-products $a_{\mu}^{(1)}b_{\mu}^{(1)}$ requires $k_{\mu}n_{\mu}$ scalar multiplications.

Now we recursively nest the $A^{(1)} \cdot B^{(1)}$ partial matrix product matrices s times, where $s > 1$. We obtain the recursive partial product $D^{(s)} = A^{(s)} \cdot B^{(s)}$ of dimension (k^s, m^s, n^s). From the partial product obtained, we select a subset of columns of matrix $A^{(s)}$ and a corresponding subset of rows of matrix $B^{(s)}$.

Now we have a discrete optimisation problem: which of the m^s available corresponding columns and rows should be selected? The selection is modelled by the following partition:

$$s = \sigma_1 + \sigma_2 + \cdots + \sigma_m \tag{7.9}$$

for non-negative integers σ_{μ}. Let us fix the variables $\overline{\sigma} = (\sigma_1, \sigma_2, \ldots, \sigma_m)$. We establish a set of multi-indices $\overline{\mu}(\sigma_1, \ldots, \sigma_m) = (\mu_1, \ldots, \mu_s)$, where σ_i of the entries μ_j are equal to i, for $1 \leq i \leq m$. This description may seem intricate, so for clarification we give two examples in Sects. 7.2.1 and 7.2.2. Let us continue with the construction of the Schonhage. From a combinatorial point of view, in a given $\overline{\mu}(\sigma_1, \ldots, \sigma_m)$ series we have s elements from m different classes. Thus they can be permuted in the following number of ways (e.g. Ryser 1963, Sect. 1.4, formula 4.18 on page 11):

$$M(\sigma_1, \ldots, \sigma_m) = \frac{s!}{\sigma_1!\sigma_2! \cdot \ldots \cdot \sigma_m!} \tag{7.10}$$

It can be deduced that for every permutation of the $\overline{\mu}(\sigma_1, \ldots, \sigma_m)$ series, induced by a fixed $(\sigma_1, \sigma_2, \ldots, \sigma_m)$ variable series, we have the same amount K of non-zero columns of $A^{(s)}$ matrix and N non-zero rows in matrix $B^{(s)}$. That amount is:

$$\begin{aligned} K(\sigma_1, \ldots, \sigma_m) &= k_1^{\sigma_1} k_2^{\sigma_2} \cdot \ldots \cdot k_m^{\sigma_m} \\ N(\sigma_1, \ldots, \sigma_m) &= n_1^{\sigma_1} n_2^{\sigma_2} \cdot \ldots \cdot n_m^{\sigma_m} \end{aligned} \tag{7.11}$$

Thus, the volume of the partial matrix product obtained is:

$$P(\sigma_1, \ldots, \sigma_m) = KMN = \frac{s!}{\sigma_1!\sigma_2! \cdot \ldots \cdot \sigma_m!}(k_1 n_1)^{\sigma_1}(k_2 n_2)^{\sigma_2} \cdot \ldots \cdot (k_m n_m)^{\sigma_m} \tag{7.12}$$

To get the most efficient matrix multiplication algorithm, we maximize the value of $P(\sigma_1, \ldots, \sigma_m)$. In practice, this means that we have to find optimal series of $(\sigma_1, \sigma_2, \ldots, \sigma_m)$ non-negative integer variables that add up to s. The problem of such partitioning is also called compositions. A well-known tool for analysing them is generating functions. A comprehensive monograph on this topic is Heubach and Mansour (2009). In particular, the number of possible partitions is of the form (7.9):

$$\binom{s + m - 1}{m - 1} \tag{7.13}$$

In practice, treating the volume $P = KMN$ as an objective function means that we want to find the optimal balance between the volume of a given column to row multiplications (the respective summands in the formula (7.8)) and the corresponding amount of that volume.

Interestingly, at the time of Schonhage's publication, there were no effective methods for finding the maximum of form (7.12). Later results in probability on the topic of multinomial distribution are useful. But in order to use them, we must first transform the objective function $P(\sigma_1, \ldots, \sigma_m)$ (7.12) into the form of this multinomial distribution. To do this, we introduce new variables $p_1, \ldots, p_m$ defined by the quotient series:

$$p_1 = \frac{k_1 n_1}{f}, p_2 = \frac{k_2 n_2}{f}, \ldots, p_m = \frac{k_m n_m}{f}$$

where $f = k_1 n_1 + \cdots + k_m n_m$ (equality (7.8)). We can observe, that the variables $p_1, \ldots, p_m$ sum up to 1 and now the problem can be treated by finding the optimal multinomial distribution problem (e.g. Feller 1968, Sect. 6.9, pages 167–168). In fact, we can rewrite the objective function as follows:

$$\begin{aligned} P(\sigma_1, \ldots, \sigma_m) &= f^{\sigma_1+\sigma_2+\ldots\sigma_m} \frac{s!}{\sigma_1!\sigma_2! \cdot \ldots \cdot \sigma_m!} \left(\frac{k_1 n_1}{f}\right)^{\sigma_1} \left(\frac{k_2 n_2}{f}\right)^{\sigma_2} \cdot \ldots \cdot \left(\frac{k_m n_m}{f}\right)^{\sigma_m} \\ &= f^s \frac{s!}{\sigma_1!\sigma_2! \cdot \ldots \cdot \sigma_m!} (p_1)^{\sigma_1} (p_2)^{\sigma_2} \cdot \ldots \cdot (p_m)^{\sigma_m} \end{aligned}$$

The first results in this area were given by Finucan (1964). However, effective algorithms were later given by Gall (2003) and White and Hendy (2010).

To sum up, the choice of $\overline{\sigma} = (\sigma_1, \sigma_2, \ldots, \sigma_m)$ series determines all the other quantities, i.e. the dimensions K, M, N, the volume P, the M-element subset of the available columns and corresponding rows of m^s that we select, and thus the efficiency of the algorithm obtained.

7.2.1 Example—Maximizing the Volume of a Matrix Product

This section shows how to maximise the volume of a matrix product using a single multiplier as an example. A multiplier of the product does not change the structure of an objective function (7.12) from an optimisation point of view. Suppose the $A^{(1)}$ matrix has the following structure:

$$A^{(1)} = \begin{bmatrix} * & * \\ 0 & * \end{bmatrix}$$

where the asterisks stand for scalar non-zero entries. Now let us find all possible partitions $s = \sigma_1 + \sigma_2 + \cdots + \sigma_m$ for triple recursive nesting of the above matrix $A^{(1)}$. We have $m = 2$ and $s = 3$, so we need to find all pairs of non-negative integers (σ_1, σ_2) that satisfy the constraint:

$$3 = \sigma_1 + \sigma_2$$

We know about possible partitions from Newton's symbol (7.13):

$$\binom{s+m-1}{m-1} = \binom{3+2-1}{2-1} = 4$$

These four possible pairs (σ_1, σ_2) are: $(0, 3)$, $(3, 0)$, $(1, 2)$ and $(2, 1)$. We have listed them in order compatible with the number of permutations (formula (7.10)). Below we have analysed each of the four pairs in the following order: a given pair (σ_1, σ_2) followed by the corresponding quantities:

- induced $\overline{\mu}(\sigma_1, \sigma_2)$ series,
- number of permutations $M = s!/(\sigma_1! \cdot \sigma_2!)$ of induced $\overline{\mu}(\sigma_1, \sigma_2)$ series,
- amount $K = k_1^{\sigma_1} k_2^{\sigma_2}$ of non-zero columns of $A^{(3)}$,
- volume $P = M \cdot k_1^{\sigma_1} k_2^{\sigma_2}$.

$$\begin{cases} \sigma_1 = 0 \\ \sigma_2 = 3 \end{cases}, \overline{\mu}(0, 3) = (2, 2, 2), M = \frac{3!}{0! \cdot 3!} = 1, K = 1^0 \cdot 2^3 = 8, P = 1 \cdot 8 = 8$$

$$\begin{cases} \sigma_1 = 3 \\ \sigma_2 = 0 \end{cases}, \overline{\mu}(3, 0) = (1, 1, 1), M = \frac{3!}{3! \cdot 0!} = 1, K = 1^3 \cdot 2^0 = 1, P = 1 \cdot 1 = 1$$

$$\begin{cases} \sigma_1 = 1 \\ \sigma_2 = 2 \end{cases}, \overline{\mu}(1, 2) = \begin{cases} (1, 2, 2) \\ or \\ (2, 1, 2) \\ or \\ (2, 2, 1) \end{cases}, M = \frac{3!}{1! \cdot 2!} = 3, K = 1^1 \cdot 2^2 = 4, P = 3 \cdot 4 = \mathbf{12}$$

$$\begin{cases} \sigma_1 = 2 \\ \sigma_2 = 1 \end{cases}, \overline{\mu}(2, 1) = \begin{cases} (1, 1, 2) \\ or \\ (1, 2, 1) \\ or \\ (2, 1, 1) \end{cases}, M = \frac{3!}{2! \cdot 1!} = 3, K = 1^2 \cdot 2^1 = 2, P = 3 \cdot 2 = 6$$

We can see, that the maximum volume obtained for $(\sigma_1, \sigma_2) = (1, 2)$ is 12. The $\overline{\mu}(\sigma_1, \sigma_2)$ vector can be interpreted as a kind of 'itinerary' in the structure of the $A^{(3)}$ matrix. For example, in the vector $\overline{\mu}_1(1, 2) = \left(\overline{\mu}_1^{(1)}, \overline{\mu}_1^{(2)}, \overline{\mu}_1^{(3)}\right) = (1, 2, 2)$ the first component $\overline{\mu}_1^{(1)} = 1$ means that we go to the left half of the columns 1..8, next

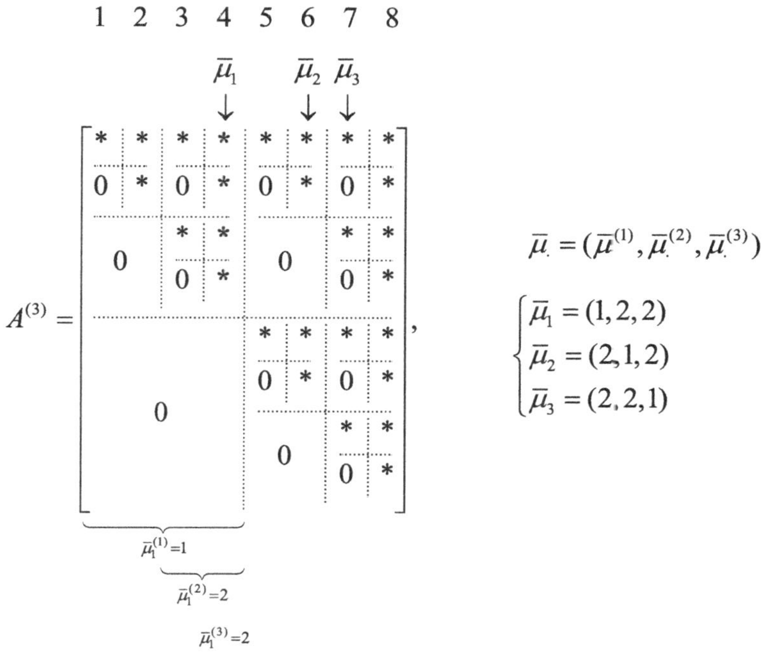

Fig. 7.1 Selection of a multiplier columns by optimal σ

$\overline{\mu}_1^{(2)} = 2$ takes us to the right half of the columns 1..4 obtained in the previous step, and finally $\overline{\mu}_1^{(3)} = 2$ means, that we go back to the right half of columns number 3..4, finally arriving at column 4; see Fig. 7.1.

We can see that in each of the three $A^{(3)}$ columns, corresponding to the $\overline{\mu}(1, 2)$ series that depend on the optimal (σ_1, σ_2), we have four non-zero entries.

Let us now continue the construction of a general method. We denote the set of all possible permutations of a given $\overline{\mu}$ series, for fixed $(\sigma_1, \sigma_2, \ldots, \sigma_m)$ variables, as $\Re(\sigma_1, \ldots, \sigma_m)$. Its cardinality is equal to $M(\sigma_1, \ldots, \sigma_m)$ (formula (7.10)). The block matrix formula (7.7) after s-fold recursion can be rewritten in the following form:

$$A^{(s)} \cdot B^{(s)} = \sum_{\mu \in \Re(\sigma_1, \ldots, \sigma_m)} a_\mu^{(s)} b_\mu^{(s)} + \sum_{\mu \notin \Re(\sigma_1, \ldots, \sigma_m)} a_\mu^{(s)} b_\mu^{(s)} \tag{7.14}$$

To obtain a partial matrix multiplication algorithm of (k^s, M, n^s) dimensionality after selecting the $A^{(s)}$ columns and corresponding $B^{(s)}$ rows, determined by permutations $M(\sigma_1, \ldots, \sigma_m)$ of $\overline{\mu}(\sigma_1, \ldots, \sigma_m)$ series, we have to nullify the second summand in (7.14). Simply, for all $\mu \notin \Re(\sigma_1, \ldots, \sigma_m)$ we assign $a_\mu^{(s)} = [0 \cdots 0]^T$ and $b_\mu^{(s)} = [0 \cdots 0]$ and the obtained product we denote as $D = AB$.

Schonhage (1981) in Sect. 4.2 shows how to adapt the (k^s, M, n^s) **partial** matrix multiplication algorithm to the **total** matrix multiplication of (K, M, N) dimensionality.

In fact, we have at our disposal a **partial** matrix multiplication algorithm $D = AB$ with K non-zero entries in each of the M columns of the multiplier (A) and N non-zero entries in each of the M rows of the multiplicand (B). Our goal is to construct on this basis an algorithm for **total** matrix multiplication, also with K entries in each of the M columns of the new multiplier matrix and the N non-zero entries in each of the M rows of the new multiplicand matrix. The description of the two matrices sounds similar, but the difference is that in the matrix A we have K non-zero entries in selected columns, while in the new matrix we have exactly K entries in the respective columns. The same applies to the B matrix with the amount equal to N. The new dense matrices will be called U, V and will be denoted by $W = UV$. They are dense in the sense that they do not contain any permanent zeros; respectively, in this section we refer to matrices as sparse in the sense that they contain such permanent zeros.

Schonhage (1981) proposed in Sect. 4.2 a kind of transition matrices, which allow to switch between sparse and dense representation of the matrices to be multiplied. As these matrices he used auxiliary matrices G and Q, which fulfil the equalities:

$$\begin{cases} U = GA \\ V = BQ \end{cases} \tag{7.15}$$

Therefore the dimensionality of G, Q matrices is $K \times k^s$ and $n^s \times N$, respectively. We also assume that all maximal order minors of G, Q matrices do not vanish. How to make sure this condition is met is shown later.

Let us denote the positions of non-zero entries in the column j^{th} of a given matrix A as integer series $(i_{j,1}, \ldots, i_{j,K})$ in ascending order. For a given column of the sparse matrix A its non-zero entries can be obtained from the dense matrix U by equality:

$$\left[a_{i_{j,1},j} \cdots a_{i_{j,K},j} \right]^T = H_j u_{\cdot j}, \quad j = 1, \ldots, M \tag{7.16}$$

where $u_{\cdot j}$ denotes the jth column of U and H_j is the inverse of such a submatrix of G, that satisfies the equality (7.15). Analogously, we fill the B's entries from the dense matrix V. How to choose suitable submatrices of G, Q is explained in the example in Sect. 7.2.2. It remains to give a structure of the transition matrices G, Q which ensures that all their submatrices of maximal size are invertible. For this purpose, Schonhage (1981) proposes to use a kind of rectangular generalisation of the Vandermonde matrices. This matrix is:

$$G = \begin{bmatrix} 1 & \cdots & 1 \\ \vdots & \ddots & \vdots \\ \alpha_1^{K-1} & \cdots & \alpha_{k^s}^{K-1} \end{bmatrix}$$

with pairwise different real α. variables. In fact, each G's submatrix constructed by K selected columns creates an $K \times K$ Vandermonde matrix that is invertible. By analogy, we construct the Q matrix.

To sum up, in order to obtain a full matrix multiplication algorithm based on a partial one, we proceed in the following stages:

- Input: U, V dense matrices to be multiplied.
- Decompose the input matrices into the partial matrices A, B by the series of inversions of properly selected submatrices of transition matrices G, Q (equality (7.16)).
- Perform the partial matrix multiplication $D = AB$.
- Obtain the dense product from equality $W = UV = G(AB)Q = GDQ$.

How to invert Vandermonde matrix we show in the Appendix, Chap. 11.

7.2.2 Example—Application of Partial Matrix Multiplication

In this example we multiply input (dense) matrices U, V of dimensions $\dim U = 3 \times 2$ and $\dim V = 2 \times 3$ using the double recursively nested particular partial matrix multiplication algorithm, which we described in Sect. 7.1 with parameters from Example 7.1.1. So we have $s = 2$ and $(k, m, n) = (3, 4, 3)$. The k_μ and n_μ parameters are: $k_1 = 1$, $n_1 = 3$; $k_2 = 1$, $n_2 = 3$; $k_3 = 3$, $n_3 = 1$ and $k_4 = 3$, $n_1 = 1$. The partition $s = \sigma_1 + \sigma_2 + \cdots + \sigma_m$ receives form:

$$2 = \sigma_1 + \sigma_2 + \sigma_3 + \sigma_4$$

However, a more optimised decision is possible, so that all the internal mechanisms of the method are revealed, we arbitrarily choose $\sigma_1 = \sigma_3 = 1$. Thus $\sigma_2 = \sigma_4 = 0$. From formula (7.11) we obtain K and N: $K = 1^1 \cdot 1^0 \cdot 3^1 \cdot 3^0 = 3, N = 3^1 \cdot 3^0 \cdot 1^1 \cdot 1^0 = 3$. Thus, the dimensions of the transition matrices are $\dim G = K \times k^s = 3 \times 3^2 = 3 \times 9$ and $\dim Q = n^s \times N = 3^2 \times 3 = 9 \times 3$. The cardinality of the $\Re(\sigma_1, \sigma_2, \sigma_3, \sigma_4)$ set is $M = 2!/(1! \cdot 0! \cdot 1! \cdot 0!) = 2$ and the two series induced by $\overline{\mu}(\sigma_1, \sigma_2, \sigma_3, \sigma_4)$ are: $\overline{\mu}_1 = (1, 3)$ or $\overline{\mu}_2 = (3, 1)$. The general mechanism of the method is illustrated in Fig. 7.2.

To decompose the input dense matrix U into a sparse matrix A, we choose the following submatrices of the transition matrix G:

$$H_1 = \begin{bmatrix} 1 & 1 & 1 \\ \alpha_7 & \alpha_8 & \alpha_9 \\ \alpha_7^2 & \alpha_8^2 & \alpha_9^2 \end{bmatrix}, \quad H_2 = \begin{bmatrix} 1 & 1 & 1 \\ \alpha_3 & \alpha_6 & \alpha_9 \\ \alpha_3^2 & \alpha_6^2 & \alpha_9^2 \end{bmatrix}$$

The integer series $\left(i_{j,1}, \ldots, i_{j,K}\right)$ for $j = 1$ is (7, 8, 9) and for $j = 2$ is (3, 6, 9), respectively. So from equality (7.16) we get:

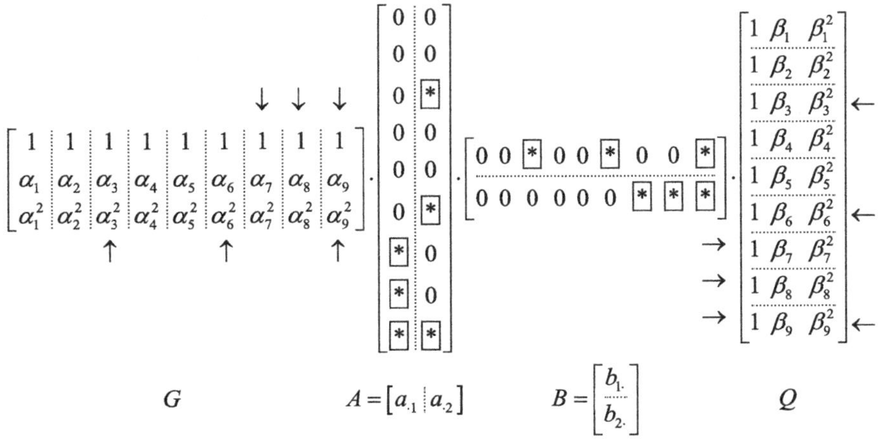

Fig. 7.2 Example construction of partial to total matrix multiplication method

$$\begin{bmatrix} a_{i_{1,1},1} \\ a_{i_{1,2},1} \\ a_{i_{1,3},1} \end{bmatrix} = \begin{bmatrix} a_{7,1} \\ a_{8,1} \\ a_{9,1} \end{bmatrix} = H_1^{-1} \cdot u_{\cdot 1}, \quad \begin{bmatrix} a_{i_{2,1},2} \\ a_{i_{2,2},2} \\ a_{i_{2,3},2} \end{bmatrix} = \begin{bmatrix} a_{3,2} \\ a_{6,2} \\ a_{9,2} \end{bmatrix} = H_2^{-1} \cdot u_{\cdot 2}$$

We construct the matrix B by analogy:

$$R_1 = \begin{bmatrix} 1 & \beta_3 & \beta_3^2 \\ 1 & \beta_6 & \beta_6^2 \\ 1 & \beta_9 & \beta_9^2 \end{bmatrix}, \qquad R_2 = \begin{bmatrix} 1 & \beta_7 & \beta_7^2 \\ 1 & \beta_8 & \beta_8^2 \\ 1 & \beta_9 & \beta_9^2 \end{bmatrix}$$

Thus:

$$\begin{bmatrix} b_{13} & b_{16} & b_{19} \end{bmatrix} = v_{1\cdot} \cdot R_1^{-1}$$
$$\begin{bmatrix} b_{27} & b_{28} & b_{29} \end{bmatrix} = v_{2\cdot} \cdot R_2^{-1}$$

At this point we have filled in the matrices A and B. Now we need to find the product $D = AB$ using the twice recursively nested partial multiplication algorithm from Sect. 7.1. Let us construct the product $D^{(2)} = A^{(2)} \cdot B^{(2)}$. From formula (7.6) we get:

$$\left[\begin{array}{cccccccc|cccccccc}
\multicolumn{4}{c}{} & \multicolumn{4}{c|}{} & 0 & 0 & a^{(1,3)}_{1,3} & a^{(1,3)}_{1,4} & 0 & 0 & a^{(1,4)}_{1,3} & a^{(1,4)}_{1,4} \\
\multicolumn{4}{c}{0} & \multicolumn{4}{c|}{0} & 0 & 0 & a^{(1,3)}_{2,3} & a^{(1,3)}_{2,4} & 0 & 0 & a^{(1,4)}_{2,3} & a^{(1,4)}_{2,4} \\
\multicolumn{4}{c}{} & \multicolumn{4}{c|}{} & a^{(1,3)}_{3,1} & a^{(1,3)}_{3,2} & a^{(1,3)}_{3,3} & a^{(1,3)}_{3,4} & a^{(1,4)}_{3,1} & a^{(1,4)}_{3,2} & a^{(1,4)}_{3,3} & a^{(1,4)}_{3,4} \\
\multicolumn{4}{c}{} & \multicolumn{4}{c|}{} & 0 & 0 & a^{(2,3)}_{1,3} & a^{(2,3)}_{1,4} & 0 & 0 & a^{(2,4)}_{1,3} & a^{(2,4)}_{1,4} \\
\multicolumn{4}{c}{0} & \multicolumn{4}{c|}{0} & 0 & 0 & a^{(2,3)}_{2,3} & a^{(2,3)}_{2,4} & 0 & 0 & a^{(2,4)}_{2,3} & a^{(2,4)}_{2,4} \\
\multicolumn{4}{c}{} & \multicolumn{4}{c|}{} & a^{(2,3)}_{3,1} & a^{(2,3)}_{3,2} & a^{(2,3)}_{3,3} & a^{(2,3)}_{3,4} & a^{(2,4)}_{3,1} & a^{(2,4)}_{3,2} & a^{(2,4)}_{3,3} & a^{(2,4)}_{3,4} \\
\hline
0 & 0 & a^{(3,1)}_{1,3} & a^{(3,1)}_{1,4} & 0 & 0 & a^{(3,2)}_{1,3} & a^{(3,2)}_{1,4} & 0 & 0 & a^{(3,3)}_{1,3} & a^{(3,3)}_{1,4} & 0 & 0 & a^{(3,4)}_{1,3} & a^{(3,4)}_{1,4} \\
0 & 0 & a^{(3,1)}_{2,3} & a^{(3,1)}_{2,4} & 0 & 0 & a^{(3,2)}_{2,3} & a^{(3,2)}_{2,4} & 0 & 0 & a^{(3,3)}_{2,3} & a^{(3,3)}_{2,4} & 0 & 0 & a^{(3,4)}_{2,3} & a^{(3,4)}_{2,4} \\
a^{(3,1)}_{3,1} & a^{(3,1)}_{3,2} & a^{(3,1)}_{3,3} & a^{(3,1)}_{3,4} & a^{(3,2)}_{3,1} & a^{(3,2)}_{3,2} & a^{(3,2)}_{3,3} & a^{(3,2)}_{3,4} & a^{(3,3)}_{3,1} & a^{(3,3)}_{3,2} & a^{(3,3)}_{3,3} & a^{(3,3)}_{3,4} & a^{(3,4)}_{3,1} & a^{(3,4)}_{3,2} & a^{(3,4)}_{3,3} & a^{(3,4)}_{3,4}
\end{array}\right]$$

$\uparrow$ $\overline{\mu}_1 = (1,3)$ $\uparrow$ $\overline{\mu}_2 = (3,1)$

$$\cdot\left[\begin{array}{cccccc|ccc}
b^{(1,1)}_{1,1} & b^{(1,1)}_{1,2} & b^{(1,1)}_{1,3} & b^{(1,2)}_{1,1} & b^{(1,2)}_{1,2} & b^{(1,2)}_{1,3} & b^{(1,3)}_{1,1} & b^{(1,3)}_{1,2} & b^{(1,3)}_{1,3} \\
b^{(1,1)}_{2,1} & b^{(1,1)}_{2,2} & b^{(1,1)}_{2,3} & b^{(1,2)}_{2,1} & b^{(1,2)}_{2,2} & b^{(1,2)}_{2,3} & b^{(1,3)}_{2,1} & b^{(1,3)}_{2,2} & b^{(1,3)}_{2,3} \\
0 & 0 & b^{(1,1)}_{3,3} & 0 & 0 & b^{(1,2)}_{3,3} & 0 & 0 & b^{(1,3)}_{3,3} \leftarrow \overline{\mu}_1 = (1,3) \\
0 & 0 & b^{(1,1)}_{4,3} & 0 & 0 & b^{(1,2)}_{4,3} & 0 & 0 & b^{(1,3)}_{4,3} \\
b^{(2,1)}_{1,1} & b^{(2,1)}_{1,2} & b^{(2,1)}_{1,3} & b^{(2,2)}_{1,1} & b^{(2,2)}_{1,2} & b^{(2,2)}_{1,3} & b^{(2,3)}_{1,1} & b^{(2,3)}_{1,2} & b^{(2,3)}_{1,3} \\
b^{(2,1)}_{2,1} & b^{(2,1)}_{2,2} & b^{(2,1)}_{2,3} & b^{(2,2)}_{2,1} & b^{(2,2)}_{2,2} & b^{(2,2)}_{2,3} & b^{(2,3)}_{2,1} & b^{(2,3)}_{2,2} & b^{(2,3)}_{2,3} \\
0 & 0 & b^{(2,1)}_{3,3} & 0 & 0 & b^{(2,2)}_{3,3} & 0 & 0 & b_{3,3} \\
0 & 0 & b^{(2,1)}_{4,3} & 0 & 0 & b^{(2,2)}_{4,3} & 0 & 0 & b_{4,3} \\
\hline
\multicolumn{3}{c}{} & \multicolumn{3}{c|}{} & b^{(3,3)}_{1,1} & b^{(3,3)}_{1,2} & b^{(3,3)}_{1,3} \leftarrow \overline{\mu}_2 = (3,1) \\
\multicolumn{3}{c}{0} & \multicolumn{3}{c|}{0} & b^{(3,3)}_{2,1} & b^{(3,3)}_{2,2} & b^{(3,3)}_{2,3} \\
\multicolumn{3}{c}{} & \multicolumn{3}{c|}{} & 0 & 0 & b^{(3,3)}_{3,3} \\
\multicolumn{3}{c}{} & \multicolumn{3}{c|}{} & 0 & 0 & b^{(3,3)}_{4,3} \\
\multicolumn{3}{c}{} & \multicolumn{3}{c|}{} & b^{(4,3)}_{1,1} & b^{(4,3)}_{1,2} & b^{(4,3)}_{1,3} \\
\multicolumn{3}{c}{0} & \multicolumn{3}{c|}{0} & b^{(4,3)}_{2,1} & b^{(4,3)}_{2,2} & b^{(4,3)}_{2,3} \\
\multicolumn{3}{c}{} & \multicolumn{3}{c|}{} & 0 & 0 & b^{(4,3)}_{3,3} \\
\multicolumn{3}{c}{} & \multicolumn{3}{c|}{} & 0 & 0 & b^{(4,3)}_{4,3}
\end{array}\right]$$

with dimensionality $(3,4,3)^2 = (9,16,9)$. We have marked with arrows the rows and columns selected by $\Re(\sigma_1 = 1, \sigma_2 = 0, \sigma_3 = 1, \sigma_4 = 0)$ set. Now we run the partial matrix multiplication algorithm, presented in detail in Sect. 7.2.1, after deleting all μth columns of $A^{(2)}$ and μth rows of $B^{(2)}$ except those for which $\mu \in \Re(\sigma_1, \sigma_2, \sigma_3, \sigma_4)$. It takes $l^s = l^2 = 9^2 = 81$ non-scalar multiplications. At this point it remains to calculate the dense product from equality $UV = GDQ$.

To sum up this chapter, it is worth saying a few words about efficiency. In the article Pan (1984) in Theorem 2.1, page 396, the efficiency exponent for a bilinear, total matrix multiplication algorithm (for dense matrices) is estimated to be $\omega \leq 3\log(r)/\log(mnp)$. Considering that the denominator contains the volume, the efficiency of the partial matrix multiplication algorithm is analogous. After s recursive nesting, we have an algorithm that works in l^s non-commutative multiplications (bilinear steps) on f^s entries. By analogy, we can estimate the efficiency of the partial algorithm as $\omega \leq 3\log(l^s)/\log(f^s) = 3 \cdot (s\log(l))/(s\log(f)) = 3\log(l)/\log(f)$. A formal proof of this fact is given by Theorem 4.1, page 443, in the article Schonhage (1981).

In the case of the partial matrix multiplication scheme we presented in Sect. 7.1, it obtains its best efficiency for $q = 3$. The volume is given by formula (7.8):

$$f = \underbrace{1 \cdot (q+1) + \cdots + 1 \cdot (q+1)}_{q} + \underbrace{(q+1) \cdot 1 + \cdots + (q+1) \cdot 1}_{q}$$
$$= (q+1)q + (q+1)q = 2q(q+1)$$

and $f(3) = 24$ and $l = (q+1)^2 = (3+1)^2 = 16$ what gives $\omega \leq 2.61725$.

After Schonhage's article, Coppersmith (1982) adapted Schonhage's method to the multiplication of partial matrices, but of rectangular shape, obtaining in this case a more favourable efficiency:

$$\omega \leq 3\tau = 3\frac{\ln 52}{\ln 110} = 2.5218127\ldots$$

References

Bini D, Capovani M, Lotti G, Romani F (1979) O(n^2.7799) Complexity for approximate matrix multiplication. Inf Process Lett 8(5):234–235
Coppersmith D (1982) Rapid multiplication of rectangular matrices. SIAM J Comput 11(3):467–471
Feller W (1968) An introduction to probability theory and its applications, vol I, 3rd edn. Wiley, New York, London, Sydney
Finucan HM (1964) The mode of a multinomial distribution. Biometrika 51(3/4):513–517
Gall F (2003) Determination of the modes of a Multinomial distribution. Statist Probab Lett 62(4):325–333
Heubach S, Mansour T (2009) Combinatorics of compositions and words. Chapman & Hall/CRC, New York
Pan V (1984) How can we speed up matrix multiplication? SIAM Rev 26(3):393–415
Ryser HJ (1963) Combinatorial mathematics. The Carus Mathematical Monographs 14. The Mathematical Association of America
Schonhage A (1981) Partial and total matrix multiplication. SIAM J Comput 10(3):434–455
White WTJ, Hendy MD (2010) A fast and simple algorithm for finding the modes of a multinomial distribution. Statist Probab Lett 80(1):63–68

Chapter 8
Summary of Fast Matrix Multiplication Algorithms

8.1 General Problems and Methods in Fast Matrix Multiplication

In general, fast matrix multiplication algorithms can be divided into the following groups:

- Commutative and non-commutative. The definitions given in Sect. 2.2. At this point we just want to remind that in a recursive way only those algorithms can be applied that are non-commutative.
- Designed for the general size of matrices and specialised for a given size.

In Table 8.1 we have summarised the selected results for the arbitrary size matrices, in particular those presented in detail in this monograph. However, some of them apply to matrices with a particular parity of the dimensions.

For $n = 3$, the best algorithm is Laderman (1976) in 23 multiplications. Rosowski (2023) arrived at 21 multiplication for this dimension, but at the cost of a commutative version. For $n = 4$, the optimal approach is to use Strassen's scheme twice recursively, which gives 49 multiplications. For $n = 5$ we multiply 97 times thanks to Kauers and Moosbauer (2023). For $n = 6$, Smirnov (2013) needs 160 multiplications. For $n = 7$, Sedoglavic (2017a) uses 250 multiplications. (There are online tables collected by various authors, but we have decided not to include a specific link as they are never guaranteed to be permanent.)

In Table 8.2 we show the pursuit for the fastest algorithm for dimension 5×5, in non-commutative version. We considered both the algorithms for arbitrary size applied to this particular dimension and those designed just for 5×5 matrices.

There have also been attempts to develop fast matrix multiplication algorithms using methods other than those discussed in this chapter.

There are papers that have successfully applied artificial intelligence to the problem. Veit (2016) studied a neural network for Strassen's algorithm. Fawzi et al. (2022) through machine learning methods proposed an algorithm for rectangular

J. S. Respondek, *Fast Matrix Multiplication with Applications*, Studies in Big Data 166,
https://doi.org/10.1007/978-3-031-76930-6_8

Table 8.1 Survey of the algorithms for arbitrary size matrices

Dimension	Commutative	Work	Number of multiplications
$a \times b \times c$	No	Definition	abc
$a = b = c = n$			n^3
n (square)	Yes	Winograd (1968)	$2n^2 + (n^2 - 2n)\lfloor (n+1)/2 \rfloor$
n (square)	No	Gastinel (1971)	$n^3 - (n-1)$
n (square)		Fiduccia (1972)	$n^3 - n(n-1)/2$
n (square)		Sykora (1977)	$n^3 - (n-1)^2$
$a \times 2 \times c$		Hopcroft and Kerr (1971)	$\lceil (3ac + \max(a, c))/2 \rceil$
n (square, even)		Pan (1980)	$(n^3 - 4n)/3 + 6n^2$
n (square, even)		Pan (1982)	$(4n^3 + 45n^2 + 128n + 108)/12$
$(a \times b) \times (b \times c) +$ $(b \times c) \times (c \times a)$		Pan (1984)	$abc + ab + bc + ac$
$(a \times b) \times (b \times c) +$ $(b \times c) \times (c \times a) +$ $(c \times a) \times (a \times b)$		Pan (1984)	$abc + 2(ab + ac + bc) +$ $4(a + b + c) + 15$

Table 8.2 Pursue for the fastest algorithm for the dimension 5×5

Publishing year	Author	Specialized for 5×5 dimension	Number of multiplications
	Definition	No	$5^3 = 125$
1971	Gastinel		121
1972	Fiduccia		115
1974	Fischer		110
1977	Sykora		109
1977		Yes	105
1978	Schachtel		103
1986 (b)	Makarow		102
1987			100
2017 (b)	Sedoglavic		99
2021			98
2023	Kauers and Moosbauer		97

problem of (4, 5, 5) dimensionality in 76 non-commutative multiplications, outperforming the previously known algorithm for this problem by 80 multiplications. The same article presents an algorithm for multiplying 4×4 matrices by 47 multiplications in modular arithmetic Z_2.

Another method is proposed in the article by Kauers and Moosbauer (2023). It used graph methods to find algorithms for fast matrix multiplication. In particular, by using these methods new algorithms were obtained for the (5, 5, 5) problem in 97 multiplications.

There are also works about hardware implementation of matrix multiplication, not only in a fast version. Cariow et al. (2012) proposed a field-programmable gate array system for computing the 3×3 matrix product in the commutative version given by Makarov (1986a), presented in Sect. 2.3. Article Bessant et al. (2023) used this hardware tool for Strassen's algorithm.

There is also a series of works on the multiplication of matrices using optical techniques. Examples in this line are Tamura and Wyant (1979) and the more recent articles by Hailong et al. (2022), Giamougiannis et al. (2023) and Peserico et al. (2023).

Article Fam (1988) analysed how to efficiently perform matrix multiplication with complex entries, using known formula for multiplying complex numbers in three instead of definitional four multiplications. In particular, a variation of this formula is given and discussed when applied to rectangular matrices.

8.2 Pursue for the Best Exponent on the Timeline

Finally, in this section we present the summary that everyone probably expected, i.e. how the time complexity exponent of the matrix multiplication algorithms has decreased on the timeline. In the case of disjoint algorithms, we present the exponent which follows from the methods of Chap. 6, i.e. which they possess in the case of their application to a single matrix multiplication. We must also remember that some algorithms are exact, while others are approximate (Table 8.3).

For the exponents of 2.479 and below, another tool is used, the so-called laser method, developed by Strassen (1986, 1987).

8.3 The Limits

There is still an open question about the minimum possible exponent in the $O(n^{\omega})$ complexity. There is a separate branch of research on the minimum number of multiplications required to perform matrix multiplication, in both non-commutative and commutative versions of the algorithms. The first estimate was given by Leeuven and Boas (1978), proposition 5.3 on page 75, that to multiply $n \times n$ square matrices we need at least $2n^2 - n$ multiplications. This results in an inequality for the exponent

Table 8.3 Pursue of the best exponent on the timeline

Publishing year	Author	Exponent
1969	Strassen	2.8074
1978	Pan	2.79512
1979	Bini et al.	2.7799
1980	Pan	2.7801419
1981	Schonhage	2.5479
1981	Schonhage	2.5218
1982	Romani	2.5117
1982	Coppersmith and Winograd (1982)	2.496
1986	Strassen	2.479
1990	Coppersmith and Winograd (1982)	2.3755
2010	Stothers (2013)[1]	2.37369
2012	Vassilevska-Williams	2.3729
2014	Le Gall	2.3728639
2020	Alman and Williams	2.3728596
2023	Duan et al.	2.371866

$\omega > 2$. In general, the following inequality applies (Blaser 2005, page 7):

$$c \leq r \leq 2c$$

where r stands for non-commutative algorithm and c for the commutative one, respectively. In particular, it is a clear fact that for a given matrix multiplication problem, if we do not allow commutativity, the number of multiplications required is either equal to or greater than in the commutative case. A sharper lower bound is given by Brockett and Dobkin (1978) for the rectangular (k, m, n) matrix multiplication problem, who obtain the following lower bound for the non-commutative algorithms:

$$r \geq km + mn - m + n - 1 = 2n^2 - 1\big|_{k=m=n}$$

Lafon and Winograd in the same 1978 proved an analogous inequality for the commutative case, i.e. proved that $c \geq km + mn - m + n - 1 = 2n^2 - 1\big|_{m=n}$. This bound was raised by Blaser (1999a) in the commutative case; he proved for the (k, m, n) problem that:

$$c \geq km + mn + k - m + n - 3 = 2n^2 + n - 3\big|_{k=m=n}, n \geq k \geq 2$$

The next result is given by Blaser (2003) for the rectangular matrix multiplication problem, in the non-commutative case, but only with 2 out of 3 allowed arbitrary

[1] The result was firstly published by Stothers in 2010 in his PhD thesis.

dimensions, i.e. for the (n, m, n) problem:

$$c \geq 2mn + 2n - m - 2 = 2n^2 + n - 2\big|_{m=n}, for\ m \geq n \geq 3$$

In particular, the last inequality gives the best lower bound so far for the 3×3 problem, with a minimum of 19 non-commutative multiplications. For the square non-commutative algorithms, even sharper bounds are given in the articles Blaser (1999b, 2005):

$$c \geq 2.5n^2 - 3n$$

Blaser (1999b) also proves that for upper triangular matrices $c \geq 1.25n^2 - 1.25n + 1$. Some results are given by Ja'Ja (1980) for commutative algorithms for selected low-dimensional problems. Namely, for $(2, n, n)$ matrix multiplication we need at least $\lceil 27n/8 \rceil$ multiplications and for $(2, 2, n)$ no less than $3n + 2$ multiplications are needed.

There are also some limits for the border rank (using ε-based approximate methods). Article Lickteig (1984) obtained lower bound for the (n, n, n) matrix problem equal to $3/2n^2 + 1/2n - 1$. Landsberg and Ottaviani (2014) provide a sharper bound on this problem equal to $2n^2 - n$. Landsberg and Michalek (2018) raise the bar to $2n^2 - \log_2 n - 1$. Article Massarenti and Raviolo (2013) gives an even sharper bound[2]:

$$\frac{11}{4}n^2 - 17n$$

In the community of scientists working with the matrix multiplication algorithm there is a popular hypothesis that it is possible to multiply the matrices in $O(n^{2+\varepsilon})$ time, for any positive ε. For example, the article Cohn et al. (2005) proposes conjecture 3.4, page 382, which—if true—implies that.

The hidden fact in this hypothesis (if true) is that to reduce ε it is necessary to increase the minimum size of the matrices to be multiplied (even to thousands and more), which leads to weaker numerical stability and a higher constant time coefficient hidden in the $O(n^{2+\varepsilon})$ complexity. The numerical stability of fast matrix multiplication algorithms is a subject of extensive research. We have already presented some work on the numerical stability of concrete algorithms. A comprehensive analysis, both theoretical and empirical, of these problems and an overview of previous results can be found in the article Ballard et al. (2016). The general conclusion of the research there is that the loss of numerical stability is not prohibitive, especially for most practical algorithms.

[2] The tile bound $3n^2 - 2\sqrt{2}n^{3/2} - 3n$ turned out to be incorrect, and next is corrected in corrigendum, LAA 445, pp. 369–371.

8.4 Rectangular Matrix Multiplication

There is also a branch of fast algorithms for multiplying non-square matrices. Examples of that are Lotti and Romani (1983), Coppersmith (1997), Huang and Pan (1998) and Le Gall (2024).

Another way to multiply rectangular matrices quickly is to use square matrix multiplication algorithms directly. The idea is to divide the rectangular matrices into square sub-matrices, then multiply them using any fast algorithm designed for square matrices. Say we have a (m, n, p) problem with $\min\{m, n, p\} = p$. Let us denote $j = m/p$ and $k = n/p$ (if necessary, apply e.g. the zero padding rule, Sect. 3.8). We obtain the following block matrix product to perform, with square blocks of equal dimension $\dim[A_{..}] = \dim[B_{.}] = p \times p$:

$$\begin{bmatrix} [A_{11}] & \cdots & [A_{1k}] \\ \vdots & \ddots & \vdots \\ [A_{j1}] & \cdots & [A_{jk}] \end{bmatrix} \cdot \begin{bmatrix} [B_1] \\ \vdots \\ [B_k] \end{bmatrix} = \begin{bmatrix} \left[\sum_{i=1}^{k} A_{1i} B_1\right] \\ \vdots \\ \left[\sum_{i=1}^{k} A_{ji} B_1\right] \end{bmatrix}$$

This way, in the best case, when p divides both m and n, instead of performing a rectangular matrix multiplication of size (m, n, p), we have to perform as many as $j \cdot k$ square multiplications of blocks of size $p \times p$. For each of the blocks we can apply the fast square matrix multiplication algorithm, which works in $O(p^{\omega})$ time. In summary, we have obtained a fast rectangular matrix multiplication algorithm with the following efficiency:

$$j \cdot k \cdot O(p^{\omega}) = O(j \cdot k \cdot p^{\omega}) = O\left[(jp) \cdot (kp) \cdot p^{\omega-2}\right] = O\left(mn \cdot p^{\omega-2}\right)$$

This efficiency is lower than the definitional $O(mnp)$, because for fast matrix multiplication algorithms we have $\omega < 3$, so the exponent $(\omega - 2) < 1$ is a root.

In general, for any order of (m, n, p) dimensions, in this method we have the following number of non-scalar products to perform:

$$O\left[\max\{m, n, p\} \cdot \mu_{1/2}\{m, n, p\} \cdot \min\{m, n, p\}^{\omega-2}\right] < mnp$$

where $\mu_{1/2}\{\cdot\}$ is the median. The idea is quite simple and has been formalised in the literature by the article Knight (1995), Sect. 2.1, p. 71.

We used the above method for multiplying rectangular matrices in Sect. 9.5 of this monograph to prove the time complexity of the LUP decomposition algorithm. Namely, it is used to perform the block rectangle multiplication (9.11).

8.5 Acceleration of Multiple Matrix Multiplication by Parenthesis Order

Another matrix multiplication problem, but one that is often encountered in practice, is to obtain the multiple matrix product:

$$A_1 A_2 \cdot \cdots \cdot A_n$$

with $\dim A_i = n_i \times n_{i+1}$, for a given $n > 2$. Let us illustrate the problem by the example of $n = 4$. Using the associativity property (1.2), we can compute it in 5 different orders:

$$(A_1A_2)(A_3A_4) = (A_1(A_2A_3))A_4 = A_1((A_2A_3)A_4) = ((A_1A_2)A_3)A_4 = A_1(A_2(A_3A_4))$$

These orders generally result in a different number of multiplications to be carried out. Note:

$$\begin{cases} (A_1A_2)(A_3A_4) : n_1n_2n_3 + n_3n_4n_5 + n_1n_3n_5 \\ (A_1(A_2A_3))A_4 : n_2n_3n_4 + n_1n_2n_4 + n_1n_4n_5 \\ A_1((A_2A_3)A_4) : n_2n_3n_4 + n_2n_4n_5 + n_1n_2n_5 \\ ((A_1A_2)A_3)A_4 : n_1n_2n_3 + n_1n_3n_4 + n_1n_4n_5 \\ A_1(A_2(A_3A_4)) : n_3n_4n_5 + n_2n_3n_5 + n_1n_2n_5 \end{cases}$$

The number of possible orders is expressed by the $(n-1)$th Catallan number:

$$C_{n-1} = \frac{1}{n}\binom{2n-2}{n-1}$$

So the brute-force algorithm has to search through an exponential number of possible decisions. The first polynomial-time approach to this problem was the dynamic programming of Godbole (1973) and Muraoka and Kuck (1973). The algorithms they present work in cubic time. In addition, the second of the two articles in Sect. 4, shows how parallel computing can be applied to this problem. Clearly, it can also be adapted to disjoint fast matrix multiplication algorithms.

Later, an $O(n \log n)$ complexity algorithm is proposed in articles by Hu and Shing (1982, 1984). Proof of that algorithm is given by Schwartz and Weiss (2019).

References

Alman J, Vassilevska-Williams V (2020) A refined laser method and faster matrix multiplication. In: 32nd annual ACM-SIAM symposium on discrete algorithms (SODA 2021), pp 522–539

Ballard G, Benson A, Druinsky A, Lipshitz B, Schwartz O (2016) Improving the numerical stability of fast matrix multiplication. Siam J Matrix Anal Appl 37(4):1382–1418

Bessant YRA, Jency JG, Sagayam KM, Jone AAA, Pandey D, Pandey BK (2023) Improved parallel matrix multiplication using Strassen and Urdhvatiryagbhyam method. CCF Trans High Perform Comput 5:102–115

Bini D, Capovani M, Lotti G, Romani F (1979) O(n^2.7799) complexity for approximate matrix multiplication. Inf Process Lett 8(5):234–235

Blaser M (1999a) Lower bounds for the multiplicative complexity of matrix multiplication. Comput Complex 8:203–226

Blaser M (1999b) A 5/2n^2-Lower bound for the rank of n×n-matrix multiplication over arbitrary fields. In: Proceedings of the 40th annual IEEE symposium on foundations of computer science (FOCS), pp 45–50

Blaser M (2003) On the complexity of the multiplication of matrices of small formats. J Complex 19(1):43–60

Blaser M (2005) Beyond the Alder-Strassen bound. Theor Comput Sci 331(1):3–21

Brockett R, Dobkin D (1978) On the optimal evaluation of a set of bilinear forms. Linear Algebra Appl 19:207–235

Cariow A, Syslo W, Cariowa G, Gliszczynski M (2012) A rationalized structure of processing unit to multiply 3×3 matrices. Pomiary Automatyka Kontrola 58(7):677–680

Cohn K, Szegedy U (2005) Group-theoretic algorithms for matrix multiplication. In: Proceedings of the 46th annual IEEE symposium on foundations of computer science (FOCS'05), Pittsburgh, PA, USA, pp 379–388

Cohn, Kleinberg, Szegedy, Umans (2005) Group-theoretic algorithms for matrix multiplication. In: Proceedings 46th annual IEEE symposium on foundations of computer science (FOCS'05), Pittsburgh, PA, USA, pp 379–388

Coppersmith D, Winograd S (1982) On the asymptotic complexity of matrix multiplication. SIAM J Comput 11(3):472–492

Coppersmith D, Winograd S (1990) Matrix multiplication via arithmetic progressions. J Symb Comput 3:251–280

Coppersmith D (1997) Rectangular Matrix Multiplication Revisited. Journal of Complexity 13:4249

Davie A, Stothers A (2013) Improved bound for complexity of matrix multiplication. Proc Royal Soc Edinb: Sect A Math 143(2):351–369

Duan R, Wu H, Zhou R (2023) Faster matrix multiplication via asymmetric hashing. In: IEEE 64th annual symposium on foundations of computer science (FOCS), Santa Cruz, CA, USA, pp 2129–2138

Fam AT (1988) Efficient complex matrix multiplication. IEEE Trans Comput 37(7):877–879

Fawzi A, Balog M, Huang A, Hubert T, Romera-Paredes B, Barekatain M, Novikov A, Ruiz FJR, Schrittwieser J, Swirszcz G, Silver D, Hassabis D, Kohli P (2022) Discovering faster matrix multiplication algorithms with reinforcement learning. Nature 610:47–53

Fiduccia CM (1972) On obtaining upper bounds on the complexity of matrix multiplication. In: Miller RE, Thatcher JW, Bohlinger JD (eds) Complexity of computer computations. The IBM research symposia series. Springer, Boston, pp 31–40

Fischer PC (1974) Further schemes for combining matrix algorithms. In: Loeck J (ed) LNCS, vol 14. Springer, pp 428–436

Gastinel N (1971) Sur le calcul des produits de matrices. Numer Math 17:222–229 (in French)

Giamougiannis G, Tsakyridis A, Moralis-Pegios M, Pappas C, Kirtas M, Passalis N, Lazovsky D, Tefas A, Pleros N (2023) Analog nanophotonic computing going practical: silicon photonic deep learning engines for tiled optical matrix multiplication with dynamic precision. Nanophotonic 12(5):963–973

Godbole SS (1973) On efficient computation of matrix chain products. IEEE Trans Comput 22(9):864–866

Hailong Z, Jianji D, Junwei C, Wenchan D, Chaoran H, Yichen S, Qiming Z, Min G, Chao Q, Hongsheng C, Zhichao R, Xinliang Z (2022) Photonic matrix multiplication lights up photonic accelerator and beyond. Light: Sci Appl 11(30):1–21

Hopcroft JE, Kerr LR (1971) On minimizing the number of multiplications necessary for matrix multiplication. SIAM J Appl Math 20(1):30–36

Hu TC, Shing MT (1982) Computation of matrix chain products. Part I. SIAM J Comput 11(2):362–373

Hu TC, Shing MT (1984) Computation of matrix chain products. Part II. SIAM J Comput 13(2):228–251

Huang X, Pan V (1998) Fast rectangular matrix multiplication and applications. J Complex 14:257–299

Ja'Ja (1980) On the complexity of bilinear forms with commutativity. SIAM J Comput 9(4):713–728

Kauers M, Moosbauer J (2023) Flip graphs for matrix multiplication. In: Proceedings of the 2023 international symposium on symbolic and algebraic computation (ISSAC 2023), pp 381–388

Knight PA (1995) Fast rectangular matrix multiplication and QR decomposition. Linear Algebra Appl 221:69–81

Laderman JD (1976) A noncommutative algorithm for multiplying 3×3 matrices using 23 multiplications. Bull Am Math Soc 82(1):126–128

Lafon JC, Winograd S (1978) A lower bound for the multiplicative complexity of the product of two matrices. Technical report, Centre de Calcul de L'Esplanade, U.E.R. de Mathematique, Univ. Louis Pasteur, Strasbourg, France

Landsberg JM, Michalek M (2018) A 2n^2-log_2(n)-1 lower bound for the border rank of matrix multiplication. Int Math Res Not 15:4722–4733

Landsberg JM, Ottaviani G (2014) New lower bounds for the rank of matrix multiplication. SIAM J Comput 43(1):126–143

Le Gall F (2014) Powers of tensors and fast matrix multiplication. In: Proceedings of the 39th international symposium on symbolic and algebraic computation, 296–303

Le Gall F (2024) Faster rectangular matrix multiplication by combination loss analysis. In: Proceedings of the 2024 annual ACM-SIAM symposium on discrete algorithms (SODA), pp 3765–3791

Leeuven J, Boas P-E (1978) Some elementary proofs of lower bounds in complexity theory. Linear Algebra Appl 19:63–80

Leeuwen J, Boas P-E (1978) Some elementary proofs of lower bounds in complexity theory. Linear Algebra Appl 19:63–80

Lickteig T (1984) A note on border rank. Inf Process Lett 18(3):173–178

Lotti G, Romani F (1983) On the asymptotic complexity of rectangular matrix multiplication. Theor Comput Sci 23(2):171–185

Makarov OM (1986a) An algorithm for multiplying 3×3 matrices. Comput Maths Math Phys 26(1):179–180

Makarov OM (1986b) A noncommutative algorithm for multiplying 5x5 matrices using 102 multiplications. Inf. Proc. Letters. 23:115–117

Makarov OM (1987) A non-commutative algorithm for multiplying 5×5 matrices using one hundred multiplications. Comput Maths Math Phys 27(1):205–207

Massarenti A, Raviolo E (2013) The rank of $n \times n$ matrix multiplication is at least. Linear Algebra Appl 438:4500–4509

Muraoka Y, Kuck DJ (1973) On the time required for a sequence of matrix products. Commun ACM 16(1):22–26

Pan V (1978) Strassen's algorithm is not optimal—trilinear technique of aggregating uniting and canceling for constructing fast algorithms for matrix operations. In: Proceedings of the 19th annual symposium on foundations of computer science, Ann Arbor, Michigan, pp 166–176

Pan V (1980) New fast algorithms for matrix operations. SIAM J Comput 9(2):321–342

Pan V (1982) Trilinear aggregating with implicit canceling for a new acceleration of matrix multiplication. Comput Math Appl 8(1):23–34
Pan V (1984) How to multiply matrices faster. Lecture notes in computer science, vol 179. Springer, New York
Peserico N, Shastri B, Sorger V (2023) Integrated photonic tensor processing unit for a matrix multiply: a review. J Lightwave Technol 41(12):3704–3716
Romani F (1982) Some properties of disjoint sums of tensors related to matrix multiplication. SIAM J Comput 11:263–267
Rosowski A (2023) Fast commutative matrix algorithms. J Symb Comput 114:302–321
Schachtel (1978) A noncommutative algorithm for multiplying 5×5 matrices using 103 multiplications. Inf Process Lett 7(4):180–182
Schonhage A (1981) Partial and total matrix multiplication. SIAM J Comput 10(3):434–455
Schwartz O, Weiss E (2019) Revisiting "Computation of matrix chain products." SIAM J Comput 48(5):1481–1486
Sedoglavic A (2017a) A non-commutative algorithm for multiplying (7×7) matrices using 250 multiplications. arXiv:1712.07935
Sedoglavic A (2017b) A non-commutative algorithm for multiplying 5×5 matrices using 99 multiplications. arXiv:1707.06860
Sedoglavic A, Smirnov AV (2021) The tensor rank of 5x5 matrices multiplication is bounded by 98 and its border rank by 89. In: Proceedings of the 2021 international symposium on on symbolic and algebraic computation, pp 345–351
Smirnov AV (2013) The bilinear complexity and practical algorithms for matrix multiplication. Comput Math Math Phys 53(12):1781–1795
Strassen V (1969) Gaussian elimination is not optimal. Numer Math 13:354–356
Strassen V (1986) The asymptotic spectrum of tensors and the exponent of matrix multiplication. In: Proceedings of the 27th annual IEEE symposium on foundations of computer science, pp 49–54
Strassen V (1987) Relative bilinear complexity and matrix multiplication. Journal Für Die Reine und Angewandte Mathematik 374:406–443
Sykora O (1977) A fast non-commutative algorithm for matrix multiplication. In: Gruska J (ed) Mathematical foundations of computer science, vol 1977. MFCS 1977. Lecture notes in computer science, vol 53. Springer, Berlin, Heidelberg
Tamura PN, Wyant JC (1979) Two-dimensional matrix multiplication using coherent optical techniques. Opt Eng 18(2):198–204
Vassilevska-Williams V (2012) Multiplying matrices faster than Coppersmith-Winograd. In: Proceedings of the 44th symposium on theory of computing. ACM, pp 887–898
Veit E (2016) A network that learns strassen multiplication. J Mach Learn Res 17(2016):1–13
Winograd S (1968) A new algorithm for inner product. IEEE Trans Comput 7:693–694

Chapter 9
Applications of the Fast Matrix Multiplication Algorithms

The approach of shifting the computational burden of various problems to matrix multiplication is known in the literature as black-box algorithms. A common goal of this paradigm is to construct an algorithm with the same time complexity as the matrix multiplication algorithm involved. This is often feasible, but for certain problems we get a complexity that is a function of the matrix multiplication complexity, usually with an additional logarithmic or linear factor, which is not a bad achievement either. Throughout the chapter, $O(M(n))$ stands for the time of the matrix multiplication algorithm involved, which is the reference point for the efficiency of the algorithms discussed in this chapter.

In this chapter we have collected algorithms that exploit the efficiency of fast matrix multiplication algorithms in other mathematical and computer science operations. In particular, these are matrix and other linear algebra operations, such as:

- Matrix multiplication by squaring a block matrix—Sect. 9.1.
- Inverting of triangular matrices—Sect. 9.3.
- Inverting arbitrary square matrices—Sect. 9.4. Some pre-processing is required here to ensure that the correct decomposition works in every case, at every recursion level.
- LU decomposition of a rectangular matrix—Sect. 9.5. We presented in detail the algorithms for matrices with full rank and surveyed work for matrices without this property.
- Solving systems of linear equations—Sect. 9.6. We show two methods.
- Finding the determinant of a matrix—Sect. 9.7. We show two methods. The first one gives the exact value, while the second one gives the absolute value of the determinant.
- Finding a value of a matrix polynomial—Sect. 9.9.
- Finding the characteristic polynomial of a matrix—Sect. 9.10. We give three algorithms of varying efficiency and generality.
- Finding successive powers of matrices—Sect. 9.11.

J. S. Respondek, *Fast Matrix Multiplication with Applications*, Studies in Big Data 166,
https://doi.org/10.1007/978-3-031-76930-6_9

Other operations of this type are described in Sect. 9.12. Surprisingly, efficient matrix multiplication can also be applied to problems in other domains, such as:

- Finding a series of polynomials at once—Sect. 9.8.1.
- Speed up the calculation of a single polynomial in a series of points—Sect. 9.8.2.

Other less common problems are listed in the Sect. 9.12. By the way, the matrix multiplication itself can also be done by inversion. We show proper construction in Sect. 9.2.

The efficiency goal mentioned in previous chapters is achieved in most matrix algorithms by using recursion as the main tool. The recursive method of algorithm construction is a very popular and powerful tool in classical computer science.

In the field of numerical algorithms, however, the meaning of the word 'recursive' is slightly different. It is understood as a series of successive calculations, each based on the results of the previous iteration. In matrix linear algebra, this is usually done by recursive formulas that express new values of the elements of the given row or column of the matrix, based on some range of previous values of the elements, rows or columns, of the given matrix.

Nevertheless, there are known matrix algorithms in the numerical algebra literature that are built using only the purely recursive paradigm, as in classical computer science algorithms. These algorithms have been presented in previous chapters, with the exception of Chap. 2, which is devoted to commutative multiplication.

Thus, another goal of this chapter is to collect—in our opinion—the most interesting recursive matrix algorithms available in the literature, which at the same time are rarely (if ever) discussed in numerical recipe monographs and academic lectures.

9.1 Matrix Multiplication by Squaring

Article Bodrato (2010) showed how to use a fast algorithm for squaring a matrix to find the matrix product. The block matrix formula presented in Sect. 2.3, page 274, makes everything clear:

$$\begin{bmatrix} 0 & A \\ B & 0 \end{bmatrix}^2 = \begin{bmatrix} AB & 0 \\ 0 & BA \end{bmatrix}$$

It should be noted that, using Bodrato's formula, we obtain both products AB and BA at the same time. Another block equality of this kind is proposed by Kalorkoti (1988) page 284, by double squaring:

$$\begin{bmatrix} 0 & A & 0 \\ 0 & 0 & B \\ I & 0 & 0 \end{bmatrix}^4 = \begin{bmatrix} 0 & ABA & 0 \\ 0 & 0 & BAB \\ AB & 0 & 0 \end{bmatrix}$$

9.2 Matrix Multiplication by the Inversion

This section goes a little beyond the main theme of the chapter, but we have included these results for completeness. Article Munro (1973) or Bunch and Hopcroft (1974), page 236, presented how to obtain the matrix product by the matrix inversion. Let A, B be the matrices of the $n \times n$ sizes and we will calculate their product. To do this, Munro cleverly suggested using block matrix algebra. Let us define the D matrix of the $3n \times 3n$ size as follows, and calculate its inversion:

$$D = \left[\begin{array}{c|c|c} I_n & A & 0 \\ \hline 0 & I_n & B \\ \hline 0 & 0 & I_n \end{array}\right], \quad D^{-1} = \left[\begin{array}{c|c|c} I_n & -A & AB \\ \hline 0 & I_n & -B \\ \hline 0 & 0 & I_n \end{array}\right]$$

So to calculate the product AB we can take the top right third of the D^{-1} matrix. Using the matrix inversion algorithm with the time complexity of the $O(n^\omega)$ algorithm, we obtain the efficiency of this method equal to $O((3n)^\omega) = O(3^\omega n^\omega) = O(n^\omega)$. Thus, the obtained efficiency of the multiplication is the same as in the applied inversion algorithm.

Another block equality of this kind is proposed by Kalorkoti (1988) page 281:

$$\begin{bmatrix} \mathbf{0} & & & I \\ & & I & C \\ & I & B & \\ I & A & & \mathbf{0} \end{bmatrix}^{-1} = \begin{bmatrix} -ABC & AB & -A & I \\ BC & -B & I & \\ -C & I & & \\ I & & & \mathbf{0} \end{bmatrix}$$

9.2.1 Example

Let us calculate by the inversion the product of the following matrices A, B:

$$A = \begin{bmatrix} 2 & -3 & 1 \\ 0 & -3 & 7 \\ 9 & 4 & 5 \end{bmatrix}, \quad B = \begin{bmatrix} 0 & 2 & 8 \\ 3 & 5 & 4 \\ 1 & -3 & 5 \end{bmatrix}$$

We use the block matrix formula for D^{-1}. Notice:

$$D = \left[\begin{array}{c|c|c} I_3 & \left.\begin{matrix} 2 & -3 & 1 \\ 0 & -3 & 7 \\ 9 & 4 & 5 \end{matrix}\right|_A & \mathbf{0} \\ \hline \mathbf{0} & I_3 & \left.\begin{matrix} 0 & 2 & 8 \\ 3 & 5 & 4 \\ 1 & -3 & 5 \end{matrix}\right|_B \\ \hline \mathbf{0} & \mathbf{0} & I_3 \end{array}\right]_{9\times 9},$$

$$D^{-1} = \left[\begin{array}{c|c|c} I_3 & \left.\begin{matrix} -2 & 3 & -1 \\ 0 & 3 & -7 \\ -9 & -4 & -5 \end{matrix}\right|_{-A} & \left.\begin{matrix} -8 & -14 & 9 \\ -2 & -36 & 23 \\ 17 & 23 & 113 \end{matrix}\right|_{AB} \\ \hline \mathbf{0} & I_3 & \left.\begin{matrix} 0 & -2 & -8 \\ -3 & -5 & -4 \\ -1 & 3 & -5 \end{matrix}\right|_{-B} \\ \hline \mathbf{0} & \mathbf{0} & I_3 \end{array}\right]_{9\times 9}$$

As we can see, the upper right third of the D^{-1} matrix is equal to AB.

9.3 Recursive Inversion of Triangular Matrices

Suppose we want to invert the following upper triangular block matrix of size $n \times n$, which is divided into four submatrices, each of size $(n/2) \times (n/2)$. We can check that the following identities are true:

$$A = \left[\begin{array}{c|c} A_{11} & A_{12} \\ \hline \mathbf{0} & A_{22} \end{array}\right], \quad A^{-1} = \left[\begin{array}{c|c} A_{11}^{-1} & -A_{11}^{-1}A_{12}A_{22}^{-1} \\ \hline \mathbf{0} & A_{22}^{-1} \end{array}\right] \tag{9.1}$$

For the downward triangular matrix, the analysis is analogous. For the block triangular matrix, its determinant is the product of the diagonal submatrices, (e.g. Gantmacher 1960, Vol. I, formula 67, page 43 or Scott 1880, page 51) i.e.:

$$\det\left|diag[A_{11},\ A_{22},\ ...,A_{pp}]\right| = \prod_{i=1}^{p} \det|A_{ii}| \tag{9.2}$$

The determinant of a non-singular matrix is not equal to zero. Thus from the equality (9.2) it follows that each of the A.'s diagonal sub-matrices is non-singular.

Now let us analyse the time complexity of inverting the triangular matrices by the equality (9.1). Suppose n is a natural power of two. At each recursion level we have to perform the following operations: two inversions of the $(n/2) \times (n/2)$ matrices, two $(n/2) \times (n/2)$ matrix multiplications and $n^2/4$ negations of the sign in the right upper quarter. So the recursive complexity equation has the form:

$$T(n) = 2T\left(\frac{n}{2}\right) + 2M\left(\frac{n}{2}\right) + \frac{n^2}{4} \tag{9.3}$$

where $T(n)$ is the unknown time complexity of the triangular matrix inversion and $M(n)$ is the complexity of the matrix multiplication algorithm involved. Considering that $M(n) = O(n^{2+\varepsilon})$ for a given $\varepsilon > 0$ we can estimate $n^2/4 \le M(n)$ and from the Eq. (9.3) it follows that:

$$T(n) \le 2\left(\frac{n}{2}\right) + 3M\left(\frac{n}{2}\right)$$

Since $M(n) = O(n^{2+\varepsilon})$, in the general recursion theorem (Cormen et al. 2001, Sect. 4.3) we have the case where the $f(n) = 3M(n/2)$ decides on the overall algorithm complexity and $T(n) = O(M(n))$.

This means that the triangular matrix algorithm, resulting from the equality (9.1) has the same time complexity as the matrix multiplication algorithm involved.

An attentive reader may ask what if the size of the matrix to be inverted is not a natural two's power. In this case, we expand the matrix by the identity matrix to the size of the next power of two, building the block matrix with the properties given by the formula:

$$\begin{bmatrix} A & 0 \\ 0 & I \end{bmatrix}^{-1} = \begin{bmatrix} A^{-1} & 0 \\ 0 & I^{-1} \end{bmatrix} = \begin{bmatrix} A^{-1} & 0 \\ 0 & I \end{bmatrix}$$

It is therefore sufficient to invert the larger matrix and to pick the principal minor of the size $n \times n$. The dimension of the extended matrix is less than twice that of the input matrix. This preliminary operation will slow down the final algorithm by a constant coefficient less than $2^k \le 8$, so this step will not affect the final time efficiency of the algorithm.

9.3.1 Example—Inversion of a Triangular Matrix

Let us calculate the inversion of the following 4×4 upper triangular matrix by the recursion:

$$A = \left[\begin{array}{cc|cc} 3 & 0 & 0 & 2 \\ 0 & 7 & 1 & 5/3 \\ \hline & & 2/7 & -53/21 \\ \multicolumn{2}{c|}{\mathbf{0}} & 0 & 119/6 \end{array}\right]_{4\times 4} = \left[\begin{array}{c|c} A_{11(1)} & A_{12(1)} \\ \hline \mathbf{0} & A_{22(1)} \end{array}\right]$$

In this section, the lower index parentheses of the form (q) we denote the q-th level of the recursion. According to formula (9.1), we need the inversions of the diagonal submatrices, i.e. $A_{11(1)}$ and $A_{22(1)}$. We show how the second level of recursion works using the matrix $A_{11(1)}$. Observe:

$$A_{11(1)} = \begin{bmatrix} 3 & 0 \\ 0 & 7 \end{bmatrix} = \left[\begin{array}{c|c} A_{11(2)} & A_{12(2)} \\ \hline \mathbf{0} & A_{22(2)} \end{array}\right]$$

$$A_{11(2)}^{-1} = \frac{1}{3}, \quad A_{2(2)}^{-1} = \frac{1}{7}, \quad -A_{11(2)}^{-1} A_{12(2)} A_{22(2)}^{-1} = 0, \quad A_{11(1)}^{-1} = \begin{bmatrix} 1/3 & 0 \\ 0 & 1/7 \end{bmatrix}$$

By analogy, we calculate $A_{22(1)}^{-1}$ and the searched inversion is:

$$A^{-1} = \left[\begin{array}{c|c} A_{11(1)}^{-1} & -A_{11(1)}^{-1} A_{12(1)} A_{22(1)}^{-1} \\ \hline \mathbf{0} & A_{22(1)}^{-1} \end{array}\right] = \left[\begin{array}{cc|cc} 1/3 & 0 & 0 & -4/119 \\ 0 & 1/7 & -1/2 & -9/119 \\ \hline & & 7/2 & 53/53 \\ \multicolumn{2}{c|}{\mathbf{0}} & 0 & 6/119 \end{array}\right]_{4\times 4}$$

In the next section we present an algorithm to invert arbitrary form matrices.

9.4 Recursive Inversion of Arbitrary Matrices

In this section we show how to invert any invertible matrix using matrix multiplication with the same efficiency as the multiplication algorithm used, i.e. with the same ω exponent in the $O(n^{\omega})$ complexity.

A method for this task is outlined in Bunch and Hopcroft (1974), Sect. 1. However, it is shown there to justify the necessity of using the LUP decomposition, since in the general case the decomposition shown there does not ensure that at each recursion level the matrices to be inverted are non-singular. A method to overcome this problem, without using the LUP decomposition proposed by Bunch and Hopcroft, was presented by Pan (1984) in Sect. 11, pages 413–414.

Let us assume that the matrix to be inverted is symmetric and positive definite, in the 2×2 block form. We can decompose it into the product of three matrices, block triangular and block diagonal[1]:

$$A = \begin{bmatrix} B & C^T \\ C & D \end{bmatrix} = \begin{bmatrix} I & 0 \\ CB^{-1} & I \end{bmatrix} \begin{bmatrix} B & 0 \\ 0 & S \end{bmatrix} \begin{bmatrix} I & B^{-1}C^T \\ 0 & I \end{bmatrix} \tag{9.4}$$

where $S = D - CB^{-1}C^T$ is the so-called Schur residual of the matrix A with respect to B. Now it is easy to check that the inversion of A has the form:

[1] Originally Pan (1984) gave this identity a in a non-symmetric version, then referred to it in the symmetric form of A. Here we use it directly in the symmetric form.

$$A^{-1} = \begin{bmatrix} I & -B^{-1}C^T \\ 0 & I \end{bmatrix} \begin{bmatrix} B^{-1} & 0 \\ 0 & S^{-1} \end{bmatrix} \begin{bmatrix} I & 0 \\ -CB^{-1} & I \end{bmatrix} =$$

$$= \begin{bmatrix} B^{-1} + B^{-1}C^T S^{-1} C B^{-1} & -B^{-1}C^T S^{-1} \\ -S^{-1}CB^{-1} & S^{-1} \end{bmatrix} \tag{9.5}$$

In fact, it is sufficient to apply the property (1.4) to the decomposition (9.4) and to use the fact that any function of the block diagonal matrix is the block diagonal matrix built up by diagonal blocks with the respective function applied to each of them.

So the inversion of the matrix can be done recursively by four matrix multiplications in the following order:

$$C \cdot B^{-1} \rightarrow (CB^{-1}) \cdot C^T \rightarrow S^{-1} \cdot (CB^{-1}) \rightarrow (C \cdot B^{-1})^T \cdot (S^{-1}CB^{-1})$$

and two inversions of the half-size matrices B and S. Since every symmetric and positive matrix is invertible, and every principal minor of such a matrix has these properties, there exists an inversion of B.

Proof of the Schur residual invertibility is more extensive. By the assumption the A matrix is positive definite, then $\det(A) > 0$. Applying property (1.5) to the matrix product (9.4) we obtain:

$$\det(A) = \det\begin{bmatrix} I & 0 \\ CB^{-1} & I \end{bmatrix} \det\begin{bmatrix} B & 0 \\ 0 & S \end{bmatrix} \det\begin{bmatrix} I & B^{-1}C^T \\ 0 & I \end{bmatrix} \tag{9.6}$$

The determinant of block triangular matrices is calculated again by the method of formula 67 on page 43 of the classical monograph Gantmacher (1960) Vol. I or even from the (much) older book Scott (1880), page 51:

$$\det\begin{bmatrix} I & 0 \\ CB^{-1} & I \end{bmatrix} = \det\begin{bmatrix} I & B^{-1}C^T \\ 0 & I \end{bmatrix} = \det(I) \cdot \det(I) = 1$$

So from equality (9.6) follows that:

$$\det(A) > 0 \quad \Leftrightarrow \quad \det\begin{bmatrix} B & 0 \\ 0 & S \end{bmatrix} > 0$$

Moreover:

$$\det\begin{bmatrix} B & 0 \\ 0 & S \end{bmatrix} > 0 \quad \Leftrightarrow \quad \det(B) \cdot \det(S) > 0 \tag{9.7}$$

By the assumption that the matrix A is positive defined, B is also [block partition (9.4)]. Thus:

$$\det(B)|_{B>0} \cdot \det(S) > 0 \quad \Leftrightarrow \quad \det(S) > 0 \tag{9.8}$$

In summary, the Schur residual is symmetric and positive, so it is also invertible. For a comprehensive monograph on Schur matrices, with over three hundred pages, see Zhang (2005).

But there is still the question of how to invert any non-singular matrix, since the formula (9.4) works in the case of symmetric and positive matrices to be inverted. The universal solution is given by Pan (1984), remark 11.2 there. By the very definition of the inverse matrix we can write:

$$\left(A^T A\right)^{-1}\left(A^T A\right) = I$$

Next, we multiply both sides by A^{-1}. We can obtain the desired inverse from the formula:

$$A^{-1} = \left(A^T A\right)^{-1} A^T \tag{9.9}$$

In summary, instead of inverting A^{-1} Pan (1984) proposes to invert an auxiliary matrix $A^T A$ that is symmetric and positive for any non-singular matrix. Next, we obtain the desired inversion from formula (9.9).

The last issue is the time complexity of the proposed algorithm. From the formula (9.5) follows the corresponding recursive equation:

$$I(n) \le 2I\left(\frac{n}{2}\right) + 4M\left(\frac{n}{2}\right) + O(n^2)$$

where $2I(n/2)$ is responsible for 2 inversions, $4M(n)$ is responsible for 4 multiplications and the last term $O(n^2)$ encapsulates all the necessary additions/subtractions and transpositions. Furthermore, $O(M(n)) > O(n^2)$ so we can rewrite the complexity inequality in the form:

$$I(n) \le 2I\left(\frac{n}{2}\right) + O(M(n))$$

From the general recursion theorem (Cormen et al. 2001, Sect. 4.3) we have $T(n) = I(M(n))$. Another, more theoretical, proof that matrix inversion can be performed in the same time as applied matrix multiplication is given by Corollary 8 in the article Winograd (1970).

9.4.1 Example—Inversion of Arbitrary Form Matrix

Let us compute the inversion of the 3×3 matrix A_0, expanded to 4×4 size in A:

$$A_0 = \begin{bmatrix} 2 & 1 & 0 \\ 6 & 3 & 1 \\ -3 & 0 & 2 \end{bmatrix}_{3\times 3}, \quad A = \left[\begin{array}{c|c} A_0 & 0 \\ \hline 0 & I_1 \end{array}\right]$$

We calculate the desired inverse in the following steps:

- calculate the $A^T A$ auxiliary matrix—to obtain a symmetric and positive one,
- invert the $A^T A$ matrix,
- transpose A, then multiply $\left(A^T A\right)^{-1} A^T$,
- pick the result as the 3×3 size principal minor of the $A^{-1} = \left(A^T A\right)^{-1} A^T$ matrix.

Observe[2]:

$$A^T A = \left[\begin{array}{cc|cc} 49 & 20 & 0 & 0 \\ 20 & 10 & 3 & 0 \\ \hline 0 & 3 & 5 & 0 \\ 0 & 0 & 0 & 1 \end{array}\right]_{4\times 4} = \left[\begin{array}{c|c} B_{(1)} & C_{(1)}^T \\ \hline C_{(1)} & D_{(1)} \end{array}\right], \quad B_{(1)}^{-1} = \begin{bmatrix} 1/9 & -2/9 \\ -2/9 & 49/90 \end{bmatrix}$$

$$S_{(1)} = D_{(1)} - C_{(1)} B_{(1)}^{-1} C_{(1)}^T = \begin{bmatrix} 1/10 & 0 \\ 0 & 1 \end{bmatrix} = \left[\begin{array}{c|c} B_{(2)} & C_{(2)}^T \\ \hline C_{(2)} & D_{(2)} \end{array}\right]$$

At the second level of recursion, we need to invert two 2×2 matrices $B_{(1)}$ and $S_{(1)}$. The recursive algorithm on this recursion level is as follows, using the latter as an example. We calculate the inverse using the general formula (9.5). The antidiagonal elements of $S_{(1)}^{-1}$ are zero, because $C_{(2)} = 0$ and it is reduced to the form:

$$S_{(1)}^{-1} = \begin{bmatrix} B_{(2)}^{-1} & 0 \\ 0 & B_{(2)}^{-1} \end{bmatrix} = \begin{bmatrix} 10 & 0 \\ 0 & 1 \end{bmatrix}$$

The auxiliary inversion $\left(A^T A\right)^{-1}$ according to the formula (9.5) is equal to:

$$\left(A^T A\right)^{-1} = \frac{1}{9} \left[\begin{array}{cc|cc} 41 & -100 & 60 & 0 \\ -100 & 245 & -147 & 0 \\ \hline 60 & -147 & 90 & 0 \\ 0 & 0 & 0 & 9 \end{array}\right]_{4\times 4}$$

And the final inversion of A_0 is equal to:

$$A^{-1} = \left(A^T A\right)^{-1} A^T = \left[\begin{array}{ccc|c} -2 & 2/3 & -1/3 & \\ 5 & -4/4 & & \mathbf{0}_{3\times 1} \\ -3 & 1 & 0 & \\ \hline & \mathbf{0}_{1\times 3} & & 1 \end{array}\right]_{4\times 4} = \left[\begin{array}{c|c} A_0^{-1} & \mathbf{0} \\ \hline \mathbf{0} & I_1 \end{array}\right]$$

[2] Again, the $(\cdot)$ subscripts denote the recursion level.

$$A = \begin{bmatrix} B \\ \cdots\cdots \\ C \end{bmatrix}$$

Fig. 9.1 The initial partition of the input matrix for the LUP decomposition algorithm

In summary, in this section we have presented a black-box recursive algorithm for inverting a non-singular matrix and proved that it performs no worse in time than the applied multiplication algorithm.

9.5 Recursive LU Decomposition of Matrices

This section shows how to decompose a non-singular **square** matrix, where the dimension is a natural power of two.

To this end, we present an algorithm to decompose a $m \times n$ matrix **with a full row rank**[3] into the LU product form:

$$A = LUP$$

where L is a $m \times m$ lower-triangular matrix with ones on the main diagonal, U is a $m \times n$ rectangular matrix with leading m columns with the rank m,[4] of the form:

$$U = \left[\, U_0 \middle| U_{res} \,\right]$$

where U_0 is a square upper triangular matrix and P is a $n \times n$ permutation matrix. The dimensions satisfy the inequality $m \leq n$ and both m and n are natural powers of two.

Similarly to other sections, our goal is to yield from the efficient matrix multiplication, obtaining algorithms of the same efficiency, measured by the big "O" notation, as that multiplication. In particular, we use the algorithm for inverting the triangular matrix from Sect. 9.3, this way indirectly using the "base" black-box matrix multiplication.

An adequate algorithm is presented and theoretically analysed in the article by Bunch and Hopcroft (1974). We also refer to the more practical argument from Sect. 6.4 of Aho et al. (1974), although. We proceed recursively as follows:

- For $m = 1$ we do the following:

[3] At the end of this section, we reviewed papers that relax this assumption.

[4] From the scalar case of the equality it follows that it is equivalent to the requirement that the main diagonal elements of the submatrix U_0, constructed from the most $m \times m$ size left elements of the U matrix, are non-zero.

Fig. 9.2 The LUP decomposition process after the first recursive call

$$A = \begin{bmatrix} L_1 & \mathbf{0} \\ \mathbf{0} & I_{m/2} \end{bmatrix} \cdot \begin{bmatrix} U_1 \\ D \end{bmatrix} \cdot P_1, \quad \begin{cases} U_1 = \begin{bmatrix} E & U_{1res} \end{bmatrix} \\ D = \begin{bmatrix} F & D_{res} \end{bmatrix} \end{cases}$$

- Assign $L = [1]$.
- Fix c as the number of the first non-zero position of A matrix.
- Build P permutation matrix that swaps the 1 and c A's columns.
- Assign $U = AP$.

- If $m > 1$ partition the input matrix A into $(m/2) \times n$-dimensional upper B and down C submatrices as depicted on Fig. 9.1.
- Repeat the algorithm recursively to decompose the upper matrix $B = L_1 U_1 P_1$.
- Permute the C matrix according to the formula.[5]

$$D = C \cdot P_1^{-1} \tag{9.10}$$

We receive form depicted on Fig. 9.2:

where E is the square upper triangular matrix, $\dim L_1 = \dim E = \dim F = (m/2) \times (m/2)$, $\dim U_{1\,res} = \dim D_{res} = (m/2) \times (n - m/2)$ and $\dim P_1 = n \times n$ (Fig. 9.2).

- Calculate the auxiliary matrix $G = D - FE^{-1}U_1$. We obtain form depicted on Fig. 9.3.
- Repeat the algorithm recursively for the decomposition $G' = L_2 U_2 P_2$
- Calculate the auxiliary permutation matrix P_3 and permute the matrix U_1 by it:

$$\begin{cases} P_3 = diag[I_{m/2}; P_2] \\ H = U_1 \cdot P_3^{-1} \end{cases}$$

- Finally, we obtain the decomposition depicted on Fig. 9.4.

Note that the algorithm calls itself recursively twice. Similarly to the other recursive matrix algorithms, if the size of the matrix is not a natural power of two, we extend the matrix by the identity matrix to the nearest power of two.

To verify the respective decompositions we present the crucial part of the algorithm, i.e. setting the left part, $(m/2) \times (m/2)$, of the D matrix to zero. Observe:

$$\begin{aligned} [F|D_{res}] &= \left[F|FE^{-1}U_{1res}\right] + \left[\mathbf{0}_{(m/2)\times(m/2)}|D_{res} - FE^{-1}U_{1res}\right] \\ &= FE^{-1}[E|U_{1res}] + I_{m/2}\left[\mathbf{0}_{(m/2)\times(m/2)}|G'\right] \end{aligned}$$

Another important tool used in the algorithm is the matrix permutation. In the operation $D = C \cdot P_1^{-1}$ we are permuting the C matrix by the inverse P_1. The

[5] For the permutation matrix holds $P^{-1} = P^T$.

aim of this step is to rearrange the C matrix in a way that allows to combine it in the common block matrix product after the triangular decomposition of B. Indeed, $DP_1 = CP_1^{-1}P_1 = C$. The same idea is used in the operation $H = U_1 \cdot P_3^{-1}$, where the order of the G' columns is also changed after their triangular decomposition $G' = L_2U_2P_2$.

To prove the correctness of the algorithm, we need to show that the decomposition assumptions are satisfied both for its output parameters and for the two input matrices $B,\ G'$ the algorithm recursively calls. Aho et al. (1974), Sect. 6.4, propose an inductive proof here.

- The base induction level is guaranteed by the algorithm for $m = 1$.
- Let us assume that the output variables of the algorithm fulfil the assumptions, for an input matrix size with a certain $m = 2^k,\ \text{for } k \geq 0,\ k \in Z$.
- It is to be proved that these conditions hold true for the input matrix size for $2m = 2^{k+1}$.
  - We call the algorithm with the input matrix A of the row rank equal to $2m$. It means that each subset of the A matrix rows is linearly independent. In particular, the matrix B has the rank m. This ensures that the $B = L_1U_1P_1$ decomposition has a correct input parameter.
  - By the base induction step the first m columns of the matrix U_1 also have the rank m, so the E matrix is invertible.
  - From the rank inequality (1.6) we can conclude that the rank of each matrix depicted in Fig. 9.3 is equal to $2m$. Thus the row rank of G' is m and the second decomposition $G' = L_2U_2P_2$ is guaranteed to work.
  - By the inductive assumption both the L_1, L_2 matrices are lower-triangular matrices with 1's on the main diagonal. Thus the L matrix, defined on the Fig. 9.4, preserves these properties.
  - Now it remains to prove that the U matrix, also defined on Fig. 9.4, is an upper-triangular matrix with leading $2m$ columns of rank $2m$. These leading $2m$ columns of the U matrix form a triangular matrix, built block-diagonally by the leading m columns of two matrices: H and U_2. The latter is the result of the $G = D - FE^{-1}U_1$ operation, which we proved above, that works correctly, thus produces an U_2 matrix with the leading columns rank equal to m.

$$A = \begin{bmatrix} L_1 & 0 \\ FE^{-1} & I_{m/2} \end{bmatrix} \cdot \begin{bmatrix} E & U_{1res} \\ 0 & G' \end{bmatrix} \cdot P_1$$

Fig. 9.3 The LUP decomposition process before the second recursive call

$$L = \begin{bmatrix} L_1 & 0 \\ FE^{-1} & L_2 \end{bmatrix}, \quad U = \begin{bmatrix} H \\ 0 \;\; U_2 \end{bmatrix}, \qquad P = P_3 \cdot P_1$$

Fig. 9.4 The result of the LUP decomposition algorithm

The case of the H matrix is a little more complicated. According to $H = U_1 \cdot P_3^{-1}$ it consists of the permuted columns of the U_1 matrix, returned by the $B = L_1 U_1 P_1$ decomposition. That permutation is defined by the matrix P_3. The U_1 matrix by the inductive assumption has a full row rank for its leading columns. The construction of P_3 reveals that those leading columns of the matrix U_1 are not affected by this permutation, since the first $m \times m$ diagonal block P_3 is just an identity matrix.

The leading m columns of the matrices H and U_2 are triangular, so by the determinant property (9.2) all their diagonal entries are non-zero. Thus the determinant of the leading $2m$ columns of the U matrix is also non-zero, so the U matrix has a full row rank.

■

The complexity of the algorithm is analysed in the article Bunch and Hopcroft (1974), originally without using the general recursion theorem. Let $T(m, n)$ denote the execution time of the algorithm.

- Recursive decompositions $B = L_1 U_1 P_1$ and $G' = L_2 U_2 P_2$ take $T(m/2, n)$ and $T(m/2, n - m/2)$ steps, respectively.
- In operations $D = C \cdot P_1^{-1}$ and $H = U_1 \cdot P_3^{-1}$ the inversions of $P_.$ take $O(n)$ steps, but the permutations of the $m \times n$ matrices C and U_1 dominate. So the complexity of those operations is $O(mn)$.
- The calculation of E^{-1} can (and should) be done by the inverting algorithm specialised in triangular matrices, presented in Sect. 9.3, in time $M(m/2)$. The calculation of the term FE^{-1} also takes $M(m/2)$ time.
- The calculation of the term $(FE^{-1}) \cdot U_1$ requires more attention because of the rectangular U_1 shape of the size $(m/2) \times n$. We should remember that both m and n are the natural powers of 2 and $m \leq n$.[6] So, to obtain the product $(FE^{-1}) \cdot U_1$, we can smartly use the block matrix algebra. Observe[7]:

$$\underbrace{\left[FE^{-1}\right]}_{m/2} \cdot \underbrace{\left[U_1^{(1)}\ U_1^{(2)}\ \cdots\ U_1^{n/(m/2)}\right]}_{n/(m/2)\ \ times}$$
$$= \underbrace{\left[FE^{-1}U_1^{(1)}\ FE^{-1}U_1^{(2)}\ \cdots\ FE^{-1}U_1^{n/(m/2)}\right]}_{n} \tag{9.11}$$

According to the above scheme we can calculate the term $(FE^{-1}) \cdot U_1$ in $n/(m/2)$ multiplications of $(m/2) \times (m/2)$ size submatrices and the total time of that operation is $O((n/m)M(m/2))$.

Based on the above analysis we can construct the recursive inequality of the complexity for the LU decomposition algorithm with respect to variables m, n, with certain positive integer constants b, c and d:

---

[6] This is ensured by the first call of the algorithm with a non-singular square matrix as argument.
[7] $\dim[U_1^{(\cdot)}] = (m/2) \times (m/2)$

$$T(m,n) \le \begin{cases} T\left(\frac{m}{2}, n\right) + T\left(\frac{m}{2}, n - \frac{m}{2}\right) + c\frac{n}{m}M\left(\frac{m}{2}\right) + dmn, & m > 1 \\ bn, & m = 1 \end{cases} \tag{9.12}$$

The second recursive call of the algorithm $G' = L_2U_2P_2$ is performed on the submatrix of size $(m/2) \times (n - m/2)$, which is smaller than the recursive call $B = L_1U_1P_1$, performed on the submatrix of size $(m/2) \times n$ dimensions. Importantly, the algorithm does not use conditional statements.[8] All its subroutines are shown on the dependency graph Fig. 9.7. Thus, the execution time of the algorithm is shorter when the dimensionality of the problem to be solved is smaller. On this basis, we can claim the inequality $T(m/2, n - m/2) < T(m/2, n)$.

The fourth factor in formula (9.12) can be rewritten as $4d\frac{n}{m} \cdot \left(\frac{m}{2}\right)^2$. For the matrix multiplication we have the inequality $M\left(\frac{m}{2}\right) > \left(\frac{m}{2}\right)^2$, so we can merge the third and the fourth factors in formula (9.12) and obtain the recursive inequality, for a certain e constant:

$$T(m,n) \le \begin{cases} 2T\left(\frac{m}{2}, n\right) + e\frac{n}{m}M\left(\frac{m}{2}\right), & m > 1 \\ bn, & m = 1 \end{cases}$$

Now in the general recursion theorem, the order of the second term determines the overall complexity of the algorithm. Thus the final complexity of the algorithm, invoked for a rectangular matrix with a $m \times n$ size, is equal to:

$$T(m,n) = O\left(\frac{n}{m}M(m)\right)$$

To obtain the time complexity for the LU decomposition of a non-singular square matrix of a $n \times n$ size, it is enough to use state $m = n$:

$$T(n) = O(M(n))$$

To summarise, in this section we have presented an LU decomposition algorithm that works in time:

- equal—in the case of a square matrix,
- proportional—for a rectangular matrix, with a ratio equal to a quotient *columns*: *rows* of a decomposed matrix,

to the involved matrix multiplication algorithm. Later than Bunch and Hopcroft's 1974 article, works appeared that did not require the matrix to be decomposed to have full row rank. The first article in this series is Ibarra, Moran et al. (1982). To distinguish it, this article called the decomposition as LSP. It also contains algorithms for finding a matrix rank and the associated maximum size square minor. Another version of a mathematically equivalent algorithm is proposed in the article Jeannerod, Pernet et al. (2013). Its authors claim that their version is the most suitable for the

---

[8] Some authors such an algorithms calls as a straight-line ones.

purposes of calculation. Article Jeannerod, Mary et al. (2019) proposes an algorithm for LU decomposition of a square matrix that can be represented by low-rank blocks.

In the next subsection we illustrated how the LU decomposition algorithm works on a 8×8 size matrix, with the entries chosen in such a way as to demonstrate how all the internals of the LU algorithm work.

9.5.1 Example—Recursive LU Decomposition

Let us compute the LU decomposition of the 7×7 matrix A_0, extended to the 8×8 size in A, as shown below:

$$A_0 = \left[\begin{array}{ccccccc} 0 & 0 & 3 & 6 & 2 & 2 & 3 \\ 1 & 7 & 2 & -9 & 3 & 4 & -5 \\ 0 & -2 & 0 & 4 & -3 & 2 & 2 \\ 2 & 3 & -1 & -6 & 6 & 7 & 2 \\ \hline 6 & 6 & 3 & 3 & 0 & 3 & 0 \\ 2 & 5 & 2 & 7 & 2 & -1 & 2 \\ 3 & -5 & 2 & 2 & 0 & 3 & 3 \end{array}\right], \quad A = \left[\begin{array}{c|c} A_0 & \mathbf{0} \\ \hline \mathbf{0} & I_1 \end{array}\right] = \left[\begin{array}{c} \left[B_{(1\div 8)}\right]_{4\times 8} \\ \hline \left[C_{(1-8)}\right]_{4\times 8} \end{array}\right] \tag{9.13}$$

In order to make the example readable enough, it is important to design a neat notation of the respective auxiliary matrices used by the algorithm 9.5, which would clearly indicate to which recursion level a given matrix belongs. The analysis of the algorithm convinces us that each matrix used by the LU algorithm, at a given recursion level, can be uniquely specified by the range of **rows** to be decomposed.

In our example, the starting matrix is an 8×8 matrix A, which is divided in the algorithm into $B_{(1\div 8)}$ and $C_{(1\div 8)}$ submatrices,[9] as shown in formula (9.13).

Now let us analyse how the algorithm 9.5 deals with the LU decomposition of the A matrix (9.13). It immediately calls the algorithm 9.5 recursively to decompose the 4×8 $B_{(1\div 8)}$ matrix (9.13) and again the same step is called to decompose the $B_{(1\div 4)}$ matrix of the form:

$$B_{(1\div 4)} = \left[\begin{array}{cccccccc} 0 & 0 & 3 & 6 & 2 & 2 & 3 & 0 \\ \hline 1 & 7 & 2 & -9 & 3 & 4 & -5 & 0 \end{array}\right]_{2\times 8} = \left[\begin{array}{c} B_{(1\div 2)} \\ \hline C_{(1\div 2)} \end{array}\right] \tag{9.14}$$

For $B_{(1\div 2)}$ we have $m = 1$. The first non-zero column of $B_{(1\div 2)}$, which is a single-row matrix, is at position $c = 3$, thus its LU decomposition has the form:

[9] The submatrices $B_{(1\div 8)}$, $C_{(1\div 8)}$ are of dimensions 4×8, but according to the notation described at the beginning of this section example, they correspond to the rows $1 \div 8$ of the A matrix to be decomposed.

$$B_{(1\div 2)}=\left[\underset{\sim}{0}\;0\;\underset{\sim}{3}\;6\;2\;2\;3\;0\right]=\underbrace{[1]}_{L_{(1\div 2)}}\underbrace{\left[\underset{\sim}{3}\;0\;\underset{\sim}{0}\;6\;2\;2\;3\middle|0\right]}_{U_{(1\div 2)}}\underbrace{\left[\begin{array}{ccc|c}0&0&\underset{\sim}{1}&\\0&1&0&\mathbf{0}_{3\times 1}\\\underset{\sim}{1}&0&0&\\\hline\multicolumn{3}{c|}{\mathbf{0}_{1\times 3}}&I_5\end{array}\right]}_{P_{(1\div 2)}}\tag{9.15}$$

Operation (9.10) swaps the 1st column with the 3rd column of the $C_{(1\div 2)}$ matrix, the result is $D_{(1\div 2)}$. Now we decompose the $D_{(1\div 2)}$ matrix. Observe how the algorithm works in the case of a single row [as the parameter $U_{1(1\div 2)}$ we assign $U_{(1\div 2)}$ from (9.15)]:

$$\begin{aligned}
&E_{(1\div 2)}=[3],\quad F_{(1\div 2)}=[2],\quad F_{(1\div 2)}E^{-1}_{(1\div 2)}=[2/3]\\
&G_{(1\div 2)}=D_{(1\div 2)}-F_{(1\div 2)}E^{-1}_{(1\div 2)}U_{1(1\div 2)}=\left[2\;7\;1\;-9\;3\;4\;-5\;0\right]+\\
&\qquad -2/3\cdot\left[3\;0\;0\;6\;2\;2\;3\;0\right]=\left[0\middle|7\;1\;-13\;5/3\;8/3\;-7\;0\right]=\left[0\middle|G'_{(1\div 2)}\right]_{1\times 8}
\end{aligned}$$

In decomposition $G'=L_2U_2P_2$. we have the second recursive call of the algorithm on the rows $1\div 2$, but now with the matrix $G_{(1\div 2)'}$ as an input parameter. It is a row vector with non-zero first left element thus its LU decomposition is straightforward: $G_{(1\div 2)'}=[1]\cdot G_{(1\div 2)'}\cdot I_7$. Now we can construct the decomposition of the $B_{(1\div 4)}$ matrix (9.14):

$$B_{(1\div 4)}=\underbrace{\begin{bmatrix}1&0\\\frac{2}{3}&1\end{bmatrix}}_{L_{1(1\div 4)}}\underbrace{\left[\begin{array}{cc|ccccc|c}\underset{\sim}{3}&0&\underset{\sim}{0}&6&2&2&3&\\\underset{\sim}{0}&7&\underset{\sim}{1}&-13&5/3&8/3&-7&\mathbf{0}_{2\times 1}\end{array}\right]}_{U_{1(1\div 4)}}\underbrace{\left[\begin{array}{ccc|c}0&0&\underset{\sim}{1}&\\0&1&0&\mathbf{0}_{3\times 1}\\\underset{\sim}{1}&0&0&\\\hline\multicolumn{3}{c|}{\mathbf{0}_1\times \mathbf{3}}&I_5\end{array}\right]}_{P_{1(1\div 4)}}\tag{9.16}$$

Let us return to the decomposition of the matrix $B_{(1\div 8)}$. Observe:

$$D_{(1\div 4)}=C_{(1\div 4)}P^{-1}_{(1\div 4)}=\left[\begin{array}{cc|cccccc}\underset{\sim}{0}&-2&\underset{\sim}{0}&4&-3&2&2&0\\-\underset{\sim}{1}&3&\underset{\sim}{2}&-6&6&7&2&0\end{array}\right],\;E_{(1\div 4)}=\begin{bmatrix}3&0\\0&7\end{bmatrix}$$

$$F_{(1\div 4)}=\begin{bmatrix}0&-2\\-1&3\end{bmatrix},\quad F_{(1\div 4)}E^{-1}_{(1\div 4)}=\begin{bmatrix}3&0\\0&7\end{bmatrix}\begin{bmatrix}0&-2\\-1&3\end{bmatrix}^{-1}=\begin{bmatrix}0&-2/7\\-1/3&3/7\end{bmatrix}$$

Operation $G=D-FE^{-1}U_1$, which sets the left part of the $D_{(1\div 4)}$ matrix to zero, is as follows:

$$G_{(1\div 4)}=D_{(1\div 4)}-F_{(1\div 4)}E^{-1}_{(1\div 4)}U_{1(1\div 4)}=\left[\begin{array}{cc|cccccc}0&0&\frac{2}{7}&\frac{2}{7}&-\frac{53}{21}&\frac{58}{21}&0&0\\0&0&\frac{11}{7}&\frac{11}{7}&\frac{125}{21}&\frac{137}{21}&6&0\end{array}\right]=\left[\mathbf{0}_{2\times 2}\middle|G'_{(1\div 4)}\right]$$

The LU decomposition of the $G_{(1\div4)}'$ matrix is similar to the $B_{(1\div4)}$ matrix decomposition, since the level of the recursion in the algorithm 9.5 is determined by the number of rows operated on, which is the same for the matrices $G_{(1\div4)}'$ and $B_{(1\div4)}$. We can get the following:

$$G'_{(1\div4)} = \underbrace{\begin{bmatrix} 1 & 0 \\ \frac{11}{2} & 1 \end{bmatrix}}_{L_{2(1\div4)}} \underbrace{\left[\begin{array}{cccccc|c} \frac{2}{7} & \overset{\rightleftarrows}{-\frac{53}{21}} & \overset{\rightleftarrows}{\frac{2}{7}} & \frac{58}{21} & 0 & & \mathbf{0}_{2\times1} \\ 0 & \frac{119}{6} & 0 & -\frac{26}{3} & 6 & & \end{array}\right]}_{U_{2(1\div4)}} \underbrace{diag\left[I_1, \begin{bmatrix} 0 & 1 \\ 1 & 0 \end{bmatrix}, I_3\right]}_{P_{2(1\div4)}} \tag{9.17}$$

Based on the decompositions of the matrices $B_{(1\div4)}$ and $G_{(1\div4)}'$, we can construct the result of the first recursive call for the initial 8×8 matrix A. In accordance with Fig. 9.4, where L_1 is $L_{1(1\div4)}$, L_2 is $L_{2(1\div4)}$, FE^{-1} is $F_{(1\div4)}E^{-1}_{(1\div4)}$, H is $U_{1(1\div4)} \cdot P^{-1}_{3(1\div4)}$, $P_{3(1\div4)}$ is $diag[I_2, P_{2(1\div4)}]$, U_2 is $U_{2(1\div4)}$, P is $P_{3(1\div4)} \cdot P_{1(1\div4)}$, we can obtain the LU decomposition of the matrix $B_{(1\div8)} = L_{1(1\div8)} \cdot U_{1(1\div8)} \cdot P_{1(1\div8)}$:

$$\left[\begin{array}{cc|cc} 1 & & & \mathbf{0} \\ \frac{2}{3} & 1 & & \\ \hline 0 & -\frac{2}{7} & 1 & \\ -\frac{1}{3} & \frac{3}{7} & \frac{11}{2} & 1 \end{array}\right] \left[\begin{array}{cc|cccccc|c} \overset{\rightarrow}{3} & 0 & \overset{\leftarrow}{0} & \overset{\downarrow}{2} & \overset{\uparrow}{6} & 2 & 3 & & \\ & 7 & 1 & \frac{5}{3} & -13 & \frac{8}{3} & -7 & & \mathbf{0}_{4\times1} \\ \hline & & \frac{2}{7} & -\frac{53}{21} & \frac{2}{7} & \frac{58}{21} & 0 & & \\ \mathbf{0} & & & \frac{119}{6} & 0 & -\frac{26}{3} & 6 & & \end{array}\right] \begin{bmatrix} \begin{bmatrix} 0 & 0 & 1 \\ 0 & 1 & 0 \\ 1 & 0 & 0 \end{bmatrix} & & \mathbf{0} \\ & \begin{bmatrix} 0 & 1 \\ 1 & 0 \end{bmatrix} & \\ \mathbf{0} & & I_3 \end{bmatrix} \tag{9.18}$$

Now we return to the first recursion level in the algorithm, where we decompose the 8×8 matrix A.

First, in operation, we permute the $C_{(1\div8)}$ columns by the permutation matrix $P^{-1}_{1(1\div8)}$, which is the third factor in the formula. It is equivalent to swapping the 1st column with the 3rd one and the 4th column with the 5th column of $C_{(1\div8)}$. The result is $D_{(1\div8)}$ below. We have highlighted the permuted columns:

$$D_{(1\div8)} = C_{(1\div8)}P^{-1}_{1(1\div8)} = \left[\begin{array}{ccccc|ccc} \underset{\smile}{3} & 6 & \underset{\smile}{6} & \underset{\sim}{0} & \underset{\sim}{3} & 3 & 0 & 0 \\ \underset{\smile}{2} & 5 & \underset{\smile}{2} & \underset{\sim}{2} & \underset{\sim}{7} & -1 & 2 & 0 \\ \underset{\smile}{2} & -5 & \underset{\smile}{3} & \underset{\sim}{0} & \underset{\sim}{2} & 3 & 3 & 0 \\ \underset{\smile}{0} & 0 & \underset{\smile}{0} & \underset{\sim}{0} & \underset{\sim}{0} & 0 & 0 & 1 \end{array}\right]$$

Now our aim is to "nullify" the left, 4×4 size, part of the matrix $D_{(1\div8)}$ obtained above. We achieve this by computing the matrix expression of a general form $G = D - FE^{-1}U_1$. In that expression the auxiliary $E_{(1\div8)}$ matrix is the left 4×4 size part of the $U_{1(1\div8)}$ matrix, given by the decomposition, and $F_{(1\div8)}$ is the 4×4 left part of the $D_{(1\div 8)}$ matrix:

$$E_{(1\div 8)} = \begin{bmatrix} 3 & 0 & 0 & 2 \\ & 7 & 1 & 5/3 \\ & & 2/7 & -53/21 \\ \mathbf{0} & & & 119/6 \end{bmatrix}, \quad F_{(1\div 8)} = \begin{bmatrix} 3 & 6 & 6 & 0 \\ 2 & 5 & 2 & 2 \\ 2 & -5 & 3 & 0 \\ 0 & 0 & 0 & 0 \end{bmatrix}$$

It should be noted that in order to keep the overall complexity of algorithm 9.5 at the same level as the matrix multiplication algorithm involved, it is important to invert the triangular matrix, generally denoted as E, by the recursive algorithm presented in Sect. 9.3. The algorithm yields:

$$G_{(1\div 8)} = D_{(1\div 8)} - F_{(1\div 8)} E_{(1\div 8)}^{-1} U_{1(1\div 8)} = \left[\begin{array}{cccc|ccc|c} 0 & 0 & 0 & 0 & 3 & -\frac{555}{17} & -\frac{165}{17} & 0 \\ 0 & 0 & 0 & 0 & 11 & -\frac{1420}{119} & \frac{205}{119} & 0 \\ 0 & 0 & 0 & 0 & -15 & -\frac{307}{17} & -\frac{236}{17} & 0 \\ \hline 0 & 0 & 0 & 0 & 0 & 0 & 0 & 1 \end{array}\right] = \left[0_{4\times 4} \middle| G'_{(1\div 8)} \right]$$

We leave it as an exercise for the interested reader to perform the LU decomposition of a 4×4 matrix $G_{(1\div 8)'}$ by the algorithm 9.5. It is simpler than the LU decomposition of the $B_{(1\div 8)}$ matrix, because in this case the permutations are not necessary. We can obtain $G_{(1\div 8)'} = L_{2(1\div 8)} \cdot U_{2(1\div 8)} \cdot P_{2(1\div 8)}$:

$$G'_{(1\div 8)} = \left[\begin{array}{ccc|c} 1 & & & \\ \frac{11}{3} & 1 & & \mathbf{0}_{3\times 1} \\ -5 & -\frac{21{,}574}{12{,}825} & 1 & \\ \hline & \mathbf{0}_{1\times 3} & & 1 \end{array}\right] \left[\begin{array}{ccc|c} 3 & -\frac{555}{17} & -\frac{165}{17} & \\ & \frac{12{,}825}{119} & \frac{4490}{119} & \mathbf{0}_{3\times 1} \\ & & \frac{301}{855} & \\ \hline & \mathbf{0}_{1\times 3} & & 1 \end{array}\right] I_4$$

The final result is constructed again according to Fig. 9.4, picking the leading 7×7 size minors of the respective matrices. The final product is

$$A_0 = L_0 U_0 P_0 \tag{9.19}$$

$$\begin{bmatrix} 1 & & & & & & \mathbf{0} \\ \frac{2}{3} & 1 & & & & & \\ 0 & -\frac{2}{7} & 1 & & & & \\ -\frac{1}{3} & \frac{3}{7} & \frac{11}{2} & 1 & & & \\ 1 & \frac{6}{7} & 18 & \frac{36}{17} & 1 & & \\ \frac{2}{3} & \frac{5}{7} & \frac{9}{2} & \frac{65}{119} & \frac{11}{3} & 1 & \\ \frac{2}{3} & -\frac{5}{7} & 13 & \frac{28}{17} & -5 & -\frac{21574}{12825} & 1 \end{bmatrix}$$

$$
\begin{bmatrix}
3 & 0 & 0 & 2 & 6 & 2 & 3 \\
 & 7 & 1 & \frac{5}{3} & -13 & \frac{8}{3} & -7 \\
 & & \frac{2}{7} & -\frac{53}{21} & \frac{2}{7} & \frac{58}{21} & 0 \\
 & & & \frac{119}{6} & 0 & -\frac{26}{3} & 6 \\
 & & & & 3 & -\frac{555}{17} & -\frac{165}{17} \\
 & & & & & \frac{12825}{119} & \frac{4440}{119} \\
\mathbf{0} & & & & & & \frac{301}{855}
\end{bmatrix}
$$

$$
\begin{bmatrix}
\begin{bmatrix} 0 & 0 & 1 \\ 0 & 1 & 0 \\ 1 & 0 & 0 \end{bmatrix} & & \mathbf{0} \\
 & \begin{bmatrix} 0 & 1 \\ 1 & 0 \end{bmatrix} & \\
\mathbf{0} & & I_3
\end{bmatrix}
$$

In the following sections we show how the LU decomposition algorithm can be used in a number of other matrix problems, such as solving a system of linear equations, finding the absolute value of a determinant, and inverting a matrix.

9.6 Solving the System of Linear Equations

We can solve a system of linear equations $Ax = b$ using two methods:

- Recursive matrix inversion algorithm

We invert the system matrix A using the algorithm presented in Sect. 9.4. The solution is straightforward: $x = A^{-1}b$.

- Matrix LU decomposition

First, we decompose the system matrix using the algorithm presented in Sect. 9.5, to obtain $A = LUP$. Next, we solve the system $LUPx = b$ in two steps. The first step is to solve the triangular system $Ly = b$ with respect to y. The next step is to solve the second triangular system of the form $UPx = y$ with respect to x. This is a classical numerical method.

It is well known that the linear triangular system can be solved in $O(n^2)$ time, so the solution of the linear system can be obtained in time determined by the LU decomposition. In Sect. 9.5 we showed that it works in time $O(M(n))$.

9.7 Finding the Determinant of a Matrix

There are two ways to find the determinant of a matrix:

9.7.1 Matrix LU Decomposition

Again, we decompose a given system matrix using the algorithm from Sect. 9.5. Next we use the well-known equality expressing the determinant of the product (property):

$$\det(A) = \det(L) \cdot \det(U) \cdot \det(P)$$

Since the matrices L, U are triangular, their determinants are equal to the product of the main diagonal elements. The matrix L contains only ones on its main diagonal, so $\det(L) = 1$.

To obtain the determinant of the permutation matrix, it is sufficient to decompose it into cycles, and then to count those cycles which are of even length. An efficient algorithm to find a parity of a permutation by this method is presented in the monograph Lipski (1989), pages 20–21. This algorithm works in linear time, so it does not affect the overall efficiency of the method.

- **Example**- finding the determinant of a matrix by LUP decomposition

Let us calculate the determinant of the 7×7 matrix A_0 from Sect. 9.5.1 (formula). Observe:

$$\det(L_0) = 1, \quad \det(U_0) = 3 \cdot 7 \cdot \frac{2}{7} \cdot \frac{119}{6} \cdot 3 \cdot \frac{12825}{119} \cdot \frac{301}{855} = 13545, \quad \det(P_0) = 1$$

So the A_0 determinant is $\det(A_0) = \det(L_0) \cdot \det(U_0) \cdot \det(P_0) = 1 \cdot 13545 \cdot 1 = 13\,545$.

9.7.2 Recursive Matrix Inversion

We can apply the recursive matrix inversion algorithm, described in Sect. 9.4, to the recursive computation of the determinant, but with the precision of the sign. In fact, it follows from the equality (9.4) that:

$$\det(A) = \det\begin{bmatrix} B & C^T \\ C & D \end{bmatrix} = \det(B) \cdot \det\left(D - CB^{-1}C^T\right) \tag{9.20}$$

It should be noted that the above equality is valid for symmetric and positively defined matrices. Nevertheless, it can be used to obtain the absolute value of the determinant for any square matrix. To do this we proceed as follows:

- For a given matrix A_0, calculate $A = A_0 \cdot A_0^T$. It is symmetric and positively defined.
- Calculate the determinant recursively by block matrix equality (9.20).
- Since $\det(A_0^T) = \det(A_0)$ from the matrix multiplication property (1.5) we deduce:

$$\det(A) = \det(A_0 \cdot A_0^T) = \det(A_0)\det(A_0^T) = \overset{2}{\det}(A_0)$$

- We can calculate the absolute value of a determinant as the square root:

$$|\det(A_0)| = \sqrt{\det(A)} \tag{9.21}$$

- **Example**—finding the determinant of a matrix by recursion

Let us calculate the square of the determinant of the 7×7 matrix A_0 from Sect. 9.5.1. We extend it to the full two's power dimension as follows:

$$A_0\prime = \mathrm{diag}[A_0;\, 1]$$

We calculate the auxiliary $A = A_0\prime \cdot (A_0\prime)^T$ matrix, and partition it into the 2×2 block matrix, according to formula (9.4).

$$A = \left[\begin{array}{cccc|cccc} 62 & -49 & 28 & -7 & 33 & 56 & 33 & 0 \\ -49 & 185 & -61 & 111 & 39 & -30 & -49 & 0 \\ 28 & -61 & 37 & -30 & 6 & 14 & 30 & 0 \\ -7 & 111 & -30 & 139 & 30 & -16 & 4 & 0 \\ \hline 33 & 39 & 6 & 30 & 99 & 66 & 9 & 0 \\ 56 & -30 & 14 & -16 & 66 & 91 & 2 & 0 \\ 33 & -49 & 30 & 4 & 9 & 2 & 60 & 0 \\ 0 & 0 & 0 & 0 & 0 & 0 & 0 & 1 \end{array}\right] = \left[\begin{array}{c|c} B & C^T \\ \hline C & D \end{array}\right]$$

Thus, from formula (9.20) we obtain:

$$\det(A) = \det(B) \cdot \det\left(D - CB^{-1}C^T\right) = \det\begin{bmatrix} 62 & -49 & 28 & -7 \\ -49 & 185 & -61 & 111 \\ 28 & -61 & 37 & -30 \\ -7 & 111 & -30 & 139 \end{bmatrix} \cdot$$

$$\cdot \det \begin{bmatrix} \frac{116{,}680{,}317}{2{,}631{,}403} & \frac{78{,}961{,}234}{2{,}631{,}403} & \frac{3{,}970{,}067}{2{,}631{,}403} & 0 \\ \frac{78{,}961{,}234}{2{,}631{,}403} & \frac{228{,}946{,}679}{7{,}894{,}209} & -\frac{106{,}120{,}898}{7{,}894{,}209} & 0 \\ \frac{3{,}970{,}067}{2{,}631{,}403} & -\frac{106{,}120{,}898}{7{,}894{,}209} & \frac{190{,}818{,}044}{7{,}894{,}209} & 0 \\ 0 & 0 & 0 & 1 \end{bmatrix}$$

The absolute values of the determinants of matrices B and $S = D - CB^{-1}C^T$ we calculate recursively by the same formula (9.20). Finally we receive:

$$\det(A) = 7{,}894{,}209 \cdot \frac{61{,}155{,}675}{2{,}631{,}403} = 183{,}467{,}025$$

And the absolute value of the determinant of A_0 we obtain from equality (9.21):

$$\left|\det(A_0)\right| = \sqrt{\det(A)} = \sqrt{183{,}467{,}025} = 13{,}545$$

9.8 Calculation of Scalar Polynomial Values

In this section we show how to represent the problem of computing scalar polynomials in the matrix domain using matrix multiplication. We present two problems. In point 9.8.1 we showed how to calculate a series of polynomials at once. In point 9.8.2 we calculated a given, single polynomial at a series of points.

9.8.1 Calculating a Series of Polynomials at Once

Article Borodin, Munro (1971) proposed to use a fast matrix multiplication algorithm to calculate a **series** of polynomials in a series of points at once. Namely, we want to evaluate n polynomials of the form $p(x) = a_0 + \sum_{k=1}^{n} a_k x^k$, for a real x, each at n points $\{x_1, ..., x_n\}$. This means, that we calculate a series of polynomials $p_i(x_j) = a_{0,i} + \sum_{k=1}^{n} a_{k,i} x_j^k$, where $i = 1, ..., n$ is the polynomial number and $j = 1, ..., n$ is the point number.

Performing such a task using Horner's rule leads to the calculation of n polynomials, each n times. Considering that a single execution of Horner's algorithm takes $O(n)$ steps, in total it requires $O(n^3)$ multiplications. Borodin and Munro (1971) proposed organising this task in a matrix form as follows:

$$\begin{bmatrix} a_{0,1} & \cdots & a_{0,1} \\ & \ddots & \\ \vdots & a_{0,i} & \vdots \\ & & \ddots \\ a_{0,n} & \cdots & a_{0,n} \end{bmatrix} + \begin{bmatrix} a_{1,1} & \cdots & a_{n,1} \\ & \ddots & \\ \cdots \cdots & a_{k,i} & \cdots \cdots \\ & & \ddots \\ a_{1,n} & \cdots & a_{n,n} \end{bmatrix} \begin{bmatrix} x_1 & \vdots & x_n \\ & \ddots \vdots & \\ \vdots & x_j^k & \vdots \\ & \vdots \ddots & \\ x_1^n & \vdots & x_n^n \end{bmatrix} =$$

$$= \begin{bmatrix} a_{0,1} + \sum_{k=1}^{n} a_{k,1} x_1^k & \cdots & a_{0,1} + \sum_{k=1}^{n} a_{k,1} x_n^k \\ & \ddots & \\ \vdots & a_{0,i} + \sum_{k=1}^{n} a_{k,i} x_j^k & \vdots \\ & \ddots & \\ a_{0,n} + \sum_{k=1}^{n} a_{k,n} x_1^k & \cdots & a_{0,n} + \sum_{k=1}^{n} a_{k,n} x_n^k \end{bmatrix}$$

We can see that in a given row we get a given ith polynomial value, while in a given column we get its value in jth point. In summary, we can calculate values of n polynomials at once, for a series of n points, in $O(n^{\omega})$ time. We have decided not to include the example in this section as the construction is quite simple.

9.8.2 Calculating a Single Polynomial at a Series of Points

Unlike the previous section, we will now show how to calculate a **single** polynomial in a series of points. More precisely, we compute a polynomial of real variable of degree n at $\sqrt{n}$ points. We assume that n is a perfect square. As Borodin and Munro (1971) point out, this task now turns out to be more complicated. In this case, they suggest constructing the coefficient matrix as follows:

$$A = \begin{bmatrix} a_1 & a_2 & \cdots & a_{\sqrt{n}-1} & a_{\sqrt{n}} \\ a_{\sqrt{n}+1} & a_{\sqrt{n}+2} & \cdots & a_{2\sqrt{n}-1} & a_{2\sqrt{n}} \\ \vdots & & \ddots & \vdots & \vdots \\ a_{n-\sqrt{n}+1} & \cdots & \cdots & a_{n-1} & a_n \end{bmatrix}$$

The matrix that contains the independent variable is:

$$X = \begin{bmatrix} x_1 & x_2 & \cdots & x_{\sqrt{n}-1} & x_{\sqrt{n}} \\ x_1^2 & & & & x_{\sqrt{n}}^2 \\ \vdots & & \ddots & & \vdots \\ x_1^{\sqrt{n}} & & \cdots & & x_{\sqrt{n}}^{\sqrt{n}} \end{bmatrix}$$

As usual, the crucial computational step, responsible for the efficiency of the resulting algorithm, is the matrix multiplication. We calculate $Y = AX = [y]_{i,k}$ and get as a result:

$$y_{i,k} = \sum_{j=1}^{\sqrt{n}} a_{(i-1)\sqrt{n}+j} x_k^j$$

Finally, we linearly combine the obtained $y_{i,k}$ values raised to successive powers of $\sqrt{n}$:

$$p(x_k) = \sum_{j=0}^{n} a_j x_k^j = a_0 + \sum_{j=1}^{\sqrt{n}} y_{j,k} \left(x_k^{\sqrt{n}}\right)^{j-1}, \quad k = 1, \ldots, \sqrt{n}$$

In the constructed algorithm, there are the following stages with the corresponding complexity:

- Filling the matrix X: a single column is calculated in $O(\sqrt{n})$ time. The matrix X contains $\sqrt{n}$ columns, so this stage takes $\sqrt{n} \cdot O(\sqrt{n}) = O(n)$ time.
- Successive powers of $x_k^{\sqrt{n}}$: we calculate it $\sqrt{n}$ times for $\sqrt{n}$ points, so this also takes $O(n)$ steps.
- Calculation of $p(x_k)$: for a given x_k we have $\sqrt{n}$ multiplications of $y_{j,k}$ by successive powers of $x_k^{\sqrt{n}}$. Number x_k of points is $\sqrt{n}$, so this step also takes $O(n)$ time.

To sum up, the most demanding stage of the algorithm is the matrix multiplication of the size $(\sqrt{n}, \sqrt{n}, \sqrt{n})$, which leads to complexity $O(\sqrt{n}^{\omega}) = O(n^{\omega/2})$. Using a Horner rule for this task, we get the $O(n\sqrt{n}) = O(n^{1.5})$ time, while Strassen's algorithm already gives a better exponent $O(n^{1.404})$. Applying the algorithm $\sqrt{n}$ times we can evaluate nth order polynomial in n points in $O(n^{(\omega+1)/2})$ time. Strassen's algorithm gives the exponent $O(n^{1.904})$, while Horner's rule gives well-known exponent $O(n^2)$.

- **Example**—Calculating the value of a ninth-order polynomial at three points.

We state $n = 9$ so $\sqrt{n} = 3$, so matrices $A,\ X$ have the following form:

$$A = \begin{bmatrix} a_1 & a_2 & a_3 \\ a_4 & a_5 & a_6 \\ a_7 & a_8 & a_9 \end{bmatrix}, \quad X = \begin{bmatrix} x_1 & x_2 & x_3 \\ x_1^2 & x_2^2 & x_3^2 \\ x_1^3 & x_2^3 & x_3^3 \end{bmatrix}$$

After multiplying them, we obtain directly:

$$Y = AX = \begin{bmatrix} a_1x_1 + a_2x_1^2 + a_3x_1^3 & a_1x_2 + a_2x_2^2 + a_3x_2^3 & a_1x_3 + a_2x_3^2 + a_3x_3^3 \\ a_4x_1 + a_5x_1^2 + a_6x_1^3 & a_4x_2 + a_5x_2^2 + a_6x_2^3 & a_4x_3 + a_5x_3^2 + a_6x_3^3 \\ a_7x_1 + a_8x_1^2 + a_9x_1^3 & a_7x_2 + a_8x_2^2 + a_9x_2^3 & a_7x_3 + a_8x_3^2 + a_9x_3^3 \end{bmatrix}$$

In point x_1, we obtain the polynomial value as follows:

$$p(x_1) = a_0 + \sum_{j=1}^{3} y_{j,1}\left(x_1^3\right)^{j-1} = a_0 + y_{11} + y_{2,1}\left(x_1^3\right)^1 + y_{3,1}\left(x_1^3\right)^2 =$$

$$= a_0 + \left(a_1x_1 + a_2x_1^2 + a_3x_1^3\right) + \left(a_4x_1 + a_5x_1^2 + a_6x_1^3\right)x_1^3 + \left(a_7x_1 + a_8x_1^2 + a_9x_1^3\right)x_1^6$$

what gives $p(x_1) = \sum_{j=0}^{9} a_j x_1^j$. The polynomial values in the remaining two points x_2, x_3 are obtained by analogy.

9.9 Calculating a Matrix Polynomial

In this section we present an algorithm optimised for efficient computation of matrix polynomials, i.e. polynomials with a matrix as independent variable. To do this, we need to minimise the number of matrix multiplications. Suppose we have a $m \times m$ square matrix and a polynomial of degree $n - 1$, where n is a perfect square of a given integer k. We want to obtain the value of $P(X) = a_0 I + \sum_{k=1}^{n-1} a_k X^k$. The corresponding algorithm is given in the article Paterson and Stockmeyer (1973) in Sect. 3.2 on pages 64–65. They proposed to perform the task in three steps:

- Calculate the powers $X^2, X^3, ..., X^k$. It involves $k - 1$ matrix multiplications.
- Using X^k now calculate the powers $X^{2k}, X^{3k}, ..., X^{n-k}$. This involves $((n-k)-k)/k = (n-2k)/3$ matrix multiplications.
- Calculate the desired matrix polynomial from the formula:

$$P(X) = \sum_{j=0}^{\frac{n}{k}-1} \left(a_{jk} I - a_{jk+1} X + a_{jk+2} X^2 + ... + a_{(j+1)k-1} X^{k-1}\right) X^{jk}$$

Each of the first two steps requires $O(k) = O(\sqrt{n})$ matrix multiplications. The last step requires $O((n - \sqrt{n})/\sqrt{n})$, thus also $O(\sqrt{n})$ matrix multiplications. In summary, we obtain an algorithm of $O(m^\omega \sqrt{n})$ efficiency, with comparison to the $O(m^3 n)$ efficiency obtained from standard matrix multiplication algorithm combined with Horner's scheme.

It is worth noting that for $m = n$ the efficiency of the matrix polynomial calculation is better than just the standard single matrix multiplication, if we use the results of this section together with the fast matrix multiplication algorithm with exponent $\omega < 2.5$.

9.9.1 Example—Calculation of Twenty Fifth Order Matrix Polynomial.

Now we state $n = 25$ thus $k = \sqrt{n} = 5$. We calculate it as follows:

$$\begin{aligned} P(X) = P(X) &= a_0 I + \sum_{k=1}^{24} a_k X^k \\ &= \left(a_0 I + a_1 X + a_2 X^2 + a_3 X^3 + a_4 X^4\right) \\ &+ \left(a_5 I + a_6 X + a_7 X^2 + a_8 X^3 + a_9 X^4\right) X^5 \\ &+ \left(a_{10} I + a_{11} X + a_{12} X^2 + a_{13} X^3 + a_{14} X^4\right)\left(X^5\right)^2 \\ &+ \left(a_{15} I + a_{16} X + a_{17} X^2 + a_{18} X^3 + a_{19} X^4\right)\left(X^5\right)^3 \\ &+ \left(a_{20} I + a_{21} X + a_{22} X^2 + a_{23} X^3 + a_{24} X^4\right)\left(X^5\right)^4 \end{aligned}$$

We think the example is so clear that it needs no further explanation.

9.10 Finding the Characteristic Polynomial of a Matrix

The problem of finding the characteristic polynomial of a matrix is a very classical one in both theoretical and numerical algebra. Gantmacher (1960) devotes the whole of Section IV of Volume I to it. By the definition it is expressed by the following determinant[10] (Faddeev and Faddeeva (1963) Chap. 4)

$$\begin{aligned} \det(A - \lambda I) &= \begin{vmatrix} a_{11} - \lambda & a_{12} & \cdots & a_{1n} \\ a_{21} & a_{22} - \lambda & \cdots & a_{2n} \\ \vdots & \vdots & \ddots & \vdots \\ a_{n1} & a_{n2} & \cdots & a_{nn} - \lambda \end{vmatrix} \\ &= (-1)^n\left[\lambda^n - p_{n-1}\lambda^{n-1} - p_{n-2}\lambda^{n-1} - \ldots - p_0\right] \end{aligned} \tag{9.22}$$

for a given square matrix A. The problem has been approached numerically many times. Pereslavtseva in articles (2008, 2011) presented a survey of algorithms for that problem. Another valuable source is monograph Faddeev and Faddeeva (1963) Chap. 4.

Those algorithms achieve—at best—cubic time complexity, while some of them are of the fourth order ones.

The question is how to use fast matrix multiplication in the title problem of finding the characteristic polynomial of a matrix. The first, and at the same time the most comprehensive and still up-to-date article on this problem is Keller-Gehrig (1985).

[10] Some authors invert the order of the matrix summands, defining it by $\det(\lambda I - A)$. It changes nothing more than the sign of coefficients by $(-1)^n$.

He proposed three algorithms for the problem. The main tools used in all three algorithms are the following:

- Matrix similarity

The two square matrices A, B are said to be similar, if there exists an invertible matrix U, such that:

$$A = U^{-1}BU \tag{9.23}$$

They have certain important properties that are important for our goal of finding the characteristic polynomial—the characteristic polynomials of two similar matrices are identical.

- Frobenius matrix[11]

Matrix of the form:

$$F = \begin{bmatrix} 0 & 0 & \cdots & 0 & p_0 \\ 1 & 0 & \cdots & 0 & p_1 \\ 0 & 1 & \cdots & 0 & p_2 \\ \vdots & \vdots & \ddots & \vdots & \vdots \\ 0 & 0 & \cdots & 1 & p_{n-1} \end{bmatrix} \tag{9.24}$$

contains the characteristic coefficients (Faddeev and Faddeeva (1963), Sect. 46) in its last column and is called a Frobenius matrix. This is an example of a structured (some authors use a 'special' name here) matrix. More information about special matrices can be found in an Appendix, page 239.

In the next sections we presented three algorithms authored by Keller-Gehrig for finding the characteristic polynomial of a matrix, which make use of fast matrix multiplication, in contrast to the algorithms examined in, for example, the articles Pereslavtseva (2008, 2011) and Monograph Faddeev and Faddeeva (1963), Chap. 4, which do not make use of fast matrix multiplication.

9.10.1 The Simple Algorithm

The first algorithm proposed by Keller-Gehrig (1985) in Sect. 3 is based on the Krylov's method (Gantmacher 1960, Vol. I, Section VII point 8). If we have a non-singular matrix of the form:

---

[11] Some authors call matrix of this form a companion matrix, leaving the name Frobenius for a block diagonal matrix only with blocks of the form. We decided to preserve the original Keller-Gehrig's (1985) article naming.

$$U = \left[e, Ae, A^2e, ..., A^{n-1}e\right] \tag{9.25}$$

then the $U^{-1}BU$ has a Frobenius form for any (!) vector $e \neq 0$. The point is how to calculate the $A, A^2, ..., A^{n-1}$ series in an efficient way. Keller-Gehrig proposed the approach of recurrent squaring the source matrix. First, we compute a series for $k = \lceil \log_2 n \rceil - 1$:

$$\left(A, A^2, A^{2^2}, ..., A^{2^k}\right)$$

We then proceed in the following, smartly designed sequence:

$$\begin{cases} Ae = Ae \\ (A^3e, A^2e) = A^2(Ae, e) \\ (A^7e, A^6e, A^5e, A^4e) = A^4(A^3e, A^2e, Ae, e) \\ \vdots \\ (A^{2\cdot 2^k-1}e, ..., A^{2^k}e) = A^{2^k}(A^{2^k-1}e, ..., e) \end{cases} \tag{9.26}$$

using in a given step all the A^ie terms calculated in previous steps. Clearly, the first step takes $O(M(n)\log n)$ time. In the second step, we need to multiply the A^i matrix by a e vector $O(n)$ times. Multiplying a matrix A^i by a vector e, for any vector e, takes n^2 scalar multiplications. Thus, in general, the calculation of the series (9.26) requires $n^2 \cdot O(n) = O(n^3)$ multiplications without the use of fast matrix multiplication. It is possible to arrange the vectors on the right side as matrices to use certain fast matrix multiplication algorithms. We deliberately use the plural, because the application of fast matrix multiplication is not straightforward here, since we are working on a series of products with right sides of rectangular matrices whose column numbers vary in a sequence $1, 2, 4, ..., 2^k$. But if we choose as e any unit vector e.g. $[\,1\ 0\ \cdots\ 0\,]^T$ the complexity of the second step is reduced to $O(n^2)$.

We compute the final Frobenius matrix in $O(M(n))$ time. We invert U in time equal to the applied matrix multiplication algorithm using the method described in Sect. 9.4. In this way we obtain the characteristic polynomial coefficients in a $O(M(n)\log n)$ time.

Unfortunately, the simple algorithm does not always work, because the matrix (9.25) turns out not to be invertible for any e, for certain A matrices. How to deal with such a matrices is shown in the next point.

- **Example**: Using the first Keller-Gehrig algorithm to find a characteristic polynomial

Let us find the a characteristic polynomial of the following 5×5 matrix:

$$A = \begin{bmatrix} 1 & 6 & 4 & -3 & 2 \\ 3 & -6 & 5 & 3 & -1 \\ 2 & 4 & 0 & 9 & 1 \\ 6 & 5 & 4 & 3 & 8 \\ -4 & 4 & 6 & 6 & 6 \end{bmatrix}$$

So $n = 5$ and $k = \lceil \log_2 5 \rceil - 1 = 2$, so we need up to the 4th A power:

$$A^2 = A \cdot A = \begin{bmatrix} 1 & -21 & 34 & 54 & -12 \\ 17 & 85 & -12 & 21 & 35 \\ 64 & 37 & 70 & 39 & 78 \\ 15 & 69 & 109 & 90 & 83 \\ 32 & 30 & 64 & 132 & 78 \end{bmatrix}$$

$$A^4 = A^2 \cdot A^2 = \begin{bmatrix} 2\,246 & 2\,818 & 7\,784 & 4\,215 & 5\,451 \\ 2\,129 & 8\,923 & 3\,247 & 8\,745 & 6\,308 \\ 8\,254 & 9\,422 & 15\,875 & 20\,769 & 15\,308 \\ 12\,170 & 18\,283 & 22\,434 & 25\,566 & 24\,681 \\ 9\,114 & 15\,694 & 24\,588 & 27\,030 & 22\,698 \end{bmatrix}$$

We define $e = [\,1\ 0\ 0\ 0\ 0\,]^T$ and calculate in a sequence (9.26):

$$Ae = Ae = \begin{bmatrix} 1 \\ 3 \\ 2 \\ 6 \\ -4 \end{bmatrix}$$

$$(A^3e, A^2e) = A^2(Ae, e) = A^2 \begin{bmatrix} 1 & 1 \\ 3 & 0 \\ 2 & 0 \\ 6 & 0 \\ -4 & 0 \end{bmatrix} = \begin{bmatrix} 378 & 1 \\ 234 & 17 \\ 237 & 64 \\ 648 & 15 \\ 730 & 32 \end{bmatrix}$$

$$(A^7e, A^6e, A^5e, A^4e) = A^4(A^3e, A^2e, Ae, e) = A^4 \begin{bmatrix} * & * & * & 1 \\ * & * & * & 0 \\ * & * & * & 0 \\ * & * & * & 0 \\ * & * & * & 0 \end{bmatrix} = \begin{bmatrix} * & * & * & 2\,246 \\ * & * & * & 2\,129 \\ * & * & * & 8\,254 \\ * & * & * & 12\,170 \\ * & * & * & 9\,114 \end{bmatrix}$$

where unneeded positions are marked with an asterisk. Now we can construct the similarity matrix (9.25):

$$U = \left[e \middle| Ae \middle| A^2e\ A^3e \middle| A^4e \right] = \left[\begin{array}{c|c|c|c} 1 & 1 & 1\ 378 & 2\,246 \\ 0 & 3 & 17\ 234 & 2\,129 \\ 0 & 2 & 64\ 237 & 8\,254 \\ 0 & 6 & 15\ 648 & 12\,170 \\ 0 & -4 & 32\ 730 & 9\,114 \end{array} \right]$$

The characteristic polynomial can be found as the last column of the Frobenius matrix $U^{-1}BU$:

$$F = U^{-1}BU = \begin{bmatrix} 0\,0\,0\,0 & -39\,030 \\ 1\,0\,0\,0 & 637 \\ 0\,1\,0\,0 & 951 \\ 0\,0\,1\,0 & 154 \\ 0\,0\,0\,1 & 4 \end{bmatrix}$$

Finally:

$$\det(A - \lambda I) = -\left[\lambda^5 - 4\lambda^4 - 154\lambda^3 - 951\lambda^2 - 637\lambda + 39\,030\right]$$

In the next section we show the second algorithm, which also uses the idea of squaring the input matrix, but works for any matrix.

9.10.2 The Branching Algorithm

In order to present the algorithm, which—most importantly—is the only one by Keller-Gehrig that works with any matrix, we first introduce the notion of a Krylov matrix generated by a pair of square $n \times n$ matrices $A,\ B$. More precisely, we construct a Krylov matrix generated by the subsequent columns of B, i.e. $B = \left[b_1 \middle| \cdots \middle| b_n \right]$, left multiplied by the following subsequent powers of A:

$$\left[b_1 \middle| \cdots \middle| A^{n-1} b_1 \middle| \cdots \middle| b_n \middle| \cdots \middle| A^{n-1} b_n \right]$$

For certain power p, the $(p + 1)$-st vector $A^p b_j$ becomes a linear combination of the previous vectors (Gantmacher 1960, Vol. I, Section VII, point 8):

$$A^p b_j = -\alpha_0 b_j - \alpha_1 A b_j - \ldots - \alpha_{p-1} A^{p-1} b_j \tag{9.27}$$

Then all the further vectors in the sequence $A^{p+1}b_j,\ A^{p+2}b_j,\ \ldots$ are also linearly dependent on the p previous vectors. Indeed, if we apply the operator A on both sides of (9.27) then $A^{p+1}b_j$ is linearly dependent on the terms $A^p b_j,\ A^{p-1}b_j,\ \ldots, Ab_j$. By induction it works for all higher A powers.

The general idea of the branching algorithm is to detect the moment when the subsequent vectors become linearly dependent, for B being the identity matrix. We proceed in an efficient way, similar to the scheme (9.26), but for all base $\left[\, e_1 \middle| \cdots \middle| e_n \,\right] = I_n$ vectors at once, where I_n is the identity matrix. The result is the so-called compressed Krylov matrix of the form

$$
\begin{aligned}
U &= \left[\, U_1 \middle| \cdots \middle| U_n \,\right], \\
U_j &= \left[\, e_j \middle| \cdots \middle| A^{m_j-1} e_j \,\right],\ j = 1, \ldots, n
\end{aligned}
\tag{9.28}
$$

for $m_1 + \ldots + m_n = n$ and $0 \le m_j \le n - 1$. It is worth noting that in particular we allow $m_i = 0$, which means that the block U_j corresponding to the vector e_j vanishes in the compressed Krylov matrix. According to Keller-Gehrig (1985), page 313, the compressed Krylov matrix U of the form (9.28) is similar to the following block triangular matrix of a special structure:

$$
F = U^{-1} A U = \left[\begin{array}{c|c|c|c|c}
F_1 & B_2 & & & \\
 & F_2 & B_3 & & \\
 & & F_3 & & B_n \\
 & & & \ddots & \\
\mathbf{0} & & & & F_n
\end{array}\right]
\tag{9.29}
$$

where each block F_j on the diagonal is in the Frobenius form, and each B_j contains only zeros except for its last column. According to Gantmacher (1960) Vol. I, formula (67) on page 43, the determinant of a block triangular matrix is the product of the determinants of the diagonal blocks. This is analogous to the well-known case of the scalar triangular matrix. Clearly, the characteristic polynomial's determinant $\det(A - \lambda I)$ is expressed by the product:

$$
\det(A - \lambda I) = \det(F - \lambda I) = \det(F_1 - \lambda I) \cdot \det(F_2 - \lambda I) \cdot \ldots \cdot \det(F_n - \lambda I)
$$

i.e. the product of the characteristic polynomials of the diagonal blocks. It can be computed e.g. using elementary symmetric polynomials (e.g. El-Mikkawy (2003)) in $O(n^2)$ time, which—importantly—does not add to the overall complexity of the Keller-Gehrig second algorithm.

Keller-Gehrig (1985) describes the algorithm in the following general steps:

- Set $V_0 = \left[\, V_{01} = e_1 \middle| V_{02} = e_2 \middle| \cdots \middle| V_{0n} = e_n \,\right] = I_n$.
- Calculate a series of block matrices V_i, *for* $i = 1, 2, \ldots, r$, as described below, where $r = \lceil \log n \rceil$.
- Each matrix V_i in the series is of the form $V_i = \left[\, V_{i1} \middle| V_{i2} \middle| \cdots \middle| V_{in} \,\right]$.
- The submatrices V_{ij} of V_i are of the form $V_{ij} = \left[\, e_j \middle| A e_j \middle| A^2 e_j \middle| \ldots \,\right]$ with possible numbers of columns from zero to 2^i.
- The next iteration of V_i is calculated by matrix squaring, multiplication and linear dependency detection:

– calculate the auxiliary matrices W_{ij} using the conditional, block matrix formula:

$$W_{ij} = \begin{cases} V_{ij}, & \text{for } cols(V_{ij}) < 2^i \\ \left[V_{ij} | A^{2^i} V_{ij} \right], & \text{otherwise} \end{cases} \tag{9.30}$$

– $V_{i+1,j}$ is such a submatrix of W_{ij}, that consists of the first $l_{ij} - 1$ columns of W_{ij}, where l_{ij}-th column of W_{ij} is the first that is linearly dependent on the **preceding** columns of W_{ij} and **all** columns of matrices $W_{i1}, \ldots, W_{i,j-1}$.
– repeat this step until the first case of formula (9.30) will occur for all $j = 1, 2, \ldots, n$.

- Finally, $U = V_i$ for $i = r$ and the similarity transformation (9.29) leads directly to the block-Frobenius form and the characteristic polynomial is the product of the characteristic polynomials of the diagonal blocks.

The algorithm requires $r - 1$ matrix multiplications to calculate the series A^{2^i} in formula (9.30), r for W_{ij} (the non-square matrix multiplication $A^{2^i} V_{ij}$ in (9.30) can be performed by blocking and then de-blocking $V_{i1}, V_{i2}, \ldots, V_{in}$, in a quadratic time) and r times of detection of the linear dependency, which takes $O(M(n))$ time. Thus, the time complexity of the second, branching Keller-Gehrig algorithm to compute the characteristic polynomial is linear-logarithmic, i.e. $O(M(n) \log n)$.

- **Example**: The second, branching Keller-Gehrig algorithm

Now we consider the 5×5 matrix of the form:

$$A = \begin{bmatrix} 0 & 2 & 1 & 0 & -1 \\ -2 & 4 & 0 & 2 & 7 \\ 0 & 0 & 1 & 0 & 0 \\ 0 & 0 & 2 & 2 & -1 \\ 0 & 0 & 0 & 1 & 0 \end{bmatrix} \tag{9.31}$$

It can be verified that the matrix U constructed by the first algorithm described in Sect. 9.10.1 is singular, so the similarity formula $U^{-1}BU$ cannot be used for U matrix constructed in that way. Instead, we use the branching algorithm. The first iteration of W_{ij} is shown in Table 9.1:

We can detect the linear dependencies[12]:

$$\begin{cases} (W_{02})_1 = -1/2 (W_{01})_1 \\ (W_{04})_1 = -1/2 (W_{01})_1 - 1/2 (W_{03})_1 + 1/2 (W_{03})_2 \\ (W_{05})_2 = -1/2 (W_{01})_1 - 7/2 (W_{01})_2 + 1/2 (W_{03})_1 - 1/2 (W_{03})_2 \end{cases}$$

[12] The notation $(W_{ij})_k$ denotes the k-th column of W_{ij} matrix.

Table 9.1 Example—first iteration of the Keller-Gehrig branching algorithm

j	1	2	3	4	5
$W_{0j} = \left[e_j \middle\vert Ae_j \right]$	$\left[\begin{array}{c\|c} \mathbf{1} & \mathbf{0} \\ \mathbf{0} & \mathbf{-2} \\ \mathbf{0} & \mathbf{0} \\ \mathbf{0} & \mathbf{0} \\ \mathbf{0} & \mathbf{0} \end{array}\right]$	$\left[\begin{array}{c\|c} 0 & 2 \\ 1 & 4 \\ 0 & 0 \\ 0 & 0 \\ 0 & 0 \end{array}\right]$	$\left[\begin{array}{c\|c} \mathbf{0} & \mathbf{1} \\ \mathbf{0} & \mathbf{0} \\ \mathbf{1} & \mathbf{1} \\ \mathbf{0} & \mathbf{2} \\ \mathbf{0} & \mathbf{0} \end{array}\right]$	$\left[\begin{array}{c\|c} 0 & 0 \\ 0 & 2 \\ 0 & 0 \\ 1 & 2 \\ 0 & 1 \end{array}\right]$	$\left[\begin{array}{c\|c} \mathbf{0} & -1 \\ \mathbf{0} & 7 \\ \mathbf{0} & 0 \\ \mathbf{0} & -1 \\ \mathbf{1} & 0 \end{array}\right]$

thus $l_{02} = 1$, $l_{04} = 1$ and $l_{05} = 2$. Then $V_{11} = W_{01}$, $V_{13} = W_{03}$, $V_{15} = (W_{05})_1$ and V_{12}, V_{14} are empty. In Table 9.1, we have bolded those matrix entries that go from W_{0j} to V_{1j}. In the second iteration we calculate the matrix series $W_{11}, \ldots, W_{15}$ in Table 9.2.

Immediately we see that column $(W_{11})_3$ is linearly dependent on columns $(W_{11})_1$, $(W_{11})_2$ and $V_{21} = \left[(W_{11})_1 \middle\vert (W_{11})_2 \right]$. Furthermore:

$$\begin{cases} (W_{13})_4 = 3(W_{11})_1 - 13(W_{11})_2 + (W_{13})_1 - 3(W_{13})_2 + 3(W_{13})_3 \\ (W_{15})_1 = (W_{11})_1 + {}^1\!/\!{}_2(W_{11})_2 + (W_{13})_1 - {}^3\!/\!{}_2(W_{13})_2 + {}^1\!/\!{}_2(W_{13})_3 \end{cases}$$

thus to V_{23} goes only the first three columns of the matrix W_{13} and the matrix V_{25} becomes empty. In summary, we have obtained the following similarity matrix according to formula (9.28):

$$U = \left[e_1 \; Ae_1 \middle\vert e_3 \; Ae_3 \; A^2e_3 \right] = \left[V_{21} \middle\vert V_{23} \right] = \left[\begin{array}{cc|ccc} 1 & 0 & 0 & 1 & 1 \\ 0 & -2 & 0 & 0 & 2 \\ 0 & 0 & 1 & 1 & 1 \\ 0 & 0 & 0 & 2 & 6 \\ 0 & 0 & 0 & 0 & 2 \end{array}\right]$$

with $m_1 = 2$, $m_3 = 3$ and $m_2 = m_4 = m_5 = 0$. The block triangular matrix $U^{-1}AU$ (9.29) is:

Table 9.2 Example—second iteration of the Keller-Gehrig branching algorithm

j	1	2	3	4	5
$W_{1j} = \left[V_{1j} \middle\vert A^2V_{1j} \right]$	$\left[\begin{array}{cc\|cc} \mathbf{1} & \mathbf{0} & -4 & -16 \\ \mathbf{0} & \mathbf{-2} & -8 & -24 \\ \mathbf{0} & \mathbf{0} & 0 & 0 \\ \mathbf{0} & \mathbf{0} & 0 & 0 \\ \mathbf{0} & \mathbf{0} & 0 & 0 \end{array}\right]$	Empty	$\left[\begin{array}{cc\|cc} \mathbf{0} & \mathbf{1} & \mathbf{1} & 3 \\ \mathbf{0} & \mathbf{0} & \mathbf{2} & 32 \\ \mathbf{1} & \mathbf{1} & \mathbf{1} & 1 \\ \mathbf{0} & \mathbf{2} & \mathbf{6} & 12 \\ \mathbf{0} & \mathbf{0} & \mathbf{2} & 6 \end{array}\right]$	Empty	$\left[\begin{array}{c\|c} 0 & 14 \\ 0 & 28 \\ 0 & 0 \\ 0 & -2 \\ 1 & -1 \end{array}\right]$

$$F = U^{-1}AU = \left[\begin{array}{cc|ccc} 0 & -4 & 0 & 0 & 3 \\ 1 & 4 & 0 & 0 & -13 \\ \hline 0 & 0 & 0 & 0 & 1 \\ 0 & 0 & 1 & 0 & -3 \\ 0 & 0 & 0 & 1 & 3 \end{array}\right] = \left[\begin{array}{c|c} F_1 & B_2 \\ \hline \mathbf{0} & F_2 \end{array}\right] \tag{9.32}$$

Linking the last columns of F_1, F_2 with the general formula for the characteristic polynomial (9.22) and its coefficient deployment in the general form of the Frobenius matrix (9.24) we can finally find the characteristic polynomial of the given matrix A (9.31). Observe:

$$\begin{aligned} \det(A - \lambda I) &= \det(F - \lambda I) = (\lambda^2 - 4\lambda + 4)(-\lambda^3 + 3\lambda^2 - 3\lambda + 1) \\ &= -(\lambda - 1)^3(\lambda - 2)^2 \end{aligned}$$

Originally Keller-Gehrig (1985) proposed to detect the linear dependencies in $W_{i1}, \ldots, W_{in}$ series by his own fast, block version of Gaussian elimination, which is outlined in Sect. 9.4 of his article. The problem of finding this was tackled once again by Dumas et al. (2005) using the Keller-Gehrig algorithm as a basis. Instead of the block Gaussian elimination they proposed to apply the LSP decomposition algorithm, we described in Sect. 9.5, with the same complexity.

9.10.3 The Fast Algorithm

The third and simultaneously the last algorithm proposed by Keller-Gehrig (1985) is the straight line algorithm presented in Sect. 6. It is the only one that works in the same time as the applied matrix multiplication algorithm, with no additional logarithmic factor, but the price is that it does not work for every matrix.

The author discloses that the fast algorithm is inspired by the Danilevsky method for finding the characteristic polynomial. That algorithm was published in the article Danilevsky (1937). A more modern version of the Danilevsky algorithm can be found, for example, in the book Faddeeva (1959), Sect. 24. The Danilevsky algorithm works on the scalar level with cubic complexity.

To get the effect of fast matrix multiplication, you have to construct an algorithm that blocks the operations in an ingenious way. In principle, the algorithm works through a series of similarity transformations that bring the source matrix into a Frobenius form. First, we generalise this form by the so-called m-Frobenius matrix. Namely, the m-Frobenius matrix is the square block matrix of the form:

$$F_m = \left[\begin{array}{c|c} 0 & \\ \hline I_{n-m} & * \end{array}\right] \tag{9.33}$$

where asterisks denote m columns with arbitrary entries. Thus 1-Frobenius matrix is the classical Frobenius matrix of the form (9.24).

The first step in presenting the fast algorithm is to design the similarity matrix. We use it in the following general form:

$$U = F_{2^i} = \left[\begin{array}{c|c} \mathbf{0} & C_1 \\ \hline I_{n-2^i} & C_2 \end{array}\right], \quad U^{-1} = \left[\begin{array}{c|c} -C_2C_1^{-1} & I_{n-2^i} \\ \hline C_1^{-1} & \mathbf{0} \end{array}\right] \tag{9.34}$$

Importantly, we require to be square only the block C_1. All the similarity transformations in a series work according to a common block matrix diagram, realising the general formula (9.23), here in a form $A_{i(j+1)} = U_{ij}^{-1}A_{ij}U_{ij}$. The meaning of the indices will be explained later. Let us first look at the first part $U_{ij}^{-1}A_{ij}$ of this formula on Fig. 9.5.

In order to make the diagrams as clear as possible, we have used the *principle of uniqueness dimensions*, familiar from mechanical engineering drawings. The independent sizes are in bold, all other sizes follow. In order to identify the structure and the dependent dimensions of the respective blocks, we should first remember that **identity and invertible matrices are square**. Furthermore:

- The vertical dimension $2 \cdot 2^i$ of the block a we obtain if we consider the dimensionality of the identity block b equal to $n - (2 \cdot 2^i + j \cdot 2^i)$ and the dimensionality of the identity matrix in the third block column.
- **The last and last but one** F_{ij} block columns should be analysed simultaneously, as they form a coherent structure in the resulting matrix.
  - To obtain the **last** column of F_{ij} blocks, we note that the multiplied block dimensions are compatible. Observe:

$$\left[\begin{array}{c|c} -C_{ij,2}C_{ij,1}^{-1} & I_{n-2^i} \\ \hline C_{ij,1}^{-1} & \mathbf{0} \end{array}\right]\left[\begin{array}{c} C_{ij,1} \\ C_{ij,2} \end{array}\right] = \left[\begin{array}{c} -C_{ij,2} + C_{ij,2} \\ I + \mathbf{0} \cdot C_{ij,2} \end{array}\right] = \left[\begin{array}{c} \mathbf{0} \\ I_{2^i} \end{array}\right]$$

  - The blocks involved in the products that create the **last but one** block column in F_{ij} are not directly outlined in Fig. 9.5. To obtain their shape, it is sufficient to note that the number of rows in block c is greater than 2^i. We can then proceed as follows:

$$\left[\begin{array}{c|c} -C_{ij,2}C_{ij,1}^{-1} & I_{n-2^i} \\ \hline C_{ij,1}^{-1} & \mathbf{0} \end{array}\right]\left[\begin{array}{c} \mathbf{0}_{2^i \times j2^i} \\ \hline \left[\begin{array}{c} \mathbf{0}_{n-(j+1)2^i \times j2^i} \\ I_{j2^i} \end{array}\right] \end{array}\right] = \left[\begin{array}{c} \mathbf{0}_{n-(j-1)2^i \times j2^i} \\ I_{j2^i} \\ \hline \mathbf{0}_{2 \times j2^i} \end{array}\right]$$

So the last but one and the last block column together now form a single block column, of the form:

$$\left[\begin{array}{c} \mathbf{0} \\ I_{(j+1)2^i} \end{array}\right]$$

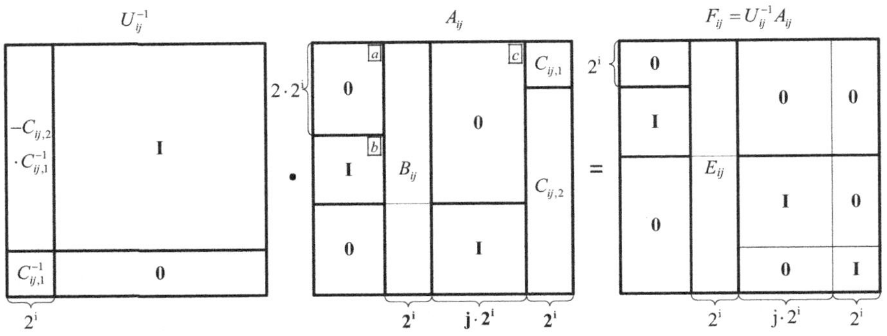

Fig. 9.5 Keller-Gehrig branching algorithm. Block matrix multiplication - diagram 1 of 2

- Similarly, we can construct **the first** F_{ij} block column. The row number of the block a is different from the corresponding dimension in the block c, but it is sufficient that the dimension in question is also greater than 2^i.

Let us now analyse the second product, shown in Fig. 9.6.

When multiplying of F_{ij} by the first block column of U_{ij} in the form $\left[\,\mathbf{0}\; I\,\right]^T$ the dimensions are not compatible. However, we can observe that the zero block $\left[\,\mathbf{0}\;\cdot\,\right]^T$ "annihilates" the first 2^i columns of the left, F_{ij} matrix. The remaining $(n-2^i)$ F_{ij}'s columns are multiplied by the identity matrix in block $\left[\,\cdot\; I\,\right]^T$. It is nothing more than shifting the F_{ij} matrix columns by 2^i entries to the left. Most importantly, the gap between B_{ij} and C_{ij} blocks, increases by 2^i entries, from $j2^i$ to $(j+1)2^i$.

Now it is time to explain how to use the matrix multiplication diagrams in Fig. 9.5 and in Fig. 9.6 to transform a square matrix into the Frobenius form (9.24). In general, they work in a double nested loop:

- In the outer loop, controlled by the i variable, we transform a given matrix A into a series of similar matrices $A \to A_r \to A_{r-1} \to A_{r-2} \to \cdots \to A_{i+1} \to A_i \to$

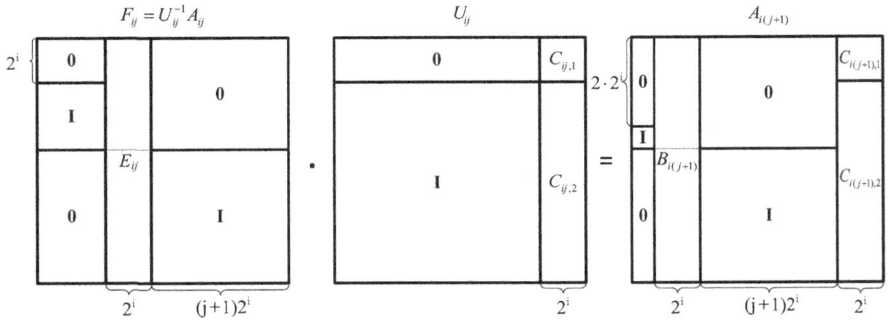

Fig. 9.6 Keller-Gehrig branching algorithm. Block matrix multiplication - diagram 2 of 2

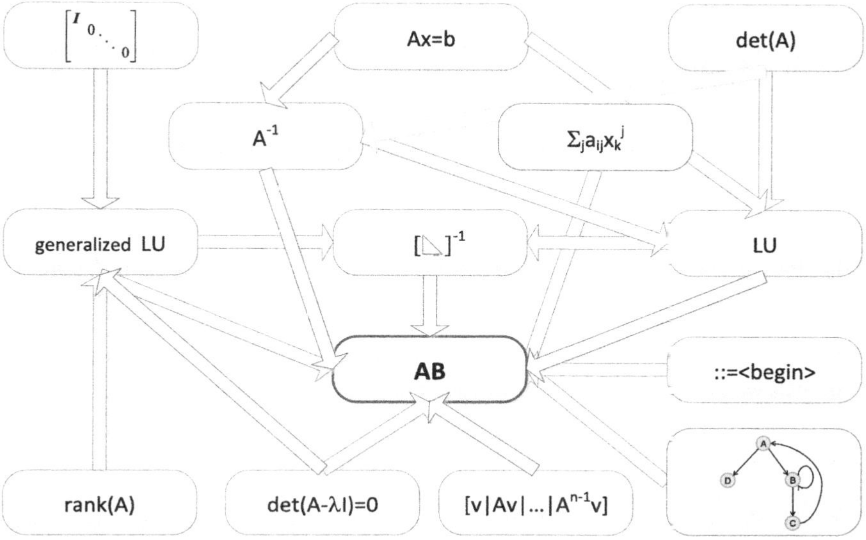

Fig. 9.7 Dependency graph of the black-boxed matrix multiplication algorithms

$\cdots \to A_1 \to A_0$, where $r = \lceil \log_2 n \rceil$ and A_i is in the 2^i-Frobenius form. Since A_0 is in the classical Frobenius form (9.24), we can directly read the characteristic polynomial (9.22).

- The internal loop is controlled by the j variable. We proceed as follows:
  - If n is an exact power of 2 assign $A_r = A$ and then go the next step. Otherwise[13] in the source matrix A extract the blocks C_1, C_2, with the highest possible $r-1$ column number that is an exact power of 2, according to the diagram:

$$A = \left[* \left| \frac{C_1}{C_2} \right. \right], \quad \dim[C_1] = 2^{r-1} \times 2^{r-1} \tag{9.35}$$

    and calculate $A_{r-1} = U^{-1} \cdot A \cdot U$, where U we construct directly by (9.34).
  - To get the matrix A_i from the matrix A_{i+1}, we write the second one in the form:

$$A_{i+1} = \left[\frac{\mathbf{0}}{I_{n-2^{i+1}}} \left| B_i \right| \frac{C_{i,1}}{C_{i,2}} \right] \tag{9.36}$$

[13] In this case from A we go directly to A_{r-1}.

dividing the last 2^{i+1} columns into two equal column blocks, each one containing 2^i columns. Next we apply $s_i = \lceil n/2^i \rceil - 1$ times formula $A_{i(j+1)} = U_{ij}^{-1} A_{ij} U_{ij}$ (diagrams Figs. 9.5 and 9.6). We obtain for a given i a series of similar matrices $A_{i+1} = A_{i,0} \rightarrow A_{i,1} \rightarrow A_{i,2} \rightarrow \cdots \rightarrow A_{i,j} \rightarrow \cdots \rightarrow A_{i,s_i} = A_i$. Note that A_i and $A_{i-1,0}$ are in the same 2^i degree of the Frobenius form.

We can now briefly summarise the working principle of the fast Keller-Gehrig algorithm:

- Each iteration of the **internal loop** shifts the—unwanted—block B_{ij} by 2^i columns to the left. This continues until the B_{ij} block vanishes and the matrix receives the 2^i-Frobenius form.
- For each iteration of the **outer loop,** the degree of the Frobenius form is halved, as long as we obtain 1-Frobenius form and thus the desired characteristic polynomial coefficients.

To prove the efficiency of the fast algorithm, as its name suggests, let us list the operations required to compute the $U_{ij}^{-1} A_{ij} U_{ij}$ term:

- An inversion of $C_{ij,1}$ matrix of $2^i \times 2^i$ dimension—by the algorithm from Sect. 9.4 feasible in $O(M(2^i))$ steps.
- Compute the term $C_{ij,2} C_{ij,1}^{-1}$. Since the dimension of the vertical block $C_{ij,1}$ is $(n - 2^i) \times 2^i$, we can divide it into $\lceil n/2^i \rceil - 1$ submatrices and calculate the product $O\big(n/2^i \cdot M(2^i)\big)$ steps.
- To compute $U_{ij}^{-1} A_{ij}$, we need to multiply block $\begin{bmatrix} -C_{ij,2} C_{ij,1}^{-1} \\ -C_{ij,1}^{-1} \end{bmatrix}$ by B_{ij} to get E_{ij}. Similarly to the previous step, after dividing it now into $\lceil n/2^i \rceil$ submatrices this can be done in $O\big(n/2^i \cdot M(2^i)\big)$ steps.
- Next, to calculate $(U_{ij}^{-1} A_{ij}) \cdot U_{ij}$ by analogy, we multiply E_{ij} by C_{ij} also by $O\big(n/2^i \cdot M(2^i)\big)$ cost.

The remaining operations are add/subtract and move blocks, which can cost up to a factor of four, so can be omitted.

In total, calculating $U_{ij}^{-1} A_{ij} U_{ij}$ term costs $O\big(n/2^i \cdot M(2^i)\big)$. For a given i we do it about $n/2^i$ times so the step $A_{i+1} \rightarrow A_i$ costs $n/2^i \cdot O\big(n/2^i \cdot M(2^i)\big)$ operations. The whole process $A \rightarrow A_0$ takes:

$$T(n) = \sum_{i=0}^{\lceil \log_2 n \rceil - 1} O\left[\frac{n^2}{2^{2i}} M(2^i)\right]$$

steps. To calculate the sum, we model the applied matrix multiplication complexity as $M(n) = C_\omega n^\omega$ and sum up the obtained geometric series. Observe:

$$T(n) = n^2 \sum_{i=0}^{\lceil \log_2 n \rceil - 1} O\left[\frac{1}{\left(2^i\right)^2} C_\omega (2^i)^\omega\right] = n^2 O\left[\sum_{i=0}^{\lceil \log_2 n \rceil - 1} (2^{\omega-2})^i\right]$$

$$= n^2 O\left(\frac{n^{\omega-2} - 1}{2^{\omega-2} - 1}\right) = n^2 O\left(n^{\omega-2}\right) = O(n^\omega) = O[M(n)]$$

■

- **Example**: Execution of the fast Keller-Gehrig algorithm

Consider the following 7×7 matrix:

$$A = \left[\begin{array}{ccc|cccc} 4 & 7 & 9 & -6 & 1 & 0 & 7 \\ 8 & 4 & -9 & 2 & 3 & 5 & 4 \\ 7 & 1 & 7 & 3 & 4 & 1 & 4 \\ 0 & -2 & 3 & 5 & 8 & 1 & 1 \\ \cline{4-7} 5 & 3 & 5 & 8 & 9 & 1 & 2 \\ 3 & -3 & 4 & 7 & 4 & 5 & -4 \\ 1 & 2 & 4 & 8 & 3 & 3 & 6 \end{array}\right]$$

with tentatively highlighted blocks to be used by the algorithm. In the example matrix we have $r = \lceil \log_2 n \rceil = 3$ and $n = 7$ is not a power of two. Thus we dispatch those blocks according to formula (9.35), which gives us the similarity matrix U leading to $A_{r-1} = A_2 = U^{-1}AU$:

$$U = \left[\begin{array}{ccc|cccc} 0 & 0 & 0 & -6 & 1 & 0 & 7 \\ 0 & 0 & 0 & 2 & 3 & 5 & 4 \\ 0 & 0 & 0 & 3 & 4 & 1 & 4 \\ 0 & 0 & 0 & 5 & 8 & 1 & 1 \\ \hline 1 & 0 & 0 & 8 & 9 & 1 & 2 \\ 0 & 1 & 0 & 7 & 4 & 5 & -4 \\ 0 & 0 & 1 & 8 & 3 & 3 & 6 \end{array}\right], \quad A_2 = \left[\begin{array}{ccc|cc|cc} 0 & 0 & 0 & 1.20 & 8.49 & 27.83 & 17.81 \\ 0 & 0 & 0 & 49.85 & 103.30 & 25.53 & 127.99 \\ \cline{6-7} 0 & 0 & 0 & 96.15 & -27.85 & 2.61 & -130.94 \\ 0 & 0 & 0 & -0.97 & 11.01 & -0.79 & 10.91 \\ \hline 1 & 0 & 0 & 14.29 & 8.07 & 0.74 & -5.21 \\ 0 & 1 & 0 & -4.92 & -8.98 & 4.24 & -14.57 \\ 0 & 0 & 1 & 4.41 & 14.42 & 7.79 & 28.67 \end{array}\right]$$

We can see that A_2 is in the 2^2-Frobenius form (9.33). From A_2 we can reach A_1 in $s_1 = \lceil n/2^1 \rceil - 1 = 3$ similarity transformations $A_{1(j+1)} = U_{1j}^{-1} A_{1j} U_{1j}$, defined by the key diagrams Figs. 9.5 and 9.6. The whole step can be modelled by a flowchart:

$$A_2 = A_{1,0} \rightarrow A_{1,1} \rightarrow A_{1,2} \rightarrow A_{1,3} = A_1$$

We now construct the similarity matrix using the second, block matrix formula (9.36).

$$U_{10} = \left[\begin{array}{ccccc|cc} 0&0&0&0&0 & 27.83 & 17.81 \\ 0&0&0&0&0 & 25.53 & 127.99 \\ \hline 1&0&0&0&0 & 2.61 & -130.94 \\ 0&1&0&0&0 & -0.79 & 10.91 \\ 0&0&1&0&0 & 0.74 & -5.21 \\ 0&0&0&1&0 & 4.24 & -14.57 \\ 0&0&0&0&1 & 7.79 & 28.67 \end{array}\right],$$

$$A_{1,1} = \left[\begin{array}{c|cc|cc|cc} 0 & 153.94 & 84.79 & 0 & 0 & -58.45 & 1\,238.16 \\ 0 & -5.92 & 1.48 & 0 & 0 & 5.74 & -72.30 \\ 0 & 16.74 & 12.70 & 0 & 0 & 24.03 & 134.36 \\ 0 & 2.45 & 4.51 & 0 & 0 & 26.94 & 131.18 \\ 1 & -6.264 & -8.21 & 0 & 0 & 1.47 & -156.54 \\ 0 & -0.24 & -0.24 & 1 & 0 & 4.24 & -15.89 \\ 0 & 0.44 & 0.86 & 0 & 1 & 8.08 & 28.98 \end{array}\right]$$

Notice that a new 2-column block $\left[\mathbf{0}\ I\right]^T$ has appeared before the last $2^i = 2$ columns. Now, in this step, we will show in detail how the diagrams Fig. 9.5 and in Fig. 9.6 work, for $i = j = 1$. First, the $F_{11} = U_{11}^{-1} A_{11}$ part is multiplied as follows:

$$\overbrace{\left[\begin{array}{cc|ccccc} -0.87 & -13.03 & 1&0&0&0&0 \\ -0.94 & -14.21 & 0&1&0&0&0 \\ 0.27 & 2.54 & 0&0&1&0&0 \\ -0.075 & -1.50 & 0&0&0&1&0 \\ -0.26 & -4.05 & 0&0&0&0&1 \\ \hline 0.025 & 0.43 & 0&0&0&0&0 \\ 0.00 & 0.02 & 0&0&0&0&0 \end{array}\right]}^{U_{11}^{-1}} \cdot \overbrace{\left[\begin{array}{c|cc|cc|cc} 0 & 153.94 & 84.79 & 0 & 0 & -58.45 & 1\,238.16 \\ 0 & -5.92 & 1.48 & 0 & 0 & 5.74 & -72.30 \\ 0 & 16.74 & 12.70 & 0 & 0 & 24.03 & 134.36 \\ 0 & 2.45 & 4.51 & 0 & 0 & 26.94 & 131.18 \\ 1 & -6.264 & -8.21 & 0 & 0 & 1.47 & -156.54 \\ 0 & -0.24 & -0.24 & 1 & 0 & 4.24 & -15.89 \\ 0 & 0.44 & 0.86 & 0 & 1 & 8.08 & 28.98 \end{array}\right]}^{A_{11}} =$$

$$= \left[\begin{array}{c|cc|cc|cc} 0 & -39.95 & -80.33 & 0 & 0 & 0 & 0 \\ 0 & -57.48 & -95.89 & 0 & 0 & 0 & 0 \\ 1 & 20.99 & 18.83 & 0 & 0 & 0 & 0 \\ 0 & -2.86 & -8.80 & 1 & 0 & 0 & 0 \\ 0 & -15.60 & -27.20 & 0 & 1 & 0 & 0 \\ 0 & 1.32 & 2.76 & 0 & 0 & 1 & 0 \\ 0 & 0.19 & 0.20 & 0 & 0 & 0 & 1 \end{array}\right] = F_{11}$$

The second multiplication $(U_{11}^{-1} A_{11}) \cdot U_{11} = F_{11} \cdot U_{11}$ of the product goes as follows:

$$
F_{11}\cdot U_1 = \overbrace{\left[\begin{array}{c|cc|cccc}
0 & -39.95 & -80.33 & 0 & 0 & 0 & 0\\
0 & -57.48 & -95.89 & 0 & 0 & 0 & 0\\
\hline
1 & 20.99 & 18.83 & 0 & 0 & 0 & 0\\
\hline
0 & -2.86 & -8.80 & 1 & 0 & 0 & 0\\
0 & -15.60 & -27.20 & 0 & 1 & 0 & 0\\
0 & 1.32 & 2.76 & 0 & 0 & 1 & 0\\
0 & 0.19 & 0.20 & 0 & 0 & 0 & 1
\end{array}\right]}^{F_{11}=U_{11}^{-1}A_{11}} \cdot \overbrace{\left[\begin{array}{ccccc|cc}
0 & 0 & 0 & 0 & 0 & -58.45 & 1\,238.16\\
0 & 0 & 0 & 0 & 0 & 5.74 & -72.30\\
\hline
1 & 0 & 0 & 0 & 0 & 24.03 & 134.36\\
0 & 1 & 0 & 0 & 0 & 26.94 & 131.18\\
0 & 0 & 1 & 0 & 0 & 1.47 & -156.54\\
0 & 0 & 0 & 1 & 0 & 4.24 & -15.89\\
0 & 0 & 0 & 0 & 1 & 8.08 & 28.98
\end{array}\right]}^{U_{11}} =
$$

$$
= \left[\begin{array}{c|cccc|cc}
-80.33 & 0 & 0 & 0 & 0 & -2\,160.07 & -7\,904.09\\
-95.89 & 0 & 0 & 0 & 0 & -2\,634.80 & -8\,728.30\\
\hline
18.83 & 0 & 0 & 0 & 0 & 514.78 & 2\,251.46\\
\hline
-8.80 & 1 & 0 & 0 & 0 & -200.93 & -843.98\\
-27.20 & 0 & 1 & 0 & 0 & -741.89 & -2\,683.02\\
2.76 & 0 & 0 & 1 & 0 & 78.13 & 259.69\\
0.20 & 0 & 0 & 0 & 1 & 13.93 & 42.20
\end{array}\right] = A_{12}
$$

The size of the internal block $\left[\mathbf{0}\ I\right]^T$ increases by $2^i = 2$ columns. Repeating the similarity transformation $A_{1(j+1)} = U_{1j}^{-1}A_{1j}U_{1j}$ once more with $j = 2$, we obtain the 2^1-Frobenius form:

$$
A_{1,3} = A_1 = \left[\begin{array}{ccccc|c|c}
0 & 0 & 0 & 0 & 0 & 2\,485.16 & 9\,093.62\\
0 & 0 & 0 & 0 & 0 & -309.77 & -220.60\\
\hline
1 & 0 & 0 & 0 & 0 & -167.55 & -245.33\\
0 & 1 & 0 & 0 & 0 & -15.35 & -164.90\\
0 & 0 & 1 & 0 & 0 & -95.76 & -318.73\\
0 & 0 & 0 & 1 & 0 & 15.89 & 31.97\\
0 & 0 & 0 & 0 & 1 & 8.98 & 24.11
\end{array}\right]
$$

To perform the last outside loop iteration $A_1 \to A_0$ we again divide the matrix $A_1 = A_{0,0}$ according to the block diagram (9.36) and apply the formula $A_{0(j+1)} = U_{0j}^{-1}A_{0j}U_{0j}$ in a loop for $j = 0$ to $j = s_0 - 1 = 5$. In this iteration of the outer loop, with $i = 0$, we have a single column just to the right of $A_{0(j+1)}$, and at the same time we are shifting the second single column to the left by one position per inner loop iteration. We have illustrated the last outside loop iteration in the following diagram:

$$A_{0,1} = \left[\begin{array}{cccc|c|c|c} & \mathbf{0} & & & * & \mathbf{0} & * \\ & & & & * & & * \\ \hline 1 & & & & * & & * \\ & 1 & & & * & & * \\ & & 1 & & * & & * \\ & & & 1 & * & & * \\ \hline & \mathbf{0} & & & * & 1 & * \end{array}\right] \rightarrow A_{0,2} = \left[\begin{array}{ccc|c|cc|c} & \mathbf{0} & & * & \mathbf{0} & & * \\ & & & * & & & * \\ \hline 1 & & & * & & & * \\ & 1 & & * & & & * \\ & & 1 & * & & & * \\ \hline & & & * & 1 & & * \\ & \mathbf{0} & & * & & 1 & * \end{array}\right] \rightarrow \ldots \rightarrow A_{0,6} = A_0 =$$

$$= \left[\begin{array}{cccccc|c} 0&0&0&0&0&0& 2\,268\,661 \\ 1&0&0&0&0&0& -309\,235 \\ 0&1&0&0&0&0& -118\,827 \\ 0&0&1&0&0&0& 24\,019 \\ 0&0&0&1&0&0& 338 \\ 0&0&0&0&1&0& -430 \\ 0&0&0&0&0&1& 40 \end{array}\right] = F_1$$

Considering the general form of the Frobenius matrix (9.24) we obtain the characteristic polynomial in accordance with (9.22):

$$\det(A - \lambda I) = -(\lambda^7 - 40\lambda^6 + 430\lambda^5 - 338\lambda^4 - 24019\lambda^3 + 118{,}827\lambda^2 + 309{,}235\lambda - 2{,}268{,}661)$$

To summarise, in Sect. 9.10 we described three algorithms for finding the characteristic polynomial that derive their efficiency from fast matrix multiplication. All those algorithms were proposed in an article by Keller-Gehrig (1985). The fastest, shown in Sect. 9.10.3 works in a complexity equal to the applied matrix multiplication algorithm, but the cost of this efficiency is that it works under the condition of invertibility of a series of matrices, encountered during the process. The branching algorithm, Sect. 9.10.2, adds a logarithmic factor to the complexity, but works for any matrix.

We can also read the matrix determinant directly from the characteristic polynomial. In notation (9.22) $\det(A) = (-1)^{n+1} p_0$. But it seems to be effective only when we need both that properties at once, since the algorithms from Sect. 9.7 work for any matrix in better time than the universal, branching algorithm for characteristic polynomial. The same is even more true for the—in linear time computable—trace, since trace(A) $= p_{n-1}$.

The branching algorithm we described in Sect. 9.10.2 leads to the block upper triangular form, with Frobenius blocks on the main diagonal. Article Storjohann (2001) proposes an algorithm that starts from this point and leads to obtain a block matrix with only Frobenius blocks on the main diagonal (some of the authors just call this matrix a Frobenius matrix, with the blocks having the name of a companion matrix). The proposed algorithm works in $O(n^{\omega}(\log n)(\log\log n))$ time.

9.11 Finding Successive Powers of Matrices

The article Galil and Pan (1989) in Sect. 4 pages 43–44 shows also how to apply an algorithm for fast matrix multiplication of rectangular matrices for finding a series of successive powers of a given square matrix. Suppose we want to calculate a series $A,\ A^2,\ A^3,\ \ldots,\ A^m$ of powers for a given A square matrix. Clearly, by the definition of matrix multiplication, and even by the fast algorithm for square matrices, it requires $m-1$ matrix multiplications (e.g. Bini and Pan 1994, Problem 2.2a, page 97). The $A,\ A^2,\ A^3,\ \ldots,\ A^m$ series can then be calculated in the following steps:

- Assign as series $A,\ A^2,\ A^3,\ \ldots,\ A^m$ a series $A,\ A^2,\ A^3,\ \ldots,\ A^{h^2-1}$. So from equality $m=h^2-1$ we get $h=\lceil\sqrt{m+1}\rceil$.
- Calculate a matrix series $A,\ A^2,\ A^3,\ \ldots,\ A^h$.
- Using the A^h obtained in the previous step, calculate a matrix series $A^h,\ (A^h)^2,\ (A^h)^3,\ \ldots\ ,\ (A^h)^{h-1}$.

Then the calculation of the $A,\ A^2,\ A^3,\ \ldots,\ A^{h^2-1}$ series can be replaced by the following $(n(h-1),\ n,\ n(h-1))$ rectangular matrix product problem (Galil and Pan 1989, page 43):

$$\begin{bmatrix} A \\ A^2 \\ \vdots \\ A^{h-1} \end{bmatrix} \begin{bmatrix} A^h & (A^h)^2 & \cdots & (A^h)^{h-1} \end{bmatrix} \tag{9.37}$$

Originally described in Galil and Pan (1989), this method is used to compute successive powers of a matrix in parallel.

9.11.1 *Example: Find First 24 Powers of a Given Square Matrix A*

We have $m=24$ thus $h=\lceil\sqrt{m+1}\rceil=\lceil\sqrt{24+1}\rceil=5$. First we calculate a series $A,\ A^2,\ A^3,\ A^4,\ A^{5(=h)}$. Next, we find a series $A^5,\ (A^5)^2,\ (A^5)^3,\ (A^5)^{4(=h-1)}$ so the rectangular matrix product (9.37) receives the $(n(5-1),\ n,\ n(5-1))=(4n, n, 4n)$ form, where $n\times n$ is the dimension of the A matrix:

$$\begin{bmatrix} A \\ A^2 \\ A^3 \\ A^4 \end{bmatrix} \begin{bmatrix} A^5 & A^{10} & A^{15} & A^{20} \end{bmatrix} = \begin{bmatrix} A^6 & A^{11} & A^{16} & A^{21} \\ A^7 & A^{12} & A^{17} & A^{22} \\ A^8 & A^{13} & A^{18} & A^{23} \\ A^9 & A^{14} & A^{19} & A^{24} \end{bmatrix}$$

To sum up, before it is possible to use a specialised rectangular fast matrix multiplication algorithm to find a series of consecutive matrix powers, we first have to use the standard or fast square matrix multiplication algorithm $2h - 3 = O(\sqrt{m})$ times.

9.12 Survey of Other Applications

Below we have reviewed other applications of fast matrix multiplication algorithms discovered so far, distinguishing those that use algorithms for square and rectangular matrices. Some of them use algorithms for fast multiplication of both square and rectangular matrices.

In general, besides the obvious major group of applications just to matrix operations, for dense matrices (interestingly, for generic, structured and indirectly to sparse), it can be noticed that a vast part of them applies to a variety of graph problems.

There are also applications to computational biology, neural networks, artificial intelligence, Boolean functions, formal grammars, pattern recognition, parallel computing, combinatorics and a range of work on linear optimisation and others.

We hope the reader will appreciate that we have grouped the applications in this order.

9.12.1 Applications of Square Matrices Multiplication

Application of fast matrix multiplication algorithms to *square* matrices are the following:

- For matrix-related problems
  - **Matrix inversion**: We can invert any non-singular matrix in a different way, than shown in Sect. 9.4. Namely, we can use the LU decomposition by the following matrix computations: $A^{-1} = (LUP)^{-1} = P^{-1}U^{-1}L^{-1} = P^TU^{-1}L^{-1}$. The inversions U^{-1} and L^{-1} can be obtained by the inversion algorithm from Sect. 9.3, specialised in triangular matrices.
  - **Trace of the inverse matrix**: Article Kalorkoti (1988) proved another method for finding trace of an inverse matrix.
  - **Problems related to Krylov matrix**: Article Dumas et al. (2005) proposes an $O(\log n \cdot M(n))$ time algorithm to find a minimal polynomial of a Krylov matrix.
  - **Special matrices**: Article Hyun et al. (2017) applies fast matrix multiplication (by its very idea for generic matrices) to special matrices,[14] i.e. Cauchy-like

[14] We devote an Appendix in the monograph to special matrices.

and Toeplitz type matrices. In particular, they use it in an iterative algorithm for solving systems of equations with these types of matrices.

- **Matrix pseudoinverse**: Article Ibarra et al. (1982) presents how to obtain the Moore-Penrose pseudoinverse in $O(M(n))$ time.
- **QR decomposition, unitary transformation and unitary triangularization of Hermitian matrices:** Article Schonhage (1972) gives algorithms for QR-decomposition and unitary transformation of arbitrary complex matrices to the upper Hessenberg form and for unitary triangularization of Hermitian matrices which work in the asymptotic time of the applied matrix multiplication algorithm. Later in 1995 Knight in Sect. 3, pages 77–80, proposed how to perform fast block QR decomposition, in general by fast matrix multiplication of rectangular matrices.
- **Parallel solving of systems of linear equations and matrix inversion**: Articles Kaltofen and Pan (1991, 1992) propose randomised algorithms for the title problems. The proposed algorithms use the results of Baur and Strassen from 1983, which will be discussed later in this section. The problem of parallel inversion of a matrix was also addressed by Galil and Pan in a 1989 article, also referred to later in this section.
- **Finding the matrix kernel basis, diagonalizing a quadratic form and matrix sparse representation**: Article Burgisser et al. (1991) showed that the three title problems, i.e. computing a basis of the kernel for a given $n \times n$ matrix, finding an invertible matrix that transforms a given quadratic form matrix (symmetric) into a diagonal form and finding a sparse representation of a given $n \times n$ matrix can be solved in a way that has the same time complexity as the applied fast matrix multiplication algorithm.
- **Maximum Subarray Problem**: Articles Takaoka (2002) and Tamaki and Tokuyama (2000) show how to find a maximum size subarray (constructed from adjacent rows and columns), with non-negative entries, in the matrix multiplication time.

* For graph-related problems

  - **Compiler construction**: For efficient memory management it is important to know which of the compiled program functions are recursive. This can be done by computing the so-called **transitive closure of a graph**, which is expressed as a certain matrix polynomial. The article Munro (1971) shows how to use matrix multiplication in a recursive way in this problem, thus obtaining an $O(M(n))$ time algorithm. The article Fisher and Meyer (1971) solves an inverse problem, i.e. how to multiply matrices using the available transitive closure algorithm.
  - **Graph shortest paths problems**: The work Seidel (1992) computes the shortest distances between all pairs of vertices of an undirected, unweighted graph in $O(\log n \cdot M(n))$ time. The article Alon et al. (1991) generalises these results to the case where the weights are between 0 and B with $O\big(B^2 \cdot \log n \cdot M(n)\big)$ time. Another work on this topic we presented in the

section dedicated to the application of fast multiplication of rectangular matrices.
  - **Maximum matching in a graph**: Another graph problem tackled by the fast matrix multiplication is finding a maximum matching in a general graph. The appropriate algorithms are given in the article Mulmuley et al. (1987).

- Other types of problems
  - **Formal grammar problems**: The black box matrix multiplication paradigm has also been applied to the formal grammar problems. The example is the article Valiant (1975) where this method allowed to achieve $O(M(n))$ time in the problem of context-free recognition.
  - **Pattern recognition tasks:** Finding and classifying similar objects using fast matrix multiplication algorithms, proposed the article Cohen and Lewis (1999).
  - **Computational biology:** Fast matrix multiplication has been applied to certain problems in gene analysis (DNA). Relevant works are Akutsu et al. (2000) and Biedl et al. (2001–14). Another work in this area, where the efficiency depends on the applied matrix multiplication algorithm, is Shabalin (2012).
  - **Boolean functions**: Work Mossel et al. (2003) applied fast matrix multiplication to the problem of learning an arbitrary Boolean function that depends on an unknown set of k out of n Boolean variables. The obtained efficiency is $(n^k)^{\omega/(\omega+1)}$ compared to the previous naive bound n^k.
  - **Partial derivatives of rational functions:** The work Baur and Strassen (1983), in particular Corollary 5 on page 328, shows how to calculate partial derivatives of a given rational function in the time of applied matrix multiplication algorithm.
  - **Generalized string matching:** The article by Bernardini et al. (2022) applied fast matrix multiplication to the problem of so-called elastic-degenerate string matching.
  - **Computation of so-called approximant bases in canonical form**: This problem is tackled by the work Jeannerod et al. (2020).
  - **Multivariate polynomials:** Article Andrews (2024) applies fast matrix multiplication to a classical problem of polynomial identity testing.
  - **Databases:** Article Xiao (2024) applies fast matrix multiplication to a fundamental problem of query processing.
  - **Artificial Intelligence**: Article Njifon and Schuhmacher (2023) applied fast matrix multiplication algorithms for graph-based neural networks.

A valuable survey of the problems of **transitive closure and context free grammars** in the context of fast matrix multiplication is given in the article Paterson (1974) in Sect. 6 pages 487–488.

9.12.2 Applications of Rectangular Matrices Multiplication

Examples of applications of fast multiplication algorithms for *rectangular* matrices are:

- For matrix-related problems
  - **Sparse matrices**: An interesting application of fast (rectangular) matrix multiplication algorithms is the work of Yuster and Zwick (2005b). They have successfully used fast matrix multiplication algorithms for dense matrices (the subject of this monograph) to construct fast algorithms for sparse matrices.
  - **Parallel computation of the determinant and inversion of a matrix**: The Article Galil and Pan (1989) applies fast algorithms for rectangular matrix multiplication in parallel computation of the determinant and inversion of a matrix. The mathematical internals for that is given in the article Preparata and Sarwate (1978), which in turn inherits and optimises results presented earlier in the work Csanky (1975).
- For graph-related problems
  - **Graph shortest paths problems**: Works on this problem uses both square and rectangular matrix multiplication algorithms. Examples of the latter are Zwick (1998), Zwick (2002), Yuster and Zwick (2005a), Alon and Yuster (2007).
  - **Finding the lowest common ancestors of a given graph class**: The corresponding algorithm for this problem is presented in an article by Czumaj et al. (2007).
  - **Subgraph problems**: Article Nesetril and Poljak (1985) constructs an algorithm that solves the problem of whether for a fixed k a given graph contains a complete subgraph with k vertices.
  - **Detecting cycles in graphs**: The article Yuster and Zwick (2004) applies fast rectangular matrix multiplication to detect certain kind of cycles in a selected type of graphs.
  - **Detecting bottlenecks in shortest graphs paths**: Work Duan and Pettie (2009) tackles this problem with the use of both square and rectangular fast matrix multiplication algorithms.
  - **Transitive closure**: Article Demetrescu and Italiano (2000) applies fast matrix multiplication algorithms to enable dynamic operations on a directed graph, such as inserting and deleting edges in a given graph in an efficient way. Importantly, its transitive closure is maintained during this dynamic process. A randomised approach to the same problem is given in article Sankowski and Mucha (2010).
  - **Other graph problems**: Bentert et al. (2023).
- Other types of problems

- **Linear programming optimization**: There is a broad literature on applications of fast rectangular matrix multiplication to the linear programming optimisation problem. Examples are the articles by Vaidya (1989), Beling and Megiddo (1998), Brand (2020) and Cohen et al. (2021).
- **Sets of permutations**: The Yuster (2009) article uses fast rectangular matrix multiplication to manipulate sets of permutations in an efficient way.
- **Artificial Intelligence**: Article Brand et al. (2021) applied fast rectangular matrix multiplication algorithms for training neural networks.
- **Modular composition of polynomials**; **formal power series**: Modular composition is the problem of computing the composition of two univariate polynomials modulo a third one. The problem is tackled using of fast matrix multiplication in the works of Brent and Kung (1978) and Hoeven and Lecerf (2018). The former of the two additionally proves that this problem is equivalent to the inversion of a formal power series.

9.13 Summary of the Application of Fast Matrix Multiplication

In this chapter we surveyed a number of applications of efficient matrix multiplication algorithms. These applications appear in the field of matrix algebra and in other types of problems, seemingly far away. The core of the black-box algorithms is always the matrix multiplication algorithm, which is often referred to indirectly. The dependencies of the procedures are summarised in Fig. 9.7.

In general, the idea of applying fast matrix multiplication algorithms to other matrix problems is to block their respective entries into submatrices, then perform the operation on those submatrices, rather than working on the scalar entries separately. In the case of non-matrix operations, we represent a given problem in matrix form, and then work on that representation. For example, in the case of graph algorithms, we do not use the list representation of graphs, but just the matrix representation.

Another important issue when constructing recursive algorithms for matrices is that we must ensure the feasibility of all operations on the respective submatrices at all levels of recursion. In particular, this concerns the non-singularity of selected submatrices during the process of inverting matrices in Sect. 9.4.

Furthermore, the implementations of matrix multiplication on modern, highly vector-oriented and multi-core-oriented hardware (incessantly trickling down to mainstream desktop computers) are the subject of work by many computer scientists, due to the essential importance of this operation in other scientific and engineering domains. An example of work in this area is the article Schwartz and Vaknin (2023). Thanks to this, the available implementations are highly optimised, so that they can be sensibly used in other matrix and non-matrix problems. In this way, we solve the other problems quickly, taking advantage of the capabilities of modern computer systems, although indirectly, by simply using the black-boxed matrix multiplication.

Surprisingly, the above applies even to implementations of the most classical $O(n^3)$ version of the algorithm, but implemented in such a way that the available hardware capabilities should be fully exploited. In particular, this applies to the Advanced Matrix Extensions (AMX) available on Intel Xeon microprocessors, which is nothing more than performing matrix multiplication at the hardware level.

In conclusion, despite the existence of many important theoretical and algorithmic solutions, fast matrix multiplication algorithms, together with their applications in other matrix and non-matrix applications, are still a promising and dynamically developing branch of numerical algorithms, involving hundreds of researchers and engineers.

References

Aho AV, Hopcroft JE, Ullman JD (1974) The design and analysis of computer algorithms. Addison-Wesley Pub., Mass. U.S.A

Akutsu T, Miyano S, Kuhara S (2000) Algorithms for identifying Boolean networks and related biological networks based on matrix multiplication and fingerprint function. J Comput Biol 7(3/4):331–343

Alon N, Galil Z, Margalit O (1991) On the exponent of the all pairs shortest path problem. In: Proceedings of 32nd annual IEEE symposium on foundations of computer science. IEEE Computer Society Press, 569–575

Alon N, Yuster R (2007) Fast algorithms for maximum subset matching and all-pairs shortest paths in graphs with a (not so) small vertex cover. Springer-Verlag Berlin Heidelberg, LNCS 4698, 175–186

Andrews R (2024) On matrix multiplication and polynomial identity testing. SIAM J Comput Special Section Prev FOCS

Baur W, Strassen V (1983) The complexity of partial derivatives. Theoret Comput Sci 22(3):317–330

Beling PA, Megiddo N (1998) Using fast matrix multiplication to find basic solutions. Theoret Comput Sci 205:307–316

Bentert M, Heeger K, Koana T (2023) Fully polynomial-time algorithms parameterized by vertex integrity using fast matrix multiplication. In: 31st annual European symposium on algorithms (ESA), Amsterdam, Netherlands 274(16), 1–16

Bernardini G, Gawrychowski P, Pisanti N, Pissis SP, Rosone G (2022) Elastic-degenerate string matching via fast matrix multiplication. SIAM J Comput 51(3):549–576

Biedl T, Brejova B, Demaine ED, Hamel AM, Vinar T (2001–14) Optimal arrangement of leaves in the tree representing hierarchical clustering of gene expression data. Tech. Report, Department of Computer Science, University of Waterloo, 12 p

Bini D, Pan V (1994) Polynomial and matrix computations. Fundamental algorithms vol. 1 progress in theoretical computer science. Birkhauser, Boston

Bodrato M (2010) A Strassen-like matrix multiplication suited for squaring and higher power computation. In: Proceedings of 2010 international symposium on symbolic and algebraic computation (ISSAC 2010), 273–280

Borodin A, Munro I (1971) Evaluating polynomials at many points. Inf Process Lett 1(2):66–68

Brand J, Peng B, Song Z, Weinstein O (2021) Training (overparametrized) neural networks in near-linear time. In: 12th innovations in theoretical computer science conference (ITCS 2021), 63, 1–15

Brand J (2020) A deterministic linear program solver in current matrix multiplication time. In: Proceedings of 2020 ACM-SIAM symposium on discrete algorithms (SODA), 259–278

Brent RP, Kung HT (1978) Fast algorithms for manipulating formal power series. J ACM 25(4):581–595
Bunch J, Hopcroft J (1974) Triangular factorization and inversion by fast matrix multiplication. Math Comput 28(125):231–236
Burgisser P, Karpinski M, Lickteig T (1991) Some computational problems in linear algebra as hard as matrix multiplication. Comput Complex 1:131–155
Cohen MB, Lee YT, Song Z (2021) Solving linear programs in the current matrix multiplication time. J ACM 68(1):1–39
Cohen E, Lewis DD (1999) Approximating matrix multiplication for pattern recognition tasks. J Algorithms 30:211–252
Cormen TH, Leiserson ChE, Rivest RL, Stein C (2001) Introduction to algorithms 2. The MIT Press, Cambridge
Csanky L (1975) Fast parallel matrix inversion algorithms. In: 16th annual symposium on foundations of computer science, USA, pp 11–12
Czumaj A, Kowaluk M, Lingas A (2007) Faster algorithms for finding lowest common ancestors in directed acyclic graphs. Theoret Comput Sci 380:37–46
Danilevskii AM (1937) The numerical solution of secular equation. Matem Sbornik 44(2):169–171 (Russian)
Demetrescu C, Italiano GF (2000) Fully dynamic transitive closure: breaking through the O(n/sup 2/) barrier. In: Proceedings of 41st annual symposium on foundations of computer science, 381–389
Duan R, Pettie S (2009) Fast algorithms for (max, min)-matrix multiplication and bottleneck shortest paths. In: Proceedings of 2009 annual ACM-SIAM symposium on discrete algorithms (SODA), 384–391
Dumas J-G, Pernet C, Wan Z (2005) Efficient computation of the characteristic polynomial. In: Proceedings of the 2005 int. symposium on symbolic and algebraic computation, 140–147
El-Mikkawy MEA (2003) Inversion of a generalized Vandermonde matrix. Int J Comput Math 80:759–765
Faddeev DK, Faddeeva VN (1963) Computational methods of linear algebra. WH Freeman and Company, San Francisco and London
Faddeeva VN (1959) Computational methods of linear algebra. Dover Publications, New York
Fisher MJ, Meyer AR (1971) Boolean matrix multiplication and transitive closure. In: 12th IEEE annual symposium on switching and automata theory, 129–131
Galil Z, Pan V (1989) Parallel evaluation of the determinant and of the inverse of a matrix. Inf Process Lett 30(1):41–45
Gantmacher FR (1960) The theory of matrices. Chelsea Publishing Company, New York
Hoeven J, Lecerf G (2018) Modular composition via factorization. J Complexity 48:36–68
Hyun SG, Lebreton R, Schost E (2017) Algorithms for structured linear systems solving and their implementation. In: ISSAC'17: proceedings of 2017 ACM on international symposium on symbolic and algebraic computation, 205–212
Ibarra OH, Moran S, Hui R (1982) A generalization of the fast LUP matrix decomposition algorithm and applications. J Algorithms 3:45–56
Jeannerod C-P, Pernet C, Storjohann A (2013) Rank-profile revealing Gaussian elimination and the CUP matrix decomposition. J Symb Comput 56:46–68
Jeannerod C-P, Mary T, Pernet C, Roche DS (2019) Improving the complexity of block low-rank factorizations with fast matrix arithmetic. SIAM Matrix Anal Appl 40(4):1478–1496
Jeannerod C-P, Neiger V, Villard G (2020) Fast computation of approximant bases in canonical form. J Symb Comput 98:192–224
Kalorkoti K (1988) The trace invariant and matrix inversion. Theoret Comput Sci 59(3):277–286
Kaltofen E, Pan V (1991) Processor efficient parallel solution of linear systems over an abstract field. In: Proceedings of third annual ACM symposium on parallel algorithms and architectures (SPAA 91), pp 180–191

Kaltofen E, Pan V (1992) Processor-efficient parallel solution of linear systems. II. The positive characteristic and singular cases. In: Proceedings of 33rd annual symposium on foundations of computer science. IEEE, 714–723
Keller-Gehrig W (1985) Fast algorithms for the characteristic polynomial. Theoret Comput Sci 36:309–317
Lipski W (1989) Combinatorics for programmers. WNT, Warsaw (in Polish)
Mossel E, O'Donnell R, Servedio R (2003) Learning juntas. In: Proceedings of 35th annual ACM symposium on theory of computing (STOC 03), 206–212
Mulmuley K, Vazirani UV, Vazirani VV (1987) Matching is as easy as matrix inversion. Combinatorica 7(1):105–113
Munro I (1971) Efficient determination of the transitive closure of a directed graph. Inf Process Lett 2:56–58
Munro I (1973) Problems related to matrix multiplication. In: Rustin R (ed) Courant institute symposium on computational complexity. Algorithmics Press, New York, 51, 137–152
Nesetril J, Poljak S (1985) On the complexity of the subgraph problem. Comment Math Univ Carol 26(2):415–419
Njifon MA, Schuhmacher D (2023) Graph convolutional networks for spatial interpolation of correlated data. Spatial Statistics 60:1–17
Pan V (1984) How can we speed up matrix multiplication? SIAM Rev 26(3):393–415
Paterson MS, Stockmeyer LJ (1973) On the number of nonscalar multiplications necessary to evaluate polynomials. SIAM J Comput 2(1):60–66
Paterson MS (1974) Complexity of product and closure algorithms for matrices. In: Proceedings of international congress of mathematicians, Vancouver, 483–489
Pereslavtseva ON (2008) On the computation of characteristic polynomial coefficients. 9(3):366–370
Pereslavtseva ON (2011) Calculation of the characteristic polynomial of a matrix. Discrete Math Appl 21(1):109–128
Preparata FP, Sarwate DV (1978) An improved parallel processor bound in fast matrix inversion. Inf Process Lett 7(3):148–150
Sankowski P, Mucha M (2010) Fast dynamic transitive closure with look ahead. Algorithmica 56:180–197
Schonhage A (1972) Unitäre Transformationen großer Matrizen. Numer Math 20:409–417
Schwartz O, Vaknin N (2023) Pebbling game and alternative basis for high performance matrix multiplication. SIAM J Sci Comput 45(6):C277–C303
Scott RF (1880) Theory of determinants. Cambridge University Press, Cambridge
Seidel R (1992) On the all pairs shortest path problem. In: Proceedings of the 24th annual ACM symposium on theory of computing. ACM Press, New York, 745–749
Shabalin AA (2012) Matrix eQTL: ultra fast eQTL analysis via large matrix operations. Bioinformatics 28(10):1353–1358
Storjohann A (2001) Deterministic computation of the Frobenius form. In: Proceedings of 42nd IEEE symposium on foundations of computer science. Newport Beach, CA, USA, pp 368–377
Takaoka T (2002) Efficient algorithms for the maximum subarray problem by distance matrix multiplication. Electron Notes Theor Comput Sci 61:191–200
Tamaki H, Tokuyama T (2000) Algorithms for the maximum subarray problem based on matrix multiplication. Interdiscip Inf Sci 6(2):99–104
Vaidya PM (1989) Speeding-up linear programming using fast matrix multiplication. In: 30th annual symposium on foundations of computer science, 332–337
Valiant LG (1975) General context free recognition in less than cubic time. J Comput Syst Sci 10:308–315
Winograd S (1970) On the algebraic complexity of functions. Actes Du Congres International des Mathematiciens 3:283–288
Xiao H (2024) Fast matrix multiplication for query processing. Proc ACM Manage Data 2(2):1–25

Yuster R, Zwick U (2004) Detecting short directed cycles using rectangular matrix multiplication and dynamic programming. In: Proceedings of fifteenth annual ACM-SIAM symposium on discrete algorithms, 254–260

Yuster R, Zwick U (2005a) Answering distance queries in directed graphs using fast matrix multiplication. In: 46th annual IEEE symposium on foundations of computer science (FOCS'05), 389–396

Yuster R, Zwick U (2005b) Fast sparse matrix multiplication. ACM Trans Algorithms 1(1):2–13

Yuster R (2009) Efficient algorithms on sets of permutations, dominance, and real-weighted APSP. In: Proceedings of 2009 annual ACM-SIAM symposium on discrete algorithms (SODA), 950–957

Zhang F (2005) The Schur complement and its applications. Numerical methods and algorithms. Springer, 4

Zwick U (2002) All pairs shortest paths using bridging sets and rectangular matrix multiplication. J ACM 49(3):289–317

Zwick U (1998) All pairs shortest paths in weighted directed graphs—exact and inexact algorithms. In: Proceedings of 39th annual symposium on foundations of computer science (FOCS'98). IEEE Computer Society Press, Los Alamos, California, 310–319

Chapter 10
Efficient Algorithms for Logical Matrices

In this chapter we give an overview of algorithms with different time complexity for matrix multiplication, for an important—especially in computer science—class of matrices with Boolean entries.

Using the example of the Strassen's algorithm, we have explained in detail how the efficient algorithm, originally designed for number matrices, can be applied to logical matrices. A special algorithm for the multiplication of logical matrices is also presented. The chapter is structured as follows:

- In Sect. 10.1 we present a tool that allows both standard and fast matrix multiplication algorithms, originally designed for matrices with number entries, to be used to multiply Boolean matrices.
- In Sect. 10.2 we present in detail a specialised algorithm for multiplying Boolean matrices. We also analyse its computational efficiency in general and bit-oriented implementations.
- In Sect. 10.3 we give an example of using Strassen's algorithm for multiplying Boolean matrices.
- In Sect. 10.4 we give examples of the execution of the algorithm from Sect. 10.2, specialised for Boolean matrices.

10.1 Multiplication of Logical Matrices by the Algorithms for Number Matrices

The operation of adding two logical matrices leads to the $O(n^2)$ time complexity algorithm; we can simply use the definition of this operation, usually used for matrices with real entries, where true is represented by 1 and false by 0.

The application of the efficient (and even definitional) multiplication algorithms for the matrices with number entries is not straightforward in this case, but we with help comes here the following observation 10.1.1.

J. S. Respondek, *Fast Matrix Multiplication with Applications*, Studies in Big Data 166,
https://doi.org/10.1007/978-3-031-76930-6_10

10.1.1 Observation

Let *us take two logical matrices A, B, both of* $n \times n$ *dimensions. Let C denote their product in the integer (number) domain and D denote their logical product. The following properties hold true:*

$$\begin{cases} D_{ij} = 0 & \Leftrightarrow & C_{ij} = 0 \\ D_{ij} = 1 & \Leftrightarrow & C_{ij} \geq 1 \end{cases} \qquad i, j = 1, 2, ..., n$$

Proof It is important that, when using algorithms originally designed for numbers, we denote true by an integer number 1 and false by the number 0. We calculate the successive entries of the product matrix using the same formula (1.1), in this case of the form $\sum_{k=1}^{n} a_{ik} b_{kj}$. The only difference is the specific operation to be performed within the sum: arithmetic product or logical product. In accordance with the well-known logical operations (Table 10.1), the result of the sum $\sum_{k=1}^{n} a_{ik} b_{kj}$ is true if and only if there exists one (or more) corresponding pair(s) of entries with true values, i.e. $\exists a_{ik} = b_{kj} = 1$ for certain $k = 1, .., n$.

If we calculate the sum $\sum_{k=1}^{n} a_{ik} b_{kj}$ using the rules of integer algebra, we also check the condition $\exists a_{ik} = b_{kj} = 1$ for certain $k = 1, .., n$; simply if there are more corresponding pairs of ones, the result in the number domain is greater than one. But equally important, if and only if there is no such a pair, i.e. $\sim \exists a_{ik} = b_{kj} = 1$ for certain $k = 1, .., n$, the value of the sum is equal to zero.

■

To summarise, in practice, to compute the logical matrix product with the algorithms originally designed for matrices with number entries, it is sufficient to change all non-zero values of the received matrix product to 1. In the next section we show a specialised algorithm for Boolean matrices.

10.2 The Four Russians Algorithm

The Four Russians algorithm was proposed in the article by Arlazarov et al. (1970). It was later rewritten in a more algorithmic spirit by Aho et al. (1974).

Suppose we want to multiply two square logical matrices A and B, each of the dimension $n \times n$. Let $m = \lfloor \log_2 n \rfloor$. The first step is to divide the A, B matrix factors into a series of submatrices A_i, B_i, for $i = 1, ..., \lceil n/m \rceil - 1$ of the sizes $n \times m$ and $m \times n$, respectively. The last two matrices in that series, i.e. $A_{\lceil n/m \rceil}$, $B_{\lceil n/m \rceil}$, contain the

Table 10.1 Tables defining the logical operations

+	0	1	*	0	1
0	0	1	0	0	0
1	1	1	1	0	1

Fig. 10.1 Division of the logical matrices in the four Russians algorithm

$$\left[A_1 \;\vdots\; A_2 \;\vdots\; \cdots \;\vdots\; A_{\lceil n/m \rceil} \right] \begin{bmatrix} B_1 \\ \cdots \\ B_2 \\ \cdots \\ \vdots \\ \cdots \\ B_{\lceil n/m \rceil} \end{bmatrix}$$

remaining - respectively - columns and rows of the matrices A, B and, if necessary, a certain number of zero columns and rows, which complement the respective matrices $A_{\lceil n/m \rceil}$, $B_{\lceil n/m \rceil}$ to the same number of m columns or rows. The idea of the block matrix division is shown in Fig. 10.1.

Next, each of the submatrices A_i and B_i are further divided into vectors in the way explained by the Eq. (10.1). We denote by C_i the product $A_i B_i$, also divided into vectors, as in (10.1):

$$A_i = \begin{bmatrix} a_1^{(i)} \\ \hline a_2^{(i)} \\ \hline \vdots \\ \hline a_n^{(i)} \end{bmatrix}, \quad B_i = \begin{bmatrix} b_1^{(i)} \\ \hline b_2^{(i)} \\ \hline \vdots \\ \hline b_m^{(i)} \end{bmatrix}, \quad C_i = \begin{bmatrix} c_1^{(i)} \\ \hline c_2^{(i)} \\ \hline \vdots \\ \hline c_n^{(i)} \end{bmatrix}, \quad i = 1, 2, \ldots, \lceil n/m \rceil \tag{10.1}$$

Now, using the partition from Fig. 10.1, we can calculate the desired product of the A, B matrices from the following sum[1]:

$$AB = \sum_{i=1}^{\lceil n/m \rceil} A_i B_i \tag{10.2}$$

The detailed code of the Four Russians algorithm Aho et al. (1974) is presented below, in Sect. 10.2.1.

[1] Obviously, in this particular application, the Σ sign used denotes the logical sum, as defined in Table 10.1.

10.2.1 The Four Russians Algorithm for the Logical Matrix Product

---

Input: Two logical matrices A, B, each of the $n \times n$ dimensions
Output: The logical product $C = AB$

---

begin

m $\leftarrow \lfloor \log_2 n \rfloor$

1. **for** i$\leftarrow$1 **to** $\lceil n/m \rceil$ **do**
2. $RowSum[0] \leftarrow \underbrace{[0,\ 0,\ \ldots,\ 0]}_{n}$
3. **for** j$\leftarrow$1 **to** 2^m-1 **do**
4. $k := \{k \in Z:\ 2^k \le j < 2^{k+1}\}$
5. $RowSum[j] \leftarrow RowSum[j - 2^k] + b_{k+1}^{(i)}$

   next j

   for j$\leftarrow$1 **to** n **do**

6. $c_j^{(i)} = RowSum[NumBitRv(a_j^{(i)})]$

   next j

   next i

7. $C \leftarrow \sum_{i=1}^{\lceil n/m \rceil} C_i$

end

The function $NumBitRv(\cdot)$ used by algorithm 10.2.1 requires additional explanation. Let $a_j^{(i)}$ be an integer number with binary representation of the form $a_j^{(i)} = [u_0 \ldots u_{p-1}\ u_p] = u_0 2^0 + \cdots + u_{p-1} 2^{p-1} + u_p 2^p$, where $u_.$ are the consecutive bits of the integer. So each $u_.$ has two possible values: 0 or 1. $NumBitRv(a_j^{(i)})$ denotes the integer number with the bits in reverse order, i.e. $NumBitRv(a_j^{(i)}) = [u_p \ldots u_1\ u_0] = u_p 2^0 + \cdots + u_1 2^{p-1} + u_0 2^p$.

The exemplary execution of the algorithm 10.2.1 is presented in detail in Sect. 10.4. Let us now analyse the time complexity of this algorithm.

10.2.2 Theorem: The Four Russians Algorithm Works in $O(n^3/\log n)$ Time

The internal loop 3–5 is the starting point for the complexity analysis of Aho et al. (1974). The assignment in line 5 requires an $O(n)$ operation because each variable $RowSum[\cdot]$ is an n-element vector. The calculation of the k variable in line 4 requires $O(m)$ steps, so less than $O(n)$. So the whole internal loop body 4–5 requires $O(n)$ operations. This loop is repeated $2^m - 1$ times, giving complexity $O(n2^m)$. Since $m \le \log_2 n$, the whole loop 3–5 requires $O(n^2)$ operations.

During the execution of line 6 we have to perform two operations: inverting the order of the integer $a_j^{(i)}$ bits and copying the logical vector $RowSum[NumBitRv(a_j^{(i)})]$ to the corresponding row of the C_i matrix. Both operations are of linear complexity, just as importantly, are performed independently (but the order is important). Considering that line 6 is executed for j varying in the range 1, .., n, its complexity is also $O(n^2)$.

Both loops in lines 3–5 and line 6 are executed $\lceil n/m \rceil$ times, by the main loop defined in line 1. Given the inequality $\lceil n/m \rceil \leq 2n/\log_2 n$, the total time complexity of the algorithm 10.2.1 is $O(n^2) \cdot O\big(n/\log_2 n\big) = O\big(n^3/\log_2 n\big)$.

■

An important advantage of the algorithm 10.2.1 is that it can be implemented in $O_{BV}(n^2/\log_2 n)$ time if we have available logical and arithmetic operations on the bit vectors ($O_{BV}(\cdot)$ denotes the number of bit-vector operations in the $O(\cdot)$ notation). The corresponding analysis is given by the Theorem 10.2.3 from Aho et al. (1974).

10.2.3 Theorem: The Four Russian Algorithm can be Implemented in $O_{BV}(n^2/\log_2 n)$ Time

With the hardware support for bit-vectors, lines 2 and 5 in the algorithm 10.2.1 are reduced to single steps, i.e. $O_{BV}(1)$. Thus the internal loop 3–5 requires $O_{BV}(n)$ operations.

To avoid calculating the $NumBitRv(\cdot)$ function, which takes linear time, we can change the 5th line of the algorithm 10.2.1 to the following form:

$$RowSum[j] \leftarrow RowSum[j - 2^k] + b_{m-k}^{(i)}$$

and—in line 6—we can insert the function $NumBitRv(\cdot)$ instead of the function $NumBit(\cdot)$. The function $NumBit(\cdot)$ also returns the integer number corresponding to the bit representation of the $a_j^{(i)}$ vector, but does not reverse the order of the bits. In other words, the expression $RowSum[NumBit(a_j^{(i)})]$ returns the logical sum of those B_i matrix rows whose numbers correspond to the bit representation of the $a_j^{(i)}$ integer, but now the row counting starts from the bottom of a given B_i matrix.

Thanks to this modification, which is easy to implement in practice, each C_i row can be calculated in $O_{BV}(1)$ time and the whole line 6 requires $O_{BV}(n)$, since we have n rows.

The whole main loop 1–6 is repeated $\lceil n/m \rceil$ times, and we obtain the final complexity of $O_{BV}\big(n^2/\log_2 n\big)$.

■

In the next sections, we will show how the Strassen and Four Russians algorithms work in practice, using the concrete example of matrices with dimensionality 8×8, chosen to ensure that the internals of these algorithms work in practice.

10.3 Example of the Strassen's Algorithms Execution

Let us calculate the square of the following 8×8 logical matrix, using the Strassen algorithm:

$$A = \begin{bmatrix} 0&0&0&0&0&0&0&0 \\ 1&0&0&0&0&0&0&0 \\ 0&1&0&1&0&0&0&0 \\ 0&0&0&1&1&0&0&1 \\ 0&0&0&0&0&1&1&0 \\ 0&0&1&0&0&0&0&0 \\ 0&0&0&0&0&0&0&0 \\ 0&1&0&0&0&0&0&0 \end{bmatrix}_{8\times 8}$$

Let us divide the matrix into four different sub-matrices of even size. The problem is represented by the following block matrix formula:

$$A = \left[\begin{array}{c|c} A_{11} & A_{12} \\ \hline A_{21} & A_{22} \end{array}\right] = \left[\begin{array}{cccc|cccc} 0&0&0&0&0&0&0&0 \\ 1&0&0&0&0&0&0&0 \\ 0&1&0&1&0&0&0&0 \\ 0&0&0&1&1&0&0&1 \\ \hline 0&0&0&0&0&1&1&0 \\ 0&0&1&0&0&0&0&0 \\ 0&0&0&0&0&0&0&0 \\ 0&1&0&0&0&0&0&0 \end{array}\right]_{8\times 8}$$

The auxiliary matrix terms $M_1 \div M_7$ by assigning $B_{ij} = A_{ij}, \ i, j = 1, 2$ we calculate by formula (3.3), obtaining the following matrix series[2]:

$$(M_1, \ldots, M_7) = \left(\begin{bmatrix} 1&1&0&1 \\ 0&1&1&0 \\ 1&0&0&1 \\ 0&0&0&1 \end{bmatrix}_{4\times 4}, \begin{bmatrix} 1&1&0&1 \\ 0&1&0&1 \\ 0&0&0&0 \\ 1&0&0&0 \end{bmatrix}_{4\times 4}, \begin{bmatrix} 0&0&0&0 \\ 0&-1&-1&0 \\ 1&0&0&1 \\ 1&0&0&1 \end{bmatrix}_{4\times 4}, \begin{bmatrix} -1&-1&1&-1 \\ 0&0&0&0 \\ 0&0&0&0 \\ 0&0&0&0 \end{bmatrix}_{4\times 4}, \right.$$
$$\left. \begin{bmatrix} 0&0&0&0 \\ 0&1&1&0 \\ 0&0&0&0 \\ 0&1&1&0 \end{bmatrix}_{4\times 4}, \begin{bmatrix} 0&0&0&0 \\ 0&1&0&1 \\ -2&0&0&-2 \\ 0&0&0&-2 \end{bmatrix}_{4\times 4}, \begin{bmatrix} 0&0&-1&0 \\ 0&0&0&0 \\ 0&0&0&0 \\ 0&2&1&0 \end{bmatrix}_{4\times 4} \right)$$

To obtain the desired product A^2, we apply the formula (3.4):

[2] In practice, to obtain the terms $M_1 \div M_7$ we have to calculate 7 matrix products, but with the source matrix reduced to half of its size, i.e. of the size 4×4, then again $7 \times 7 = 49$ matrix products of the size 2×2, in accordance with the recursive formula (3.4).

$$A^2 = \left[\begin{array}{c|c} M_1 + M_4 - M_5 + M_7 & M_3 + M_5 \\ \hline M_2 + M_4 & M_1 + M_3 - M_2 + M_6 \end{array}\right] = \left[\begin{array}{cccc|cccc} 0&0&0&0&0&0&0&0 \\ 0&0&0&0&0&0&0&0 \\ 1&0&0&1&1&0&0&1 \\ 0&1&0&1&1&1&1&1 \\ \hline 0&0&1&0&0&0&0&0 \\ 0&1&0&1&0&0&0&0 \\ 0&0&0&0&0&0&0&0 \\ 1&0&0&0&0&0&0&0 \end{array}\right]_{8\times 8}$$

10.4 The Four Russians Algorithm Execution Example

Let us find the logical square of the same matrix using the Four Russians algorithm. The first step in calculating the $C = A \cdot A$ matrix is to divide it as shown in Fig. 10.1. We have $m = \log_2 8 = 3$ and $\lceil n/m \rceil = \lceil 8/3 \rceil = \lceil 2.(6) \rceil = 3$, so the sizes of the respective matrices are $\dim\,[A_i] = 8 \times 3$ and $\dim\,[B_i] = 3 \times 8$ for $i = 1, ..., \lceil n/m \rceil = 1, .., 3$. Given the special form of the A matrix, which in our example is both the multiplicand and the multiplier, the matrix division has the following form:

$$\left[A_1 \middle| A_2 \middle| A_3\right] = \left[\begin{array}{ccc|ccc|cc|c} 0&0&0&0&0&0&0&0& \\ 1&0&0&0&0&0&0&1& \\ 0&1&0&1&0&0&0&0& \\ 0&0&0&1&1&0&0&1& \\ 0&0&0&0&0&1&1&0& \mathbf{0}_{8\times 1} \\ 0&0&1&0&0&0&0&0& \\ 0&0&0&0&0&0&0&0& \\ 0&1&0&0&0&0&0&0& \end{array}\right]_{8\times 9}, \quad \left[\begin{array}{c} B_1 \\ \hline B_2 \\ \hline B_3 \end{array}\right] = \left[\begin{array}{cccccccc} 0&0&0&0&0&0&0&0 \\ 1&0&0&0&0&0&0&0 \\ 0&1&0&1&0&0&0&0 \\ \hline 0&0&0&1&1&0&0&1 \\ 0&0&0&0&0&1&1&0 \\ 0&0&1&0&0&0&0&0 \\ \hline 0&0&0&0&0&0&0&0 \\ 0&1&0&0&0&0&0&0 \\ \hline \multicolumn{8}{c}{\mathbf{0}_{1\times 8}} \end{array}\right]_{9\times 8}$$

As we can see, the matrices A_3 and B_3 must be completed with the column and row zero, respectively, in order to extend them to the desired number of columns or rows equal to 3.

In our case, in the main loop (steps 1÷6), the variable i of the algorithm 10.2.1 changes in the range from 1 to 3. To illustrate how the algorithm 10.2.1 works in practice, let us analyse that loop step by step for $i = 2$. In this particular loop cycle, find the logical product of the following matrices:

$$A_2 = \begin{bmatrix} 0\,0\,0 \\ 0\,0\,0 \\ 1\,0\,0 \\ 1\,1\,0 \\ 0\,0\,1 \\ 0\,0\,0 \\ 0\,0\,0 \\ 0\,0\,0 \end{bmatrix}_{8\times 3}, \quad B_2 = \begin{bmatrix} 0\,0\,0\,1\,1\,0\,0\,1 \\ 0\,0\,0\,0\,0\,1\,1\,0 \\ 0\,0\,1\,0\,0\,0\,0\,0 \end{bmatrix}_{3\times 8} \tag{10.3}$$

The variable j of the loop executed in steps 3÷5 varies from 1 to $2^3 - 1 = 7$. The successive iterations of the 3÷5 loop are shown in Table 10.2. Before that, in step 2, the assignment $RowSum[0] \leftarrow [0 \;\cdots\; 0]$ is performed.

We can see that after the execution of the 3÷5 loop, the $RowSum[\cdot]$ variable contains <u>all</u> the possible combinations of the logical sums of the rows of the B_2 matrix (10.3), for each iteration of the 1÷6 main loop.

Now we move on to the next internal loop, executed in step 6. Now the variable j varies from 1 to 8. In this loop we use the $RowSum[j]$ values, calculated in the first loop, for $j = 1, .., 8$. The successive iterations are shown in Table 10.3.

Finally, the desired matrix product A_2B_2 has the form:

$$A_2B_2 = \begin{bmatrix} a_1^{(2)} \\ \vdots \\ a_8^{(2)} \end{bmatrix} = \begin{bmatrix} 0\,0\,0\,0\,0\,0\,0\,0 \\ 0\,0\,0\,0\,0\,0\,0\,0 \\ 0\,0\,0\,1\,1\,0\,0\,1 \\ 0\,0\,0\,1\,1\,1\,1\,1 \\ 0\,0\,1\,0\,0\,0\,0\,0 \\ 0\,0\,0\,0\,0\,0\,0\,0 \\ 0\,0\,0\,0\,0\,0\,0\,0 \\ 0\,0\,0\,0\,0\,0\,0\,0 \end{bmatrix}$$

Table 10.2 Consecutive iterations of the first internal loop

j	k	$j-2^k$	$RowSum[j]$	$[b_3\ b_2\ b_1]$
1	0	0	$RowSum[0] + b_1^{(2)} = b_1^{(2)} = [0\,0\,0\,1\,1\,0\,0\,1]$	[0 0 1]
2	1	0	$RowSum[0] + b_2^{(2)} = b_2^{(2)} = [0\,0\,0\,0\,0\,1\,1\,0]$	[0 1 0]
3		1	$RowSum[1] + b_2^{(2)} = b_1^{(2)} + b_2^{(2)} = [0\,0\,0\,1\,1\,1\,1\,1]$	[0 1 1]
4	2	0	$RowSum[0] + b_3^{(2)} = b_3^{(2)} = [0\,0\,1\,0\,0\,0\,0\,0]$	[1 0 0]
5		1	$RowSum[1] + b_3^{(2)} = b_1^{(2)} + b_3^{(2)} = [0\,0\,1\,1\,1\,0\,0\,1]$	[1 0 1]
6		2	$RowSum[2] + b_3^{(2)} = b_2^{(2)} + b_3^{(2)} = [0\,0\,1\,0\,0\,1\,1\,0]$	[1 1 0]
7		3	$RowSum[3] + b_3^{(2)} = b_1^{(2)} + b_2^{(2)} + b_3^{(2)} = [0\,0\,1\,1\,1\,1\,1\,1]$	[1 1 1]

Table 10.3 Consecutive iterations of the second internal loop

j	$a_j^{(2)}$	$NumBitRv(a_j^{(2)})$ *(binary)*	$NumBitRv(a_j^{(2)})$ *(decimally)*	$RowSum[NumBitRv(a_j^{(2)})]$
1	[0 0 0]	[0 0 0]	0	[0 0 0 0 0 0 0 0]
2	[0 0 0]	[0 0 0]	0	[0 0 0 0 0 0 0 0]
3	[1 0 0]	[0 0 1]	1	[0 0 0 1 1 0 0 1]
4	[1 1 0]	[0 1 1]	3	[0 0 0 1 1 1 1 1]
5	[0 0 1]	[1 0 0]	4	[0 0 1 0 0 0 0 0]
6	[0 0 0]	[0 0 0]	0	[0 0 0 0 0 0 0 0]
7	[0 0 0]	[0 0 0]	0	[0 0 0 0 0 0 0 0]
8	[0 0 0]	[0 0 0]	0	[0 0 0 0 0 0 0 0]

Analogously, the Four Russians algorithm goes through the remaining iterations of the main loop, which gives as output the products A_1B_1 and A_3B_3. Finally, in step 7 of the algorithm 10.2.1, the desired logical matrix product $A \cdot A$ is calculated according to formula (10.2):

$$C \leftarrow \sum_{i=1}^{3} A_i B_i = \begin{bmatrix} 0&0&0&0&0&0&0&0 \\ 0&0&0&0&0&0&0&0 \\ 1&0&0&0&0&0&0&0 \\ 0&0&0&0&0&0&0&0 \\ 0&0&0&0&0&0&0&0 \\ 0&1&0&1&0&0&0&0 \\ 0&0&0&0&0&0&0&0 \\ 1&0&0&0&0&0&0&0 \end{bmatrix} + \begin{bmatrix} 0&0&0&0&0&0&0&0 \\ 0&0&0&0&0&0&0&0 \\ 0&0&0&1&1&0&0&1 \\ 0&0&0&1&1&1&1&1 \\ 0&0&1&0&0&0&0&0 \\ 0&0&0&0&0&0&0&0 \\ 0&0&0&0&0&0&0&0 \\ 0&0&0&0&0&0&0&0 \end{bmatrix} + \begin{bmatrix} 0&0&0&0&0&0&0&0 \\ 0&0&0&0&0&0&0&0 \\ 0&0&0&0&0&0&0&0 \\ 0&1&0&0&0&0&0&0 \\ 0&0&0&0&0&0&0&0 \\ 0&0&0&0&0&0&0&0 \\ 0&0&0&0&0&0&0&0 \\ 0&0&0&0&0&0&0&0 \end{bmatrix}$$

$$= \begin{bmatrix} 0&0&0&0&0&0&0&0 \\ 0&0&0&0&0&0&0&0 \\ 1&0&0&1&1&0&0&1 \\ 0&1&0&1&1&1&1&1 \\ 0&0&1&0&0&0&0&0 \\ 0&1&0&1&0&0&0&0 \\ 0&0&0&0&0&0&0&0 \\ 1&0&0&0&0&0&0&0 \end{bmatrix}$$

To sum up, the general idea of the Four Russians algorithm is to generate a so-called look-up table that stores the possible combinations of the logical matrix multiplier rows (steps 3÷5 of algorithm 10.2.1). Next, we use the lookup table to select the desired logical vector, depending on the values of the matrix multiplicand row. The use of lookup tables is possible thanks to the fact that the counter range of the logical products of the vector is limited; each entry of the vector, once it has obtained the value 1, cannot change this value in further calculations, being limited to logical sums.

It is interesting to note that not all of the data in the lookup table is usually used. In our example, the data for $i = 2$ the *RowSum*[j] values are not used for $j > 4$. Nevertheless, the efficiency remains better than in the case of the multiplication algorithm based directly on its mathematical definition (Theorem 10.2.2).

The decision on which algorithm to use should be made on a case-by-case basis. In addition, the hardware capabilities should be taken into account. For these reasons it is not possible to judge which algorithm is the best; non-commutative algorithms designed for matrices with real input time complexities are better, but on the other hand it cannot be deduced directly from the bit vector capabilities of the system.

References

Arlazarov V, Dinic E, Kronrod M, Faradzev I (1970) On economical construction of the transitive closure of a directed graph. Dokl Akad Nauk SSSR 194(3):487–488

Aho AV, Hopcroft JE, Ullman JD (1974) The design and analysis of computer algorithms. Addison-Wesley Pub., Mass. U.S.A

Chapter 11
Appendix—Algorithms for (Confluent) Vandermonde Matrices

Throughout the lecture on algorithms for fast matrix multiplication, well-known Vandermonde matrices appeared as an auxiliary tool in several places. Namely, in Sect. 5.2 they are used to transform an approximate fast matrix multiplication algorithm into an exact one, and in Sect. 7.2 they are used to transform given partial matrix multiplication algorithm to a total one. More precisely, in the first of these two tasks we have to solve a linear system with a Vandermonde matrix, and in the second we have to invert it. Obviously, we cannot perform these tasks using generic numerical recipes, because what we have gained by fast matrix multiplication is lost by the cubic execution time of these methods.

Fortunately, with help comes another branch of acceleration methods for operations on matrices. Namely, the advantage of generic numerical methods for matrices and fast matrix multiplication algorithms is their generality. Their input parameters are matrices with allowed arbitrary value of any entry. But the cost of this is that they are time-consuming. For example, in the case of the fast matrix multiplication algorithms mentioned above, we can see that since 1990 the ω coefficient has improved by only $\sim$ 0.003. One can ask whether it is possible to further improve their time complexity. Today, we do not know of any universal algorithm for matrix multiplication that is more efficient than $O(n^{2.37})$.

On the opposite side, further significant efficiency improvements are possible thanks to specialised algorithms designed for selected classes of special[1] matrices. Special matrices are characterised by a particular structure of their entries. So we can distinguish the following matrices: band, block-diagonal, block-triangular, Pascal, Frobenius, companion, Jordan, Toeplitz, Sylwester, Hilbert, bool matrices, just (confluent) Vandermonde matrices and others. From the point of view of algorithm construction, their important property is that it is often possible to achieve $O(n^2)$ time complexity by the algorithms for certain parameter values of a given

[1] We call them also 'structured'.

J. S. Respondek, *Fast Matrix Multiplication with Applications*, Studies in Big Data 166,
https://doi.org/10.1007/978-3-031-76930-6_11

special matrix type, or even in the general case. A comprehensive monograph on fast algorithms that operate on various types of special matrices is Pan (2001).

We decided to show how to invert the Vandermonde matrix in the case of its most important generalisation, i.e. confluent Vandermonde matrices. There are specialised algorithms for inverting a classical Vandermonde matrix that work in quadratic time, but there are also such algorithms for its generalised version. Obviously, we get a quadratic time algorithm for the classical case as a special case. We write more about this later in this section, along with its formal definition. In general, compared to the classical Vandermonde matrix, its confluent generalisation also contains derivatives of the columns multiplied by some constant factor.

Confluent Vandermonde matrices first appeared in **1901—in a form almost identical to the one used today**.[2] Article Vogt (1901) on page 350 gave a 6×6 instance of a determinant, with one of the eigenvalues multiplicity equal to three. We can read there (in French): "*Ce déterminant est la **généralisation** de celui de **Vandermonde***".

In a classic monograph from 1923, Muir's "*History of Determinants*", the confluent Vandermonde matrix is explicitly presented on pages 178–180 exactly in the form we use today, with decent dimensionality 6 by 6. Three different eigenvalues are considered, including one triple.

The next time the confluent Vandermonde matrix appears is in the monograph Turnbull (1928), page 28, example 4. However, here we have a kind of regression compared to Muir (1923) or even Vogt (1901). It now appears as a determinant of rather timid parameters; we have a 4×4 dimension with, what is even worse, a maximum multiplicity of entries equal to two.

The Vandermonde matrices, both in their classical and confluent versions, are mainly associated with their two main applications: in the so-called classical canonical form and in interpolation. Namely, the confluent Vandermonde matrix is the change-of-base matrix that transforms the companion matrix into the Jordan canonical form, according to the matrix equation:

$$F \cdot V = V \cdot J$$

where V stands for just a (generally confluent) Vandermonde matrix, F for a companion matrix and J for Jordan canonical form.

This fact was noted in two works in 1932. The first is the article Aitken (1932), Sect. 6 "*Relation between Two Forms of Canonical Matrix*", pages 88–90. On page 89 we have the relationship in question both in the form of a matrix equation and as 4×4 example in formula (IV). The second work is a classic but still valuable monograph Turnbull, Aitken (1932), Sect. VI.1: "*The Classical Canonical Form deduced from the Rational Form*", pages 58-62. On page 60 we have an 6×6 instance of this type of special matrix. For a more recent study of this idea, see also the article by Kalman (1984).

---

[2] Some authors consider the article Schendel (1891) in German to be the first article in which this matrix appeared.

Then the confluent Vandermonde matrix appears in the article Aitken (1938), formula 5 on page 288, as a tool for finding matrix eigenvalues.[3]

In addition to the special matrix in question, a separate section VI.50 is devoted to it in the monograph Aitken (1944), pages 119–121.

Furthermore, the determinant of the confluent Vandermonde matrix was found in 1932 in the general case. A suitable formula can be found in the monograph Turnbull, Aitken (1932), page 63, example 5 or in Aitken (1944), formula 7 on page 121.

As already mentioned, the classical Vandermonde matrix is often associated with polynomial interpolation. More precisely, this is the case when we prescribe the values of the polynomial at given points. This type of interpolation was generalised by Hermite in (1878), in an article by this author in French. Namely, he not only considered the sole values of an interpolation polynomial, but also allows to predetermine the values of the derivative of a polynomial of any order at any point in the series, starting their order from zero to a given, with a step of one.[4] Nevertheless, Hermite formulated his generalised interpolation on the basis of the integral calculus in the complex domain. Aitken (1944) gives an example of how this type of polynomial interpolation leads to the confluent Vandermonde matrix (formula 2, page 119, section 50). Next, Spitzbart (1960) formulated Hermite interpolation in a general case using this class of special matrices.

Another branch here is its generalisation to a 2D case, but already without the Vandermonde matrix. Ahlin (1964) constructed a bivariate osculating interpolation polynomial $f(x, y)$, in which at each node both the values and its partial and mixed partial derivatives match prescribed values. However, in Ahlin (1964) there is an additional restriction that at each node of a 2D grid we consider the same total order of derivatives or partial derivatives. This restriction removes the article Chawla and Jayarajan (1974).

In addition to the two classical applications discussed above, the confluent Vandermonde matrix is used in the following problems involving mathematics, physics and control theory:

- Computing exponential of a matrix [Moler and V-Loan (2003)—methods 11 and 13, Harris et al. (2001)].
- Calculation of matrix functions: Hermite interpolation for matrices (Higham 2008, Sect. 1.2.2, especially note 1.9 page 7).
- Calculation of the minimum polynomial of a matrix (Halidias 2024).
- Coding information in the Hermitian code (Lee and O'Sullivan 2010).
- Optimization of the non-homogeneous differential equations (Gorecki 1968).
- Control theory: controllability analysis of linear systems (Respondek 2008), quantised control of sampled-data systems (Shen et al. 2017).
- Calculation of Schur polynomials [Serrano, Maximenko (2023)].

---

[3] At the time of the article appeared, the eigenvalues were referred to as '*latent roots*'.

[4] Without this additional assumption we face a Birkhoff interpolation, which is in general ambiguous.

Let us now return to the operations that need to be performed on the Vandermonde matrix in fast matrix multiplication problems, i.e. inverting it and solving linear systems of equations with this matrix.

We have many algorithms available for **solving linear systems** with such a structure. Moreover, some of them work in quadratic (or even faster) time for any input parameter allowed by the definition of this matrix. The algorithms presented in the article Galimberti and Pereyra (1971) solve the linear system in quadratic time. Articles Lu (1994, 1995) presents algorithms that work in linear-polylogarithmic time. Another article on that problem is Tang and Golub (1981). The article by Bjorck and Pereyra (1970) is often cited, but it does not give an explicit formula in the general case; it only discusses it descriptively.

Nevertheless, the problem of solving linear systems of equations (for any kind of system matrix) is a simpler problem than inverting the system matrix, in the sense that in the first problem we have n unknowns, while in the second the number of unknowns is n^2. In the case of the matrix in question, this is manifested in the fact that after the Second World War a number of algorithms for **inverting** the confluent Vandermonde matrix were published. However, unlike in the case of solving a system, none of them preserved efficiency for arbitrary values of the input parameters. We can distinguish the following main research lines, both of which have not been fully successful:

- **Algorithms that always work in quadratic time, but only in special cases.** For example, Gautschi's (1963) algorithm only allows matrix parameters with multiplicities equal to two. If some of the eigenvalues have a higher multiplicity, the algorithm does not apply.
- **Algorithms that work for arbitrary values of input parameters, but are only efficient in special cases.** The articles Zhong and Zhaoyong (1998) and Luther and Rost (2004) achieve quadratic efficiency only for small values of all eigenvalue multiplicities (compared to the matrix dimension). Otherwise, the algorithms presented become algorithms with cubic efficiency.

Until 2011, there was no algorithm in the literature for the inversion of the confluent Vandermonde matrix that simultaneously had the following capabilities:

- Works with all values of the input parameter series allowed by the definition of the particular special matrix.
- Always works in quadratic time, even for large multiplicities of one or a few eigenvalues.
- Maintains high accuracy for any input parameter series. This problem is presented in e.g. Pan (2016).

The breakthrough came only with the articles by Respondek (2011a, b),[5] together with the previous work by Hou and Pang (2002), which met all three of the

[5] The problem of solving a system of equations $Vx = b$ can be solved using matrix inversion by equality $x = V^{-1}b$.

above requirements. Moreover, article Hou and Hou (2009) makes the triangular factorisation of the inverse.

Let us present the definition of the confluent Vandermonde matrix (Turnbull, Aitken (1932)), Chap. VI page 60, Kalman (1984), pages 18–19). It is determined by the characteristic polynomial:

$$p(s) = (s - \lambda_1)^{n_1}(s - \lambda_2)^{n_2} \cdot \ldots \cdot (s - \lambda_r)^{n_r}$$

where $\lambda_1, \lambda_2, \ldots\lambda_r$ are given pair wise distinct real zeros with $n_1 + \ldots + n_r = n$. The confluent Vandermonde matrix V related to the zeros of $p(s)$ is defined to be the $n \times n$ block matrix $V = \left[V_1 \middle| V_2 \middle| \cdots \middle| V_r \right]$. Let us denote as $\overline{f}(\lambda)$ the column $\left[1\ \lambda\ \cdots\ \lambda^{n-1} \right]^T$ and by $\overline{f}^{(j)}(\lambda)$ the jth derivative of this column. Now V_k block is $n \times n_k$ matrix with columns $\overline{f}^{(j)}(\lambda_k)/j!$ for $j = 0, 1, \ldots, n_j - 1$. We can also present the block in the form after differentiation as follows:

$$(V_k)_{ij} = \begin{cases} \binom{i-1}{i-j}\lambda_k^{i-j}, & \text{for } i \geq j \\ 0, & \text{otherwise} \end{cases}$$

for $k = 1, 2, \ldots, r$; $i = 1, 2, \ldots, n$ and $j = 1, 2, \ldots, n_k$. The works of Respondek (2011a, b) and Hou, Pang (2002) present the following theorem for inverting the confluent Vandermonde matrices:

11.1 Theorem—Inversion of the Confluent Vandermonde Matrix

The inverse of the confluent Vandermonde matrix V has the form $V^{-1} = \left[W_1^T \middle| W_2^T \middle| \cdots \middle| W_r^T \right]^T$. The column vectors $h_{k\cdot}$ of the block matrix $W_k = \left[h_{kn} \middle| h_{k(n-1)} \middle| \cdots \middle| h_{k1} \right]$ in the inverse confluent Vandermonde matrix V^{-1} may be recursively computed by the following scheme:

$$\begin{cases} h_{k1} = \left[K_{k,1} \cdots K_{k,n_k} \right]^T \\ h_{k2} = J_k(\lambda_k, n_k)h_{k1} + a_1 h_{k1} \\ \vdots \\ h_{kn} = J_k(\lambda_k, n_k)h_{k(n-1)} + a_{n-1}h_{k1} \end{cases}, \quad k = 1, 2, \ldots, r$$

where a_k are the coefficients of the characteristic polynomial $p(s) = s^n + \sum_{i=0}^{n-1} a_{n-i}s^i$, $J_k(\lambda_k, n_k)$ is the elementary Jordan block and $K_{k,j}$ are the auxiliary coefficients which may be computed by the following recursive scheme:

$$\begin{cases} L_{ki}^{(q+1)}(\lambda_k) = q!(\lambda_k - \lambda_i)^q \cdot K_{k,n_k-q} - q \cdot L_{ki}^{(q)}(\lambda_k), & i = 1, .., k-1, k+1, .., r \\ K_{k,n_k-q-1} = -\frac{1}{(q+1)!} \sum\limits_{i=1,\ i \neq k}^{r} n_i \frac{L_{ki}^{(q+1)}(\lambda_k)}{(\lambda_k - \lambda_i)^{q+1}} & \end{cases} \tag{11.1}$$

for $q = 0, 1, .., n_k - 2$ *and* $k = 1, 2, ..., r$. *The* K_{k,n_k} *coefficients may be computed directly from the formula:*

$$K_{k,n_k} = \frac{1}{(\lambda_k - \lambda_1)^{n_1} \cdot ... \cdot (\lambda_k - \lambda_{k-1})^{n_{k-1}} (\lambda_k - \lambda_{k+1})^{n_{k+1}} \cdot ... \cdot (\lambda_k - \lambda_r)^{n_r}}, \quad k = 1, 2, ..., r \tag{11.2}$$

Theorem 11.1 allows to invert the confluent Vandermonde matrix without the need to perform symbolic computations. The article Respondek (2011b) provides a ready-to-use C/C++ implementation of the algorithm.

The determination of the auxiliary coefficients $K_{k,j}$ by Theorem 11.1 deserves a closer look because of its sophisticated form. First we have to calculate the K_{k,n_k} coefficient directly from the formula (11.2). Next for each $k = 1, 2, ..., r$ and $q = 0, 1, .., n_k - 2$ we have to perform the following steps:

- Calculate the series $L_{k1}^{(q)}, L_{k2}^{(q)}, ..., L_{k(k-1)}^{(q)}, L_{k(k+1)}^{(q)}, ..., L_{kr}^{(q)}$ by the formula (11.1).
- Calculate the coefficient K_{k,n_k-q-1} from the formula (11.1).

The order of the necessary calculations is shown in detail in Fig. 11.1. The following should be noted:

- There is no need to initialise the $L_{ki}^{(0)}(\lambda_k)$, because for $q = 0$, due to the formula (11.1) they vanish.
- The calculation scheme shown in Fig. 11.1 must be performed separately for each $k = 1, 2, ..., r$. However, the order of the k variable does not matter.
- In each iteration of the q variable the previous $L_{ki}^{(q)}(\lambda_k)$ values may be overwritten by the new $L_{ki}^{(q+1)}(\lambda_k)$.

11.2 Example—Inversion of the Confluent Vandermonde Matrix with Four Nodes

Let us invert the matrix defined by the following characteristic polynomial:

$$p(s) = (s + 0.5)(s + 3)^2(s + 2)^3(s + 1)^4$$

We can rewrite its data in a table form (Table 11.1).

The expansion of the $p(s)$ polynomial can be done in quadratic time. We can receive (Table 11.2).

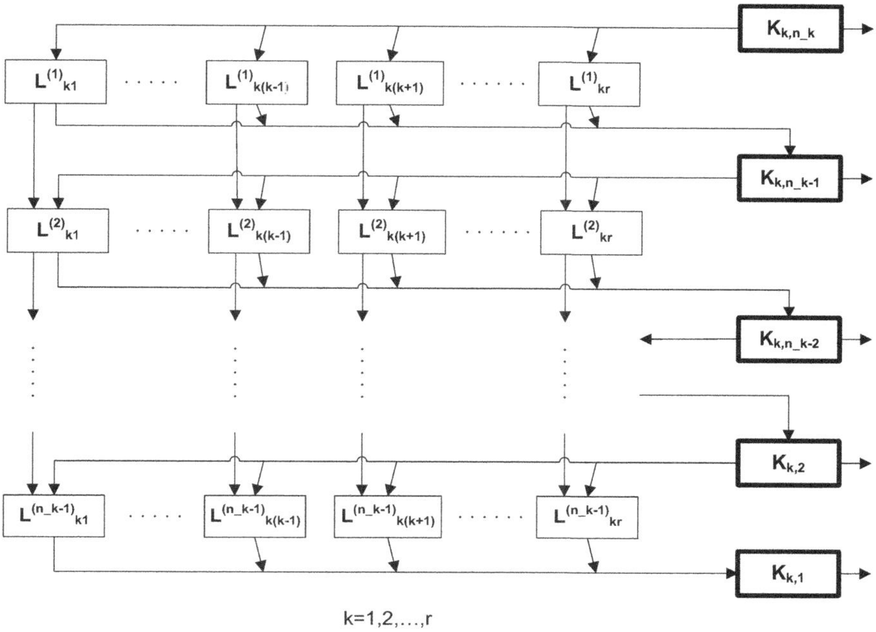

Fig. 11.1 Scheme for the auxiliary coefficients determination algorithm

Table 11.1 Parameters of confluent Vandermonde matrix to invert

i	1	2	3	4
λ_i	– 0.5	– 3.0	– 2.0	– 1.0
n_i	1	2	3	4

Table 11.2 Characteristic polynomial coefficients after expansion

i	1	2	3	4	5	6	7	8	9	10
a_i	16.5	119.0	493.5	1302.0	2281.5	2687.0	2098.5	1039.0	294.0	36.0

The matrix to be inverted has the following form:

$$V = \left[V_1 \middle| V_2 \middle| V_3 \middle| V_4 \right]$$

$$
= \left[\begin{array}{c|cc|ccc|cccc}
1 & 1 & 0 & 1 & 0 & 0 & 1 & 0 & 0 & 0 \\
-0.5 & -3 & 1 & -2 & 1 & 0 & -1 & 1 & 0 & 0 \\
0.25 & 9 & -6 & 4 & -4 & 1 & 1 & -2 & 1 & 0 \\
-0.125 & -27 & 27 & -8 & 12 & -6 & -1 & 3 & -3 & 1 \\
0.0625 & 81 & -108 & 16 & -32 & 24 & 1 & -4 & 6 & -4 \\
-0.0312 & -243 & 405 & -32 & 8 & -80 & -1 & 5 & -10 & 10 \\
0.01562 & 729 & -1458 & 64 & -192 & 240 & 1 & -6 & 15 & -20 \\
-0.0078 & -2187 & 5103 & -128 & 448 & -672 & -1 & 7 & -21 & 35 \\
0.0039 & 6561 & -17496 & 256 & -1024 & 1792 & 1 & -8 & 28 & -56 \\
-0.0019 & -19683 & 59049 & -512 & 2304 & -4608 & -1 & 9 & -36 & 84
\end{array}\right]
$$

Table 11.3 presents the execution of Theorem 11.1.

We can now obtain the following inversion of the confluent Vandermonde matrix:

$$
V^{-1} = \begin{bmatrix} W_1 \\ \hline W_2 \\ \hline W_3 \\ \hline W_4 \end{bmatrix} = \left[\begin{array}{ccccccccc|c}
h^{(1)}_{1(10)} & h^{(1)}_{19} & h^{(1)}_{18} & h^{(1)}_{17} & h^{(1)}_{16} & h^{(1)}_{15} & h^{(1)}_{14} & h^{(1)}_{13} & h^{(1)}_{12} & K_{1,1} \\ \hline
h^{(1)}_{2(10)} & h^{(1)}_{29} & h^{(1)}_{28} & h^{(1)}_{27} & h^{(1)}_{26} & h^{(1)}_{25} & h^{(1)}_{24} & h^{(1)}_{23} & h^{(1)}_{22} & K_{2,1} \\
h^{(2)}_{2(10)} & h^{(2)}_{29} & h^{(2)}_{28} & h^{(2)}_{27} & h^{(2)}_{26} & h^{(2)}_{25} & h^{(2)}_{24} & h^{(2)}_{23} & h^{(2)}_{22} & K_{2,2} \\ \hline
h^{(1)}_{3(10)} & h^{(1)}_{39} & h^{(1)}_{38} & h^{(1)}_{37} & h^{(1)}_{36} & h^{(1)}_{35} & h^{(1)}_{34} & h^{(1)}_{33} & h^{(1)}_{32} & K_{3,1} \\
h^{(2)}_{3(10)} & h^{(2)}_{39} & h^{(2)}_{38} & h^{(2)}_{37} & h^{(2)}_{36} & h^{(2)}_{35} & h^{(2)}_{34} & h^{(2)}_{33} & h^{(2)}_{32} & K_{3,2} \\
h^{(3)}_{3(10)} & h^{(3)}_{39} & h^{(3)}_{38} & h^{(3)}_{37} & h^{(3)}_{36} & h^{(3)}_{35} & h^{(3)}_{34} & h^{(3)}_{33} & h^{(3)}_{32} & K_{3,3} \\ \hline
h^{(1)}_{4(10)} & h^{(1)}_{49} & h^{(1)}_{48} & h^{(1)}_{47} & h^{(1)}_{46} & h^{(1)}_{45} & h^{(1)}_{44} & h^{(1)}_{43} & h^{(1)}_{42} & K_{4,1} \\
h^{(2)}_{4(10)} & h^{(2)}_{49} & h^{(2)}_{48} & h^{(2)}_{47} & h^{(2)}_{46} & h^{(2)}_{45} & h^{(2)}_{44} & h^{(2)}_{43} & h^{(2)}_{42} & K_{4,2} \\
h^{(3)}_{4(10)} & h^{(3)}_{49} & h^{(3)}_{48} & h^{(3)}_{47} & h^{(3)}_{46} & h^{(3)}_{45} & h^{(3)}_{44} & h^{(3)}_{43} & h^{(3)}_{42} & K_{4,3} \\
h^{(4)}_{4(10)} & h^{(4)}_{49} & h^{(4)}_{48} & h^{(4)}_{47} & h^{(4)}_{46} & h^{(4)}_{45} & h^{(4)}_{44} & h^{(4)}_{43} & h^{(4)}_{42} & K_{4,4}
\end{array}\right]
$$

$$
= \left[\begin{array}{ccccccccc|c}
54.61 & 336.78 & 902.64 & 1378.23 & 1319.82 & 821.475 & 332.23 & 84.196 & 12.132 & 0.7585 \\ \hline
1.72 & 13.440 & 44.90 & 84.4200 & 98.5875 & 74.205 & 36.005 & 10.860 & 1.8475 & 0.1350 \\
0.300 & 2.350 & 7.875 & 14.8625 & 17.4375 & 13.200 & 6.4500 & 1.9625 & 0.3375 & 0.0250 \\ \hline
-100.33 & -758.22 & -2433.0 & -4362.15 & -4820.22 & -3405.6 & -1539.1 & -429.6 & -67.29 & -4.518 \\
-38.00 & -288.33 & -929.55 & -1675.22 & -1861.33 & -1322.3 & -600.67 & -168.3 & -26.44 & -1.777 \\
-12.00 & -92.00 & -300.33 & -549.333 & -621.00 & -450.00 & -209.00 & -60.00 & -9.666 & -0.666 \\ \hline
45.00 & 408.00 & 1485.50 & 2899.50 & 3401.81 & 2509.87 & 1170.86 & 334.50 & 53.31 & 3.6250 \\
-85.50 & -612.75 & -1872.8 & -3222.06 & -3439.1 & -2358.7 & -1039.0 & -283.6 & -43.56 & -2.875 \\
18.00 & 147.00 & 501.50 & 938.250 & 1064.00 & 761.50 & 345.50 & 96.250 & 15.00 & 1.000 \\
-18.00 & -129.00 & -390.50 & -658.750 & -684.75 & -456.00 & -195.00 & -51.75 & -7.750 & -0.500
\end{array}\right]
$$

To complete the example, let us calculate the determinant. Turnbull and Aitken (1932), page 63, example 5, gives the following formula:

$$
\det(V) = \prod_{1 \le i < j \le r} (\lambda_j - \lambda_i)^{n_i n_j} = \prod_{j=2}^{r=4} \prod_{i=1}^{j-1} (\lambda_j - \lambda_i)^{n_i n_j}
$$

Thus:

Table 11.3 Consecutive iterations of Theorem 11.1

k	λ_k	n_k	q	i	$L_{ki}^{(q+1)}(\lambda_k)$	K_{k,n_k-q-1}
1	− 0.5	1	–	–	–	$K_{1,1} = 0.75851$
2	− 3	2	–	–	–	$K_{2,2} = 0.02500$
			0	1	$L_{21}^{(1)}(\lambda_2) = 0.02500$	$K_{2,1} = 0.13500$
				3	$L_{23}^{(1)}(\lambda_2) = 0.02500$	
				4	$L_{24}^{(1)}(\lambda_2) = 0.02500$	
3	− 2	3	–	–	–	$K_{3,3} = -0.66666$
			0	1	$L_{31}^{(1)}(\lambda_3) = -0.66666$	$K_{3,2} = -1.77777$
				2	$L_{32}^{(1)}(\lambda_3) = -0.66666$	
				4	$L_{34}^{(1)}(\lambda_3) = -0.66666$	
			1	1	$L_{31}^{(2)}(\lambda_3) = 3.33333$	$K_{3,1} = -4.51851$
				2	$L_{32}^{(2)}(\lambda_3) = -1.11111$	
				4	$L_{34}^{(2)}(\lambda_3) = 2.44444$	
4	− 1	4	–	–	–	$K_{4,4} = -0.50000$
			0	1	$L_{41}^{(1)}(\lambda_4) = -0.50000$	$K_{4,3} = 1.00000$
				2	$L_{42}^{(1)}(\lambda_4) = -0.50000$	
				3	$L_{43}^{(1)}(\lambda_4) = -0.50000$	
			1	1	$L_{41}^{(2)}(\lambda_4) = 0.00000$	$K_{4,2} = -2.87500$
				2	$L_{42}^{(2)}(\lambda_4) = 2.50000$	
				3	$L_{43}^{(2)}(\lambda_4) = 1.50000$	
			2	1	$L_{41}^{(3)}(\lambda_4) = -1.43750$	$K_{4,1} = 3.62500$
				2	$L_{42}^{(3)}(\lambda_4) = -28.0000$	
				3	$L_{43}^{(3)}(\lambda_4) = -8.75000$	

$$\det(V) = \left[\prod_{i=1}^{1}(\lambda_2 - \lambda_i)^{n_i n_2}\right] \cdot \left[\prod_{i=1}^{2}(\lambda_3 - \lambda_i)^{n_i n_3}\right] \cdot \left[\prod_{i=1}^{3}(\lambda_4 - \lambda_i)^{n_i n_4}\right] =$$

$$= \left[(\lambda_2 - \lambda_1)^{n_1 n_2}\right] \cdot$$

$$\cdot \left[(\lambda_3 - \lambda_1)^{n_1 n_3}(\lambda_3 - \lambda_2)^{n_2 n_3}\right] \cdot$$

$$\cdot \left[(\lambda_4 - \lambda_1)^{n_1 n_4}(\lambda_4 - \lambda_2)^{n_2 n_4}(\lambda_4 - \lambda_3)^{n_3 n_4}\right] = -337.5$$

References

Aitken AC (1932) XII—further numerical studies in algebraic equations and matrices. Proc R Soc Edinburgh 51:80–90

Aitken AC (1938) XX—studies in practical mathematics. II. The evaluation of the latent roots and latent vectors of a matrix. Proc R Soc Edinburgh 57:269–304

Aitken AC (1944) Determinants and matrices, 3rd edn. Interscience Publishers, New York

Ahlin AC (1964) A bivariate generalization of Hermite's interpolation formula. Math Comput 18:264–273

Bjorck A, Pereyra V (1970) Solution of Vandermonde systems of equations. Math Comp 24:893–903

Chawla MM, Jayarajan N (1974) A generalization of Hermite's interpolation formula in two variables. J Austr Math Soc 18(4):402–410

Gorecki H (1968) On switching instants in minimum-time control problem. One-dimensional case n-tuple eigenvalue. Bull De L'acad Pol des Sci 16:23–30

Galimberti G, Pereyra V (1971) Solving confluent Vandermonde systems of Hermite type. Numer Math 18:44–60

Gautschi W (1963) On inverses of Vandermonde and confluent Vandermonde matrices. II. Numer Math 5:425–430

Gonzalez-Serrano La, Maximenko EA (2023) Bialternant formula for Schur polynomials with repeating variables. arXiv:2312.15680

Harris WA, Fillmore JP, Smith DR (2001) Matrix exponentials-another approach. SIAM Rev 43(4):694–706

Hermite MCh, Borchardt M (1878) Sur la formule d'interpolation de Lagrange. Journal Für Die Reine und Angewandte Mathematik 84:70–79. https://doi.org/10.1515/crelle-1878-18788405

Higham NJ (2008) Functions of matrices. Theory and computation. SIAM, Philadelphia, USA

Halidias N (2024) Computing the minimum polynomial, the function and the drazin inverse of a matrix with matlab. Asian J Res Comput Sci 17(5):1–9

Hou S-H, Pang W-K (2002) Inversion of confluent Vandermonde matrices. Comput Math Appl 43(12):1539–1547

Hou S-H, Hou ES-H (2009) A recursive algorithm for triangular factorization of inverse of confluent Vandermonde matrices. AIP Conf Proc 1089(1):277–288

Kalman D (1984) The generalized vandermonde matrix. Math Mag 57(1):15–21

Lee K, O'Sullivan ME (2010) Algebraic soft-decision decoding of Hermitian codes. IEEE Trans Inform Theor 56(6):2587–2600

Lu H (1994) Fast solution of confluent Vandermonde linear systems. SIAM J Matrix Anal Appl 15(4):1277–1289

Lu H (1995) Fast algorithms for confluent Vandermonde linear systems and generalized trummers problem. SIAM J Matrix Anal Appl 16(2):655–674

Luther U, Rost K (2004) Matrix exponentials and inversion of confluent Vandermonde matrices. Electron Trans Numer Anal 18:91–100

Muir T (1923) History of determinants, vol 4. The period 1880 to 1900. Dover Publications, New York

Moler C, V-Loan C (2003) Nineteen dubious ways to compute the exponential of a matrix, twenty-five years later. SIAM Rev 45(1):3–49

Pan V (2001) Structured matrices and polynomials unified superfast algorithms. Springer Science+Business Media, LLC

Pan V (2016) How bad are vandermonde matrices? SIAM J Matrix Anal Appl 37(2):676–694

Respondek JS (2008) Approximate controllability of the nth order infinite dimensional systems with controls delayed by the control devices. Int J Syst Sci 39(8):765–782

Respondek JS (2011a) On the confluent Vandermonde matrix calculation algorithm. Appl Math Lett 24:103–106

Respondek JS (2011b) Numerical recipes for the high efficient inverse of the confluent Vandermonde matrices. Appl Math Comput 218:2044–2054

Schendel L (1891) Das alternirende Exponentialdijferenzenproduct. Zeitschrift Math Phys: 84–87
Spitzbart A (1960) A generalization of Hermite's interpolation formula. Am Math Monthly 67(1):42–46
Shen B, Tan H, Wang Z, Huang T (2017) Quantized/saturated control for sampled-data systems under noisy sampling intervals: a confluent vandermonde matrix approach. IEEE Trans Autom Control 62(9):4753–4759
Tang WP, Golub GH (1981) The block decomposition of a Vandermonde matrix and its applications. BIT Numer Math 21:505–517
Turnbull HW (1928) The theory of determinants, matrices and invariants. Blackie & Son, Londong & Glasgow
Turnbull HW, Aitken AC (1932) An introduction to the theory of canonical matrices. Blackie and Son, London, Glasgow and Bombay
Vogt M (1901) Sur l'apolarité des formes binaires. Nouvelles Annales De Mathématiques 4(1):337–365
Zhong X, Zhaoyong Y (1998) A fast algorithm for inversion of confluent vandermonde-like matrices involving polynomials that satisfy a three-term recurrence relation. SIAM J Matrix Anal Appl 19(3):797–806

Uncited References

Bini D, Lotti G, Romani F (1980) Approximate solutions for the bilinear form computational problem. SIAM J Comput 9(4):692–697
Cariow A, Syslo W, Cariowa G, Gliszczynski M (2012) A rationalized structure of processing unit to multiply 3×3 matrices. Pomiary Automatyka Kontrola 58(7):677–680
Coppersmith D (1997) Rectangular matrix multiplication revisited. J Complex 13:42–49
Coppersmith D, Winograd S (1982) On the asymptotic complexity of matrix multiplication. SIAM J Comput 11(3):472–492
Egervary E (1954) On hypermatrices whose blocks are commutable in pairs and their application in lattice-dynamics. Acta Sci Math 15:211–222
Fam AT (1988) Efficient complex matrix multiplication. IEEE Trans Comput 37(7):877–879
G-Serrano LA, Maximenko EA (2024) Bialternant formula for Schur polynomials with repeating variables. Linear and Multilinear Algebra (to appear)
Ja'Ja (1980) On the complexity of bilinear forms with commutativity. SIAM J Comput 9(4):713–728
Le Gall F (2014) Powers of tensors and fast matrix multiplication. In: Proceedings of the 39th international symposium on symbolic and algebraic computation, 296–303
Makarov OM (1987) A non-commutative algorithm for multiplying 5 × 5 matrices using one hundred multiplications. Comput Math Math Phys 27(1):205–207
Romani F (1980) Complexity measures for matrix multiplication algorithms. Calcolo 17:77–86
Schumer PD (2004) Mathematical journeys. Wiley, Hoboken, New Jersey
Sedoglavic A (2017) Laderman matrix multiplication algorithm can be constructed using Strassen algorithm and related tensor's isotropies. arXiv:1703.08298
Shabalin AA (2012) Matrix eQTL: ultra fast eQTL analysis via large matrix operations. Bioinformatics 28(10):1353–1358

The manufacturer's authorised representative in the EU is Springer Nature Customer Service Centre GmbH, Europaplatz 3, 69115 Heidelberg, Germany. If you have any concerns regarding our products, please contact ProductSafety@springernature.com

Printed and bound by CPI Group (UK) Ltd, Croydon, CR0 4YY
30/09/2026
02225599-0004